Nothing Natural Is Shameful

THE MIDDLE AGES SERIES

A complete list of books in the series
is available from the publisher.

NOTHING NATURAL IS SHAMEFUL

Sodomy and Science in Late Medieval Europe

JOAN CADDEN

PENN

UNIVERSITY OF PENNSYLVANIA PRESS

PHILADELPHIA

Published by
University of Pennsylvania Press
Philadelphia, Pennsylvania 19104-4112
www.upenn.edu/pennpress

Printed in the United States of America on acid-free paper
10 9 8 7 6 5 4 3 2 1

Library of Congress Cataloging-in-Publication Data
Cadden, Joan.
Nothing natural is shameful : sodomy and science in late medieval Europe / Joan Cadden. — 1st ed.
p. cm. — (The Middle Ages series)
Includes bibliographical references and index.
ISBN 978-0-8122-4537-0 (hardcover : alk. paper)
1. Sodomy—Europe—History—To 1500. 2. Male homosexuality—Europe—History—To 1500. 3. Science, Medieval—Europe—History—To 1500. 4. Philosophy, Medieval—Europe—History—To 1500. I. Title. II. Series: Middle Ages series.
HQ76.2.E9.C33 2013
306.77—dc23

2013005800

For Paul and Carol

Contents

Introduction

The Natural Philosophy of Sodomites and Their Kind

As he prepared to discuss a series of questions about sexual desires and pleasures, the fourteenth-century natural philosopher Walter Burley issued this "apology to the reader for the indecency of the subject": "Nothing natural is shameful but all things [that] exist in the world [are] clean. For contemplating base things should not be spurned, since in anything, no matter how base, something wonderful and divine is contained, according to the Philosopher [Aristotle], *On Animals*, Book 13. We are not made base by contemplating base things but by doing or willing them, [according to his] *Ethics*."[1] It is the spirit behind Burley's declaration that makes this book about sodomy and science in the Middle Ages possible. The inclusion of sexual subjects within the ambit of the natural made them, by definition, the proper domain of natural philosophers, whose role it was to explain the causes of objects and events in nature. Burley's outspoken confidence was not typical, but he was not alone in his willingness to address questions ranging from the causes of erections to the reasons it is difficult have sex in the water. And among the phenomena Burley and others were challenged to explain was the existence of men susceptible to anal sexual stimulation, that is, the existence of what some of them called "sodomites."

That challenge came from an Aristotelian text. Without it these medieval scholars might not have chosen to address such a subject—and even so some demurred. Aristotle provided the occasion for scientific discussions of men construed as sexually "passive" by inquiring into what causes them, in a work known as the *Problems* (*Problemata*), which Burley was summarizing. And, as his citation of two other Aristotelian works suggests, the revered ancient thinker also provided a philosophical justification, a moral shield, and an authoritative ally for late medieval intellectuals prepared to consider in terms of natural processes not only sex in general but also this widely stigmatized group. The authors, scribes, and readers willing to "contemplate base things" never argued that they were not vile, but most did share the conviction that they could be explained. Thus fourteenth- and fifteenth-century texts associated with the Aristotelian *Problemata* contain answers to questions of causation. But they also contain hints about the moral and religious tensions surrounding the pursuit of science that are reflected in Burley's apology and his appeal to a work of moral philosophy, Aristotle's *Ethics*, as he prepared to expound on such questions as why horseback riding causes sexual arousal and why men hate the first person they have sex with. In his juxtaposition of the natural and the moral an anonymous indexer summed up the predicament faced by those who attempted to reject shame in favor of the search for causes. His entry for the twenty-sixth chapter of the fourth book of the *Problemata* reads in full: "Some people, moved by nature, commit the sodomitical sin [*sodomiticum peccatum*]."[2]

This book is the history of a family of texts clustered around a single question among hundreds in an ancient Greek work: the causes of some men's susceptibility to anal sexual stimulation and their consequent enjoyment of the "passive" role in coitus.[3] From the shards of evidence that have survived in manuscripts of and related to the *Problemata*, two themes emerge concerning medieval medical theorists' and natural philosophers' approaches to that question. The first is the story of their earnest attempts to understand the causes of these desires and pleasures in terms of natural processes; the second is the story of their concerns about whether their disciplines were equipped or permitted to deal with such subjects at all. Although those who participated in this conversation in the fourteenth and fifteenth centuries belonged to a small social elite—men with university training in the natural sciences and medicine—they disagreed both about the specific causal explanations and even about the appropriateness of acknowledging natural causes at all.

The "Science" of "Sodomy"

Mining a group of texts related to the *Problemata* and the more than one hundred manuscripts of them that have survived, this book shows how European scholars deployed a standard set of philosophical tools and a variety of rhetorical strategies to produce scientific approaches to sodomy in the period from the late thirteenth through the fifteenth century. Although medieval scholastics used the term "science [*scientia*]" in connection with all systematic academic disciplines, the pull of its modern connotations is too strong to allow it to be useful, so the medieval phrase "natural philosophy" will serve better to convey the specifically medieval intellectual practices involved. But that substitution gives rise to a set of issues surrounding the object of natural philosophy, that is, "nature." Questions about the functions and meanings of the natural are thus an essential element of this book.

The term "sodomy" presents a somewhat different problem.[4] Although it is a thoroughly medieval term (some of whose connotations persist) and appears along with "sodomite" in many of the texts in question, it is at once too narrow and too vague for the subject treated here. A fuller elaboration of what it meant to writers and readers interested in medicine and natural philosophy will emerge only in the course of a close examination of the texts. In the subtitle of this book, it serves as a recognizable, if imperfect, stand-in for a cluster of overlapping subjects that the medieval works treat. Among these are sex acts between males (in connection with which I will sometimes use the modern adjective "homosexual," but not the noun); improper relations between the "active" and the "passive" (terms I use because they recur in the sources); the anus as a site of desire and pleasure; associations with effeminacy, impotence, and eunuchs; and various layers of moral and religious disapprobation.

The texts at the center of this investigation address one of about eight hundred questions posed in the ancient Greek *Problemata*: Why do some men enjoy the passive role in sex and some enjoy both the active and passive roles? Chapters 1 and 2 deal with medieval authors' two basic answers: some men are born with anatomical defects that divert the semen from its proper path; others develop their sexual appetites as habits formed from experiences early in life. Blunt discussions of friction between body parts make clear that the sexual behaviors involved are almost all unambiguously homosexual, in the sense that they involve two male participants, but both of the causal explanations reduce individuals' desires and pleasures to a susceptibility to anal sexual stimulation. That is, the texts seldom betray an interest in the men's whole bodies, much less in their whole persons, and still less in their relationships. One writer remarked that some of these men also have sex with women, and another declared that they ought to.[5] But, in general, partner preference per se was not the authors' concern. Similarly, these sources reveal nothing about an internally experienced self. Yet they do not permit the application of the common modern opposition between acts and identities, for there was broad agreement that those whose disposition was innate did represent a fixed category of the human male. Reconstructing late medieval reasoning on these questions will reveal a concern with much more than behavior. For natural philosophers there existed a group (often referred to evasively as "such men") that constituted a recognized natural phenomenon, susceptible to causal analysis of the same kind as they applied to other topics treated in the *Problemata*, such as shrubs, winds, and sweat.[6]

The two lines of causation—birth defects and bad habits based on early experience—offer different perspectives on late medieval understandings of "these men." The explanation centering on birth defects presents most dramatically the persistent occurrence of an effect—homosexual pleasure—that deviated from the regular order of things. That is, it confronted the indubitable presence of the unnatural in nature. The dominant framework of medieval natural philosophy was teleological. In this view, each individual object or creature existed in order to perfect the essence of its kind. Laid out systematically by Aristotle, incorporated by other ancient authorities such as Galen, and elaborated by Arabic scholars widely read in the Latin West, this metaphysical principle meant that nature did everything for a purpose and nothing in vain. And by the late thirteenth century, when the story chronicled here begins, it had converged with Christian theological notions about divine wisdom and its manifestation in Creation. Stones, essentially heavy in nature, were supposed to fall to earth, and acorns were destined to become oak trees. Teeth existed to chew food; hunger existed so that creatures would seek food, eat, and grow to mature perfection. Sexual desire and pleasure likewise had their proper place in the teleological and providential scheme of things. From the perspective of the individual (female or male), they promoted the excretion of physiological superfluities that might be harmful if allowed to accumulate in the body. And from the perspective of the species, they promoted reproduction. For if the successful maturation of a single creature is a good thing, a cycle of successive creatures is even better—and nature always prefers what is better.

How, then, and why did natural processes repeatedly produce individuals whose bodies and appetites were out of conformity with the proper forms and functions of their kind? In part the answer will be seen to lie in medieval concepts of nature, which, though imbued with purpose, was not expected to operate like clockwork and was not understood in modern terms of natural law. In part, at least for some authors, the explanation also involved making a distinction between the particular nature of an individual (a man given to specific pleasures) and the general nature of the category (men) to which he belonged. This more sophisticated approach was not only philosophically tricky but also doctrinally suspect. The difficulties it raised are manifest in the pronouncement of a late thirteenth-century bishop of Paris: it was an excommunicable offense to argue that "the sin against nature," which betrayed the true nature of the human male, could be regarded as consistent with the real nature of an individual man.[7]

When men's irregular desires and pleasures were explained by the formation of a habit (the subject of Chapter 2), a different set of issues emerged. First, the meaning of "nature" was further expanded or, the bishop might say, distorted. For many natural philosophers, a deeply ingrained habit functioned rather like an individual's nature. Although for some authors the creation of bad habits was associated with a corporeal change, for most it involved a sequence of pleasure, memory, and desire. In either case, a consequent question was how a person could be trained to betray his essential nature, to which the answers were both metaphysical and psychological. Also, it was especially in the context of "certain men in whom this condition occurs from a habit"[8] that medieval authors wondered whether the group whose desires were shaped by experience was congruent with the group that enjoyed both the active and the passive roles. Their reflections led them to ponder active sodomites and passive effeminates; men who also like sex with women and men who do not. In the course of discussing who does what to whom and evaluating the strength of habits, the medieval authors considered those labeled as "sodomites" and those labeled as "soft"; anal penetration and intercrural friction; the age of those who are "subjected"; and the influence of the stars. In the case of some of these themes, clear historical conclusions are possible. For example, penetration did not define the sex acts involved, nor was it required for categorizing those who engaged in them. In the case of other topics, individual authors left matters ambiguous, and readers' notes in the margins of manuscripts record differences of interpretation. For example, although youth was clearly regarded as a criterion for habit formation, either discord or confusion surrounded the relative susceptibility of adolescents and younger boys.

Dealing with sexual dispositions based on early experiences appears to have required of scholars much greater effort than was needed to contemplate explanations based on innate corporeal factors. But, aside from those who produced highly abbreviated summaries, medieval natural philosophers followed the lead of Aristotle in accepting that either nature or habit could cause men to desire and enjoy perianal stimulation by other men. So, whereas modern polemics have occasionally erupted around concepts like "the gay gene" and "the homosexual lifestyle," most of the medieval authors treated here thought that both "nature" and "nurture" made sense. And although they differentiated between two groups of men prone to irregular appetites, they attributed similar effects to the two distinct lines of causation. In general, learned authors and readers applied scholastic methods, familiar concepts, and revered authorities to carry out the task implied in Walter Burley's defense of their discipline: to discern and lay out the reasons for male homosexual inclinations in the very limited terms in which Aristotle had presented them. They manifested considerable resourcefulness in devising plausible causal explanations for irregular desires within the constraints of the Aristotelian text, academic conventions, and the social burdens of the subject matter.

As they did so, they revealed two salient concerns that recurred, explicitly and implicitly, when they addressed the text of *Problemata*, Book IV, Problem 26 (the one that called for an explanation of these passive pleasures): the relevance of women to the subject, and the relationship between natural philosophy and moral philosophy. The two issues are distinct, but they have in common that they commanded the interest of medieval readers. The most influential commentator on the Aristotelian text, Pietro d'Abano, drew attention to a connection with the feminine, but individual scholars have left credible signs of their independent judgment. In one fifteenth-century manuscript, for example, the only marginal mark in this chapter occurs at the point of a comparison between the men in question and young women.[9] The fruits of that juxtaposition are the subject of my Chapter 3, which argues that women served two contradictory functions in this context. On the one hand, they provided a resource for the elaboration of a certain type of man, based on similarity. Both groups are in some sense defective, and thus are examples of the recurrent failure of nature to fulfill its goals in every individual case. And, more particularly, both groups are associated at once with passivity and sexually insatiability. On the other hand, women are that which the men in question (in spite of their effeminacy) are not. That is, women's specific nature clarifies what ought to be a bright line between the male and the female, the masculine and the feminine. Defective though she may be as a human being, a woman who is passive and insatiable is still true to her type, in a way that a man who is passive and insatiable is not.

Unlike their explorations of gender, which are at least cued by the ancient text they are reading, medieval natural philosophers' attempts to square their project with moral philosophy were a spontaneous response to the problem of justifying the application of their discipline to a subject so laden with contemporary social and religious significance. Their efforts are the subject of Chapter 4. Burley, who felt called upon to acknowledge the difficulties of speaking about sex in general, had recourse to the standard university textbook on moral philosophy, Aristotle's *Nicomachean Ethics*. But establishing the right to speak was not the only problem natural philosophers faced when they undertook to explain the particular phenomena of always passive men and sometimes passive men. The *Ethics* helped them to confront certain specific conceptual difficulties, such as providing an explanation for the formation of habits that seem to contravene nature. How could a man learn by experience to do something that is unnatural for a man? But the most urgent need for engagement with moral philosophy arose from the issue of one's responsibility for one's acts. There were legal and moral dispensations for children and madmen. To what extent was a person with a defective nature accountable? Few natural philosophers made reference to Christian theology or doctrine in this context, but religious tenets (themselves by no means simple) cast a shadow over their efforts. The *Ethics* provided a safer textual site at which to locate discussions of such issues, and the fact that it happened to mention in passing the case of men who have sex with boys sealed its connection.[10] Like their contemporaries who wrote commentaries on the *Ethics* itself, those who dealt with the *Problemata* perceived tensions between the powers and the limitations of the philosophically defined rational soul (and thus, implicitly, theologically defined free will).

Arriving at a philosophical posture that, on the one hand, allowed for the constraints imposed by a defective nature and that, on the other hand, gave due weight to self-determination involved something beside philosophical skill. It also had to do with the reactions of individual scholars to what Burley called "the shamefulness of the subject matter." Pietro d'Abano, the early Italian commentator, expressed scorn for men who engaged in sex with men, but he made no apologies for discussing them and even carved out a little moral space for them. Evrart de Conty, a later French commentator, thought the subject itself was very rude and ruled out any moral dispensation based on the men's nature, even if it was innate. The final chapter of *Nothing Natural Is Shameful* deals with the reactions of a wide range of mostly anonymous scholars to the texts that dealt with sexual subjects in general and with the susceptibility of some men to anal stimulation in particular. One group of scholars composed expositions or summaries of the *Problemata*. How they treated the contents of Book IV, Problem 26, as well as what they decided to include and exclude, provides evidence of their attitudes. Another group undertook, either for themselves or for others, to make copies of the works in question and sometimes indexes for them. Their decisions and their deviations from the texts they were copying are also telling. A final group, readers, have left marks in the margins of the surviving manuscripts. Even the signs that are inscrutable, such as a squiggle or a "nota bene," nevertheless record a level of attention; occasionally the presence of a word or two will reveal a little more.

In spite of significant differences in the attitudes expressed, those who studied the natural philosophy of sex shared certain sensibilities. If questions like "Why is sexual intercourse the source of the greatest pleasure?" might be cause for levity among the university men who encountered it in the *Problemata*, the question about men who enjoy the passive role was not. It is a measure of the challenge it posed to earnest practitioners of natural philosophy that even Walter Burley, that stalwart defender of the prerogatives of his discipline, was far from open when he came to deal with the accepted fact that some men enjoyed the passive role in sex.

Questions That "Battle in the Mind": Aristotle's *Problemata*

The work known as the *Problemata*, or *Problems*, was a compilation of questions (*problemata*), mainly about natural phenomena, loosely organized by subject into thirty-eight books (*particule*).[11] It originated with Aristotle and underwent editing and accretion over several centuries, arriving in late antiquity at the Greek form in which it has come down to us and from which it was translated into Latin.[12] Whether or not it is designated "pseudo-Aristotelian" depends in part on the concept of authorship applied; here, in deference to its medieval readers, its author will be called "Aristotle." Groups of questions addressed the winds, the senses, oily plants, salt water, and exercise. Book IV contained thirty two questions on "venereal matters," ranging from why eunuchs and those who have too much sex suffer from bad vision (IV.1) to why birds and hairy men are lustful (IV.31). Although some topics treated in the work, such as the shape of Ethiopians' legs (XIV.4), were exotic, and others, such as whirlpools (XXIII.5), were marvelous, the work dealt neither with singular facts nor with objects and events whose occurrence was itself in doubt. It was not concerned with prodigies or omens that might be suspected of supernatural origins. Rather the subjects of the *Problemata* were recognized as "commonly knowable," and only the explanations for them were open to question.[13] Thus, the men of Book IV, *problema* 26, were understood to exist—occurring and recurring in the ordinary course of things. The question posed by each brief chapter was, "Why?" Feeling dizzy when you were drunk (III.20) and falling asleep while reading (XVIII.1) were not necessarily good things, but they were ordinary and recognizable phenomena that called for explanations. Male homosexual desire was another such fact.

Diverse as they were, the questions nonetheless had an essential trait in common: they could not be easily and definitively answered by the universal principles of Aristotelian science. Answers to questions like "Why is the earth in the center of the cosmos?" and "Why are the male and female separate in higher animals?" could be established on the basis of specific and well-developed doctrines, such as the notion of natural place and the distinction between necessity and the good. Those, in turn, formed part of a teleological natural philosophy that emphasized the order and purpose of the components and processes in the natural world. The proper subject matter of demonstrative science, *scientia*, complied with these principles. In contrast, the problems posed in the *Problemata*—whether about male homosexual practices or about the existence of huge vegetables—appeared to occur outside or even contrary to the harmonious, universal, and purposeful norms of the Aristotelian world. They could not easily be tamed by systematic natural philosophy. For this reason, the brief responses provided by Aristotle to his questions in the *Problemata* are not really answers but merely suggestions—themselves mainly in interrogative form. Indeed, an individual *problema* often offered more than one possible avenue of reasoning: Why is the abdomen the fattest area of the body? Is it because it is closest to the stomach? Or is it because it gets less exercise than other parts?[14] In keeping with accepted scholarly practice, medieval commentators paused to consider the nature of the work as a whole and reflected on the concept of a *problema* itself. Pietro d'Abano, the earliest known Latin expositor of the *Problemata*, explained in his prologue: "*Problema* is Greek, meaning 'test' [*probatio*] in Latin. For it is a difficult question containing something that must be resolved by disputation and that seems to indicate an abyss—in the sense that a *problema* is a proposition [that] may devour and ensnare."[15] Or, as a later author put it, these are questions that "battle in the mind."[16]

At the behest of the king of Sicily, the *Problemata* was translated into Latin in the mid-thirteenth century by Bartholomeus de Messina.[17] Perhaps because it entered Western Europe later than most Aristotelian texts, it never achieved the status of canonical Aristotelian works, such as *Physics* or *On the Heavens*, which formed part of the core curriculum at medieval universities.

Its format may also have made it difficult to fit into the tree of knowledge: a compilation of miscellaneous topics, its contents belonged to no particular discipline. The work is also very long, which might have discouraged scholars. And, finally, medieval readers were aware that it might not really have been written by Aristotle. In spite of these drawbacks, the *Problemata* attracted considerable attention from the late thirteenth through the fifteenth century and beyond. "The assertion of some others to the contrary notwithstanding," insisted one fourteenth-century Italian scribe, adducing specific textual evidence, "this book was compiled by Aristotle."[18]

The Latin text itself and commentaries, abbreviations, and indices based upon it were produced and disseminated all over Europe, from Oxford to Milan, from Paris to Prague. More than a hundred manuscripts of these works have survived, many containing marks made by readers, suggesting a significant level of interest. In the mid-fifteenth century George of Trebizond, an Italian humanist, declared hyperbolically in a promotional letter that the *Problemata* contained something for everyone—old and young; educated and ignorant; rich and poor; physicians, women, and farmers—and that it served the purposes of study, crafts, and business.[19] And a deluxe manuscript produced for a wealthy abbot in the Low Countries at about the same time echoes that opinion (Figure 1). It depicts Aristotle with his book, lecturing to a group of academics ranging from young pupils to sober, book-owning masters, while to his right the merchants, patricians, and officeholders of the town and to his left its masons, carpenters, farmers, and other artisans look on as auditors.[20] Indeed, the history of the work's medieval commentaries and readership confirms that there were many reasons for interest in this sprawling compendium of phenomena that were common, knowable, and subject to debate.

"Few Study It, Fewer Understand It": The Cast of Characters

At the turn of the fourteenth century, Jean de Jandun, a rising star among natural philosophers at the University of Paris, articulated the frustrations as well as the temptations of the *Problemata* as perceived by medieval readers. In a preface to his study of Aristotle's *Physics* he mapped out the relations among the subjects and texts of the sciences. In a final, catch-all category, he listed the works of secondary importance that served to supplement the major ones. Aristotle's *Physiognomy*, for example, is ancillary to his *On the Soul*. The *Problemata* comes at the end of his list, and he suggested that it is

Figure 1. Aristotle teaching students, scholars, and townsmen. Opening illumination of a fifteenth-century Flemish manuscript of Pietro d'Abano's *Exposition of Aristotle's "Problemata."* Ghent, Bibliotheek van de Rijksuniversiteit te Gent 72, 1r. Reproduced with permission of the Bibliotheek van de Rijksuniversiteit te Gent.

"in large part compiled of many natural accidents manifest to the senses, the causes of many of which are difficult and hidden, and pertain to various parts of the aforementioned sciences." He gave some examples of how a couple of its chapters are relevant to the subject of this or that canonical text, and went on to bemoan the abysmal quality of the Latin translation from the Greek. In addition, he complained that there was no good commentary on it: "It is not much explicated by anyone known or famous, and thus few people study it and fewer understand it enough." The state of affairs was clearly disappointing to Jean, who claimed that "many and beautiful propositions of wonderful delight are brought together in it."[21]

We know the names and careers of four medieval authors who took up the challenge presented by Bartholomeus's translation of the Aristotelian text in the fourteenth century: Pietro d'Abano (ca. 1250–1315), Jean de Jandun himself (ca. 1285–1338), Walter Burley (ca. 1275–1345), and Evrart de Conty (ca. 1330–1405). All were men whose careers and thinking owed much to long ties with universities, and all spent considerable time in Paris. Their engagement with a work attributed to Aristotle and their participation in explicating and disseminating its contents reflected the contemporary academic curriculum and practices, with which they were intimately familiar. In addition, all were deeply interested in the substance and approaches of natural philosophy. Thus, although they had various other intellectual commitments—whether medicine, theology, or poetry—each produced an extensive body of writing devoted to natural phenomena and their causes.

Yet differences of circumstance, experience, and personal outlook have also left traces in their versions of the *Problemata*. Pietro's exuberant pursuit of natural causes was undoubtedly due in part to the intellectual atmosphere in northern Italy where he was educated. He treated the *Problemata* as a quarry for advanced medical and natural philosophical research, and thus aimed for thoroughness of exposition. But while Jean was delighted with Pietro's commentary, his main research interests (as represented by his extant writing) lay elsewhere, and he scarcely intervened in its text. He saw it not as an opportunity for further research but as a textbook that might be useful to him as a lecturer in the Paris arts faculty, and did not trouble himself to reflect in depth upon any part of it. Burley anticipated a broader audience, and he produced an abbreviated reworking of Pietro's exposition to be read independently of university lectures. He pictured theologians as well as natural philosophers among his readers, and his posture reflected a consciousness of disciplinary boundaries. Emphasizing the marvelous and useful aspects of the text, and having rendered it into the vernacular, Evrart's version presents a substantive orthodoxy and rhetorical delicacy appropriate to a constituency that bridged university and court. Pietro and Evrart, the two physicians in the group, placed greater proportional emphasis on medicine in their prologues than did the others; Evrart and Burley expressed more explicitly a consciousness of the moral and doctrinal issues raised by the text than did Pietro and Jean. These authors were reflecting their own social positions, beliefs, and sensibilities when they confronted Book IV and its *problema* 26, and so, undoubtedly, were their anonymous colleagues.

Although the specific identities of most of the other fourteenth- and fifteenth-century commentators, copyists, and owners are beyond reach, some can be linked to a place (Bavaria, Flanders, Krakow, Seville), an institution (the monastery of St. Victor in Paris, the University of Erfurt, the court of

Cesena's ruling family), or a profession (physician, abbot, ducal secretary). The evidence, albeit fragmentary, thus makes possible a rough mapping of medieval approaches to the *Problemata*. It reveals some patterns that divide North from South and physicians from philosophers, but, at least as much, it suggests that individuals with similar backgrounds operating in similar environments could have very different reactions to the same intellectual and moral predicaments.

Finally, the cast of characters in the story of *problema* IV.26 includes three other figures: one Arabic medical writer and two humanist translators. Avicenna (980–1037), who was both an Aristotelian and a medical authority, wrote no commentary on the *Problemata*, but he exercised a profound influence on Latin opinion concerning men suffering from the condition he called *ubna*. George of Trebizond and Theodore Gaza produced new Latin translations of the work in 1452 and 1454, respectively. These Renaissance alternatives to Bartholomeus's translation provide different perspectives on what Aristotle had said about the men described in *problema* IV.26.

Pietro d'Abano and Jean de Jandun

Pietro d'Abano, the first Latin commentator on the *Problemata*, bridged a number of significant divides in the late medieval European intellectual scene.[22] A northern Italian whose intellectual appetites reflected the naturalistic and rationalistic tendencies of thirteenth-century Padua, he was trained in both natural philosophy and medicine. There is no evidence that he practiced as a physician, though he may have done so.[23] In any case, he devoted a significant part of his career to exploring the relationship between the two disciplines, an interest reflected not only in his *Problemata* commentary but also in his most successful work, the *Conciliator*, which directly addressed the disagreements between medical and natural philosophical experts. His medical writing is theoretical in nature, and he tended to side with Aristotle over the corresponding medical authority, Avicenna.[24]

Pietro's career also connected his native Veneto and its prominent university at Padua with Paris, where he spent perhaps a decade from at least as early as 1293, when he would have been around forty. He probably studied and possibly taught there, though the ample documents of the university reveal no official connection. In any case, it was in Paris that he began work on his exposition of the *Problemata*, a text that suited his intellectual style. Whether writing about astrology or poisons, he was a master of causal explanations, and the invitation extended by the *Problemata* to consider likely answers in terms of proximate causes (as distinguished from demonstrable answers in terms of final causes) may have particularly appealed to him.[25] In the course of writing and revising his commentary he traveled to Constantinople, learned Greek, and took an interest in other collections of natural questions, some of which he translated into Latin.[26] Although he occasionally had recourse to Greek in his commentary on the *Problemata*, he does not appear to have had access to a copy of the Greek text and relied instead on the Latin translation of Bartholomeus de Messina.[27]

During the same period, while he was also working on the *Conciliator* and his major astrological treatise, the *Lucidator*, religious authorities in Paris censured him.[28] He himself tells us that he was cited for fifty-five errors, some of which had to do with the powers of celestial bodies upon human souls and human history.[29] Like most of his contemporaries, Pietro was convinced that the heavens influenced terrestrial events, and he was well aware of the dangers associated with astrological determinism. Although he tells us that the Dominicans who harassed him were simply incapable of understanding such matters and that he was freed from his troubles with divine and papal assistance, he took the precaution of reassuring the readers of his works that he knew perfectly well the limits of astrological science. He was not alone among scholars of his time in entertaining ideas about nature that were officially proscribed. In his commentary on the *Problemata*, for example, he allowed for the possibility that (from a rational perspective) the world might be thought to be eternal, and it is conceivable that his views on sex also attracted attention.[30] But the subjects of the other "errors" for which Pietro was attacked, along with details of the story, remain undocumented. He left Paris in the early years of the fourteenth century and returned to Padua, where he held a position at the university until his death in 1315, earning a salary from the state of Venice. He is the only one of the commentators about whom we know anything that could be characterized as personal: his will indicated that he had children.[31]

Dated manuscripts of Pietro's *Conciliator*, *Problemata*, and *Lucidator* indicate that they were all completed in 1310. All three are lengthy and complex works, and the date clearly refers to the publication of his final revisions. In the case of the *Problemata*, he had in fact produced two distinct, though not radically different, versions by that date.[32] The massive, comprehensive commentary occupies more than two hundred folios (four hundred pages) in the manuscripts—four times the length of the already long Aristotelian text itself. He alluded to students in his prologue, but he seems to have regarded the

Problemata and his exposition of it as an advanced work, saying, "This book cannot be fully understood except by someone who has examined philosophy in all its parts."[33] It influenced most, though not all, of those who produced *Problemata*-related works in the fourteenth and fifteenth centuries, many aimed at making it shorter and more accessible. Two dozen manuscripts of Pietro's *Exposition of Aristotle's "Problemata"* survive. It appeared early in print (Mantua, 1475), and by 1520 it had been published four more times in northern Italy and Paris. Pietro's commentary reached Jean de Jandun by the hand of a friend, Marsilius of Padua (1280–ca. 1343), perhaps sent by Pietro himself, and Jean was delighted to declare himself the first among the Paris professors to get his hands on a copy.[34] The manuscript he received contained the first of Pietro's two versions. As a result, that was the form in which it became established in Paris, while the revised form was known mainly in Italy.

Jean produced a slightly emended edition of Pietro's commentary, adding a new prologue in which he called Pietro a "most diligent man and most excellent doctor."[35] More than half the surviving manuscripts of Pietro's work, especially those associated with northern Europe, are derived from the redaction Jean completed in 1318.[36] Since he did not intervene a great deal in Pietro's text,[37] it is sometimes difficult to detect his attitudes toward its more inflammatory elements, and he seems not to have deemed any of the sexual topics worthy of special note. Nevertheless, his occasional interventions, including several alterations in *problema* IV.26, are significant because of the influence his redaction exercised. In addition, his role in the textual history illustrates the networks of social and intellectual exchange through which the *Problemata* traveled. His prologue makes clear that he intended to use the work as a basis for lectures at the university; whether or not he actually did so, students at Paris were the main vectors by which Pietro's commentary spread through northern and eastern Europe.

Throughout his career, Jean was involved in both friendly and bitter debates on subjects ranging from sense perception to the infinity of God. He exhibited an awareness of the political nature of philosophical orthodoxy and the role of intellectual fads in the determination of truth.[38] He held and defended many controversial positions and is often numbered among the "Latin Averroists" by historians. In the context of his interest in the *Problemata*, this questionable historical category has relevance insofar as it associates him with a strong commitment to philosophical explanations and arguments.[39] Unlike Pietro d'Abano, Walter Burley, Evrart de Conty, or even his friend and colleague Marsilius of Padua, Jean de Jandun's academic training and career were limited to the faculty of arts, that is, to Aristotelian philosophy. He was not reluctant to challenge a theological position with a philosophical argument. Indeed, in a work in praise of the city of Paris, he represents philosophy as the most important discipline and makes fun of the theology faculty.[40] Nevertheless, although charges of erroneous and heretical teachings were hardly unknown during this period, there is no evidence that Jean's teaching caused him any trouble with the authorities, until his involvement in political conflicts during the last years of his life.[41]

How we identify and gauge Jean's interventions in Pietro's work and Pietro's own changes from one version to the other depends on how we interpret the relationship among the manuscripts that have come down to us. At what point, for example, and by whom was "sodomy" with one's self added to or subtracted from consideration? The works of Pieter De Leemans and Gijs Coucke provide details;[42] here some specific points will be useful for our understanding of the *problema* in question and the men it describes. The texts can be grouped on the basis of a comparison of the prologues. Eleven manuscripts contain some version of the one Jean added. Pietro's own two versions can likewise be distinguished by differences in their prologues, as well as on the basis of other textual comparisons. The first survives in three manuscripts, the second in six.[43] But each copy of each version is unique, and at many points in the present book the similarities and differences among manuscripts will provide evidence about ideas and debates concerning men who enjoy the passive role in coitus. For that reason, the notes here record readings from specific manuscripts rather than from Gijs Coucke's edition of Pietro's treatment of the questions on sexual intercourse (Book IV). For while Coucke's goal was to produce a text as close to Pietro's "authentic revised version" of the *Exposition of Aristotle's "Problemata"* as possible, free from "unauthorized emendation and contamination," mine is to take advantage of the instability of the text as manifested in the manuscripts.[44]

Many of the variants take the form of omissions and (apparent) misreadings; sometimes the writer's intention is apparent, often it is not. Taking advantage of the opportunities afforded by manuscript culture is complicated by the practices of manuscript culture itself. In a system of learning and research that was grounded in texts (starting with the literal interpretation of the words of an authority) scholars took a particular interest in finding a good version of a work. When Jean remarked on the lack of respectable commentaries on the *Problemata*, he also complained that the work "is usually found corrupt and uncorrected."[45] Not surprisingly, then, manuscripts of the Aristotelian work itself and of those based on it often show signs of scribes' or readers' attempts to establish the correct wording by comparing two or more copies. The practice produced hybrid versions of Pietro's commentary.[46] Sometimes copyists intervened on their own authority.

Walter Burley

The significance of the malleability of medieval texts is manifest in the case of Walter Burley's treatment of the *Problemata*. Originally it was a summary that followed the order of the original, but before long Burley or someone else had rearranged it, putting the problems in alphabetical order by topic and thus transforming it into a reference work.[47] Burley was at the University of Paris from about 1310 to 1326, and it may have been there that he obtained a copy of Jean's redaction of Pietro's commentary.[48] He was well versed in Aristotelian philosophy, having studied at Oxford, where he was associated with Merton College, a cradle of intellectual innovation in the late Middle Ages. As a master of arts, he had begun to write expositions of Aristotle's logical and natural philosophical works. Apparently to pursue the study of theology, he left for Paris, arriving soon after Pietro's departure and while Jean was still there. They were both in the city in 1310, when the university's faculty of theology was involved in the prosecution of the Order of Templars for heresy, including the closely associated crime against nature—sodomy.[49] Burley became a doctor of theology, but to judge from his writings, his main interest remained Aristotelian philosophy. He almost certainly knew Jean, and agreed with his views on some important academic issues of the time.[50] Burley's encounter with Pietro's *Problemata* commentary fits this picture, supporting the idea that he was the author of the epitome attributed to him in two manuscripts. Burley may still have been in Paris when he produced his own version of that work, but it is more likely that he composed it later, when he had left the university setting for a courtly one.[51] He occupied various offices for the English royal household, and later he served in the entourage of Richard de Bury, bishop of Durham. The bishop possessed a significant library and supported a distinguished group of scholars, in whose company Burley wrote and revised works on logic, natural philosophy, and ethics. Throughout his career Burley enjoyed increasingly lucrative benefices (ecclesiastical appointments from which scholars commonly derived their income) and spent significant periods of time at the papal court in Avignon. His philosophical views were not regarded as suspect, but a combative exchange with the Oxford theologian William of

Ockham accords with his vigorous defense of the prerogatives of philosophy with which this chapter began.

His scholarly production included summary expositions of several Aristotelian works, and the *Problemata*, given its great length, especially invited such treatment. Although based on Pietro's exposition (as he announces), rather than on the text of the *Problemata* itself, Burley's text is so abbreviated—about one-fourth the length of the commentary—that in some respects it resembles the original Aristotelian work, giving questions and listing possible answers with a minimum of explanation. Pietro had regarded prior comprehensive knowledge of philosophy as a prerequisite for approaching the work, and Jean, less prohibitively, had thought in terms of a student audience. But Burley saw his shorter, more accessible work in a different light. In his own prologue he did include (as Pietro and Jean had done) some formulaic language about the subject matter, format, and goals of the work, but he also reminded his readers that Aristotle had been the tutor to King Alexander and made reference to a well-known work on the education of princes.[52] And while Pietro and Jean introduced a *problema* as something "knowable," Burley emphasized that it was something "knowable and wonderful." These adaptations suggest that he imagined a somewhat broader or different audience for his work, including members of the literate elite beyond academic circles. The book is not dedicated to the bishop of Durham (as some of his late works were), and details of the extant manuscripts suggest that its initial audience consisted of students and masters at Oxford, but because of its brevity, his *Problemata* was more accessible than Pietro's. The alphabetical version rendered it less of a text and more of a handbook. In both formats it has survived in smaller, more informal manuscripts, often apparently produced for personal use rather than by professional scribes. In all these ways, Burley's intervention in Pietro's text was far more profound than Jean's edition of it, with its substitute prologue and its minimal revisions. Burley's was a different, if derivative work, marked among other things by his repeated reference to the marvelous and by the care he took to defend not just the study of the *Problemata* as a whole but the chapters on coitus in particular. His justification of Book IV, "Nothing natural is shameful," may reflect not only his personal sensibilities but also his view of his imagined audience.

Evrart de Conty

Writing half a century after Burley, Evrart de Conty, the last medieval *Problemata* commentator who can be identified, shared many affiliations and experiences with his predecessors, notably a long and fruitful relationship with the University of Paris. But, like Burley's, his work transformed and redirected the Aristotelian text and Pietro's commentary on it.[53] Evrart came from a prominent family in Picardy, and his status as a medical master is first documented in 1357. He was thus intimately acquainted with the natural philosophy that formed the basis of the university curriculum and with standard medical texts like Avicenna's *Canon of Medicine*. In spite of his half-century affiliation with the medical faculty, there is no evidence about his teaching. But his university position would have provided an income as well as a source of prestige upon which to base his success as a court physician, educator, and counselor. Like many professors of the medical faculty, he took minor orders and enjoyed the financial benefits of holding a nominal ecclesiastical office.[54] But the primary focus of his career was the French royal court, where Charles V was sponsoring a program of translation that included major works of natural philosophy as well as Aristotle's books on ethics and politics.[55]

Working from a copy of Pietro's work (probably Jean's version), and with copies of other *Problemata*-related texts, Evart rendered into French the questions posed in the Aristotelian text (often as rephrased by Pietro d'Abano) and Pietro's commentary, making a few cuts, offering supplemental explanations, providing references to additional sources, and interjecting comments in the form of "glosses."[56] The resulting work, completed shortly after 1380, was so long that (like Pietro's) it sometimes circulated in two volumes. Evrart was working at a moment when a vernacular vocabulary for scientific terms was just being formed, and he was conscious of building a new language.[57] At *problema* IV.26, as well as elsewhere, he commented on the task's awkwardness—social as well as linguistic.[58] He rendered the subject of Book IV as a whole (*venerea* in Bartholomeus's Latin and *coitus* in Pietro's) as "*oeuvre de generation*"—"act of reproduction."

His frequent and learned references to Latin authorities as well as certain aspects of his glosses indicate that Evart anticipated scholars among his audience. But, like Burley, he also seems to have envisaged a more diverse public.[59] His other surviving work exhibits a similar mix of the courtly and the learned. (Whereas his *Problemes* rendered an academic work more accessible, his elaboration of *The Chess Game of Love* and its French prose commentary made a courtly work more learned by adding a series of encyclopedic disquisitions on topics, such as music and astrology, to the vernacular allegorical poem.)[60] Evart did not deem it necessary to shorten the work: his royal patrons had the resources to underwrite large folio volumes. Following Pietro, he remarked that the long first book of the *Problemata*, containing questions about medicine, was useful in a practical sense.[61] But, like Burley and far more often, he referred to the phenomena discussed in the *Problemata* as "marvelous"—not in the sense that they could not be accounted for, but in the sense that they could delight the reader or listener. In his prologue he alludes to the well-known Aristotelian idea, cited by Pietro, that humans by nature desire to know, but he follows Burley in emphasizing happiness and in quoting classical poetry:

> The Prince of philosophers, Aristotle, wishes to extract and choose from numerous subjects and various sciences the marvelous and delightful ones from among the questions that can be posed about the things one sees in nature. . . . And [the books of the *Problemata*] are desirable and agreeable because human understanding desires to know and be acquainted with the causes of the marvels of nature. And it takes pleasure in this above all, and, in truth, this knowledge of causes contributes much to human perfection and happiness. [Adding the Latin of Virgil in the margin:] "Happy is he who can know the causes of things."[62]

In the same spirit, when he reaches the group of thirty-two questions about sex, he announces that "in this subject there are many issues [*doubtes*] and many great marvels."[63]

Linking the delights of contemplating the puzzling phenomena of nature with the fulfillment of humanity's highest natural capacity gave the vernacular *Problemes* at once an allure and a justification. The combination of leisured pleasure and edification was familiar and appropriate to Evrart's aristocratic audience. Scholars are fortunate to have a copy of the work written on paper in Evrart's inelegant hand,[64] but the physical form of the other surviving manuscripts of his book confirms its relation to his patrons' cultural and social milieu. They are beautifully produced large codices, some of vellum, the most expensive material.[65] Graceful illuminations depict Aristotle as the master of all philosophy. In one, for example, he is both teaching and, apparently, collecting information from figures representing the various branches of knowledge contained in the book, such as medicine, sound, and winds (Figure 2).

Figure 2. Aristotle teaching, with figures representing various topics treated in the *Problemata*, such as medicine, music, justice, astronomy, and winds. Incongruous, but not disruptive, a naked man with a golden bowl is engaged in conversation with a scholar. He may be the artist's representation of occult sciences, which are not in fact included in the work. Opening illumination of a fifteenth-century French manuscript of Evrart de Conty's *Les Problemes d'Aristote*, Jena, Thüringer Universitäts- und Landesbibliothek El.f.81, 1r. Reproduced with permission of the Thüringer Universitäts- und Landesbibliothek.

The artist has included some generic scholars and has omitted the subjects of many of the thirty-eight books, including sexual intercourse, though a mysterious naked man occupies the position farthest from Aristotle on the page. As the substance and language of Evrart's approach to Book IV, *problema* 26, reflect his sense of his audience, the physical presentation of the manuscripts reflects his audience's sense of his work.

In a similar way, the material and configuration of other *Problemata*-related works resonated with the social and institutional environments in which they circulated. The more than fifty extant copies of Bartholomeus's Latin translation vary in size and quality, but most are dignified, if small.[66] Colleges and religious houses associated with urban schools as well as some academics and physicians with more than a passing interest in natural philosophy would have possessed copies, and the ownership of some can be specifically traced. With a single beautiful exception (the one depicting Aristotle flanked by the patricians and artisans of Ghent, Figure 1), the manuscripts of Pietro's commentary in all its versions are appropriate for academic libraries and the collections of serious philosophers and physicians or their patrons—large and professionally produced, handsome but not luxurious. Burley's epitome, small in format and modest in production, was copied and bound together with various other short philosophical or medical texts. In its alphabetically arranged form it bore a general resemblance to some brief encyclopedic works and medical handbooks. It suited its owners who were individuals rather than institutions as well as men of no particular prominence with some university connections and undoubtedly small collections of books.

"The Expositors"

Evrart de Conty declined to provide a full treatment of Book IV, *problema* 26: "I will touch on it a little below, in general and rather lightly. And anyone who wishes to see it more in full may read the Latin, both the text of Aristotle and of the expositors."[67] He nowhere names these expositors other than Pietro, but that reference and other evidence indicate that the work was being discussed and even taught.[68] And the same is suggested by a handful of short, anonymous treatments of the *Problemata* that have survived, four of them in multiple copies.[69] They are all much briefer than Pietro's work, and some are even shorter than the Aristotelian text itself. Referred to as "commentaries" here, some are abbreviations or summaries. For the most part, the manuscripts in which they have been preserved are of the small, informal sort made for personal use.

A few of these texts are of interest because, in their brevity, they omit or obscure the question about sexually passive men.[70] Others make a more substantive contribution to its discussion among medieval scholars. Some have acquired modern labels, often based on very slim evidence, that are useful for keeping track of them. These include the "Bavarian" *Problemata*, the "Aachen" *Problemata*, the "Erfurt" *Problemata*, and the "Vatican" *Problemata*.[71] The author of each of these depended heavily on Pietro's commentary (mainly Jean's edition),[72] as did Evrart and Burley; each, in his choices and rephrasings, conveys his own ideas about the text and its subject matter; and the manuscripts of each sometimes reveal the thinking of the individuals who made them. For example, the briefest, the Erfurt text, declares that the involvement of the buttocks in coitus (of any kind) is confirmed by experience (*experimentaliter*), and one person who copied the work labeled IV.26 simply "On the Act [*De actu*]." The reach of ideas and students from Paris is evident in the history of these texts and their manuscripts—north to the Low Countries, south to Bavaria, and east to Krakow. In Prague a scholar produced a brief but energetic exposition (called below, the "Prague" *Problemata*) apparently without relying on the family of works descended from Pietro's. He phrased the *problema* in question this way: "Why are there some men who want only to be rubbed and certain ones [who want] only to rub and certain ones [who] want to be rubbed and to rub in intercourse?"[73] This author, as will be seen, spoke more clearly and concretely than some of those working in the Paris tradition.

The final group of anonymous "expositors" are the scribes and readers who left their marks, and sometimes their words, in the manuscripts. They remind us, first of all, that the *Problemata* was a compilation of phenomena relating to all manner of subjects, and not everyone in the fourteenth and fifteenth century was as interested in reading about sex as are historians of sexuality. The part of the work most often marked up with inarticulate lines or an only slightly more articulate "nota bene" was Book I, containing medical questions. These, Pietro and others noted, were the ones most necessary to human well-being. But some readers did give special attention to the *problema* under consideration here. In one manuscript an academic owner marked lemmas—the divisions of the text into the sections used in lectures or commentaries—in the first dozen questions about medicine, and then skipped over more than a hundred chapters. When he reached *problema* IV.26, he again marked out the lemmas.[74] There is no way of knowing what he would have said if he had given a lecture on it, but the evidence of his interest is part of the story here.

Avicenna

Western scholars faced with the challenges of interpreting a work of Aristotelian natural philosophy were accustomed to turning to Latin translations of Arabic commentaries on it for help. In the case of the *Problemata*, however, no such assistance was available. Although the Greek had been translated early into Arabic, it had not commanded significant attention from Aristotelians in the Arabic-speaking world.[75] It was not, for example, among the commentaries of the Cordovan scholar Averroes (Ibn Rushd, 1126–98) whose works were so influential in the Latin West that he was sometimes referred to simply as "the Commentator." But the great synthesis of Galenic and Arabic medicine by Avicenna (Ibn Sīnā, 981–1037) had been translated into Latin in the twelfth century, and by the time Pietro and Evrart were studying medicine, it was firmly established as the basis for advanced learning in theory and practice.[76] Far too long to fit into one manuscript (the early printed editions run to more than a thousand pages) and far too comprehensive to be required reading, Avicenna's *Canon of Medicine* was known to most learned physicians only through a few of its sections, most notably its first book. The subject of many commentaries, that section lays out the definition and components of medicine, the types of diseases that affect the whole body, the basic methods of maintaining health, and the ways of treating diseases that affect the whole body. Book III deals with illnesses that affect specific parts of the body, organized from head to foot. A number of other, widely available texts with similar content were more practical (lighter, cheaper) and less burdened with transliterated Arabic terminology, and perhaps for that reason Part III of the *Canon* was less widely circulated and commented upon.[77] It is there, in the section on genital ailments, that Avicenna compares the effects on men of intercourse with women and intercourse with boys. And it is there, in a chapter on the "illness" of men given to being anally penetrated by other men, that Western scholars found authoritative help.[78]

The Latin translation of Avicenna retains the Arabic term "*al ubna*," which might be rendered into English as "passive sodomy."[79] It appears as a chapter heading in various barely Latinized forms, such as "*halubnathi*," but even without help from the translator the *Problemata*'s Latin audience had no difficulty associating it with the subject of Book IV, *problema* 26. One reader, writing in the margin of Pietro's commentary, put it succinctly: "'On Alubnati' according to Avicenna. And it is the sodomitical act."[80] Avicenna first described the "sickness [*egritudo*]" that happens to those who are accustomed to having men lie on top of them, and then discussed its etiology and the possibility of curing them. Both their hearts and their erections are weak, hence they experience a lot of desire; their semen, though plentiful, is inert, hence they have difficulty performing active intercourse. On the other hand, their minds (*potentia imaginativa*) can help, hence they like to see two people having intercourse.[81] Given this condition, these men only get erections if they are penetrated. For some of them, being penetrated itself induces the pleasure of ejaculation (whether or not semen is emitted). Others ejaculate not then but later, taking the active role with someone else. Such is their illness, and Avicenna is unambiguous, that—whatever their underlying physiological weaknesses may be—these men suffer from a fundamentally psychological condition that medical intervention cannot remedy: "In truth they are [men] of a debased soul, an evil nature, a bad habit, and a womanly constitution. . . . And the people who want to cure them are stupid. For the origin of their sickness is mental, not natural. In truth, if a cure of it is applicable, then it is that which breaks their desire by sadness and hunger and staying awake and confinement and beating."[82] This position may have been related to treatment within Arabic ethical literature about lovesickness and moral responsibility, some elements of which filtered into the West within translations of other medical material.[83] In any case, Avicenna's clear rejection of natural causes and medical solutions played a significant role in Western discussions of *Problemata* IV.26.

Having concluded his account of the symptoms, origins, and treatments of this condition, Avicenna briefly outlined an alternative view, which he rejected. According to some unnamed individuals, anal sexual sensitivity arises from the fact that, in such men, the sensitive pathway leading to the penis has two branches, a major one to the tip of the penis (presumably the only one found in unaffected men) and a finer one to its base, that is, at the anus. In the case of the latter, a man requires vigorous rubbing to produce sensation, and so he will achieve full arousal only if someone lies (or, perhaps, ejaculates) on him.[84] It is possible that this explanation, which Avicenna dismisses as highly unlikely, itself derives directly or indirectly from *Problemata* IV.26, which, as will be seen shortly, involved the vessels leading to both the base and end of the penis. In any case, Pietro d'Abano took Avicenna to be targeting Aristotle. Conversely, in his glosses on Avicenna's *Canon* the Italian physician Gentile da Foligno (d. 1348), although he declined to say much about the subject, made a connection between this passage and that *problema*.[85] Some medieval authors who commented on the Aristotelian work, notably Pietro d'Abano and Evrart de Conty, cited Avicenna directly. Others encountered indirectly the implications of the views reflected in the passage just quoted as well as in an earlier chapter of the *Canon* which says that men having intercourse with boys is regarded in many cultures as disgusting and is therefore prohibited by law. Avicenna was not the only Arabic medical writer to address homosexual desires, pleasures, and acts,[86] but his ideas and authority were the ones that influenced the story of *problema* IV.26. His voice is heard when authors weigh the causal explanations for the condition of *halubnathi* and when they address the question of its cure. Even the untranslated, transliterated term itself gets some play, as scribes struggle to copy it.

George of Trebizond and Theodore Gaza

Late medieval scholars were conscious that they were reading many of the great works of the past in translation. Thus, for example, they often compared two translations of a work they were interested in.[87] Even so, a manuscript made by a Venetian physician in 1477 is unusual. Perhaps influenced by the humanist culture of northern Italy and certainly motivated by an interest in the subject matter, Jacobo Suriano acquired three different translations of the *Problemata*.[88] Using different colored inks, he collated the texts, so that the resulting manuscript gave all three Latin translations in sequence for each *problema*.[89] Bartholomeus's translation had pride of place, next came the one George of Trebizond had completed in 1452, and finally that of Theodore Gaza, completed about a year later.[90] Like many Renaissance scholars, those two fifteenth-century translators found the medieval versions of classical texts unsatisfactory, though Gaza was the more scornful. Both were native Greek speakers; both were polished Latin rhetoricians, though Gaza tended toward the pretentious. Both versions underwent revisions, and George wrote occasional glosses or "scholia" to his.[91] In addition, he was responsible for the translation and exposition of an astrological work containing material that, in his view as well as in Pietro's, related to the men of *Problema* IV.26.

The new translations of the *Problemata* did not influence the late medieval treatments of the work that are the subject of this study, although Suriano's composite manuscript is a sign that they were becoming known. Gaza's became the dominant Latin version from its first appearance in print in 1473, but thanks to the importance of Pietro's commentary, which was based on Bartholomeus's translation, the latter continued to be printed as well.[92] No commentaries comparable to those that resulted from the Parisian dissemination of the work appear to have originated from the fertile intellectual soil of Italy, although copies of the *Problemata* itself (sometimes with annotations) and of Pietro's exposition were to be found there. So the significance of the new translations for the present book is that they offer yet another perspective on the contents of and reactions to *problema* IV.26. Like those who commented on and annotated the Aristotelian work, they have left us evidence about their understanding of why "some people with whom intercourse is had experience enjoyment" (Bartholomeus); or why "some people experience pleasure when being subjected to venery" (George); or why "some people may be subjected to the venereal act with enjoyment" (Gaza).[93]

"What Many Do Today": Historical Perspectives on Homosexual Desire

Pietro d'Abano, whose commentary is featured in this book, was enough of a historian to make note of differences between Aristotle's time and his own. In that spirit some preliminary distinctions between medieval and modern interests and language are in order. Medieval natural philosophers did not seek out opportunities to investigate the causes of men who enjoy the passive role in sex. Thus when Pietro composed a compilation of material on physiognomy, he did not raise the subject, even though that science was devoted to moral types and the correspondence between physical and behavioral traits.[94] And, although he and others interested in the *Problemata* expressed discomfort and disgust, there is no evidence in their texts that they sensed in the subject a particular social or cultural salience specific to their environment.[95] They did not write separate tracts about it, and had they not encountered it in a work by Aristotle, they probably would not have addressed it at all. Therefore the isolation of this question from the textual contexts in which it resided—indeed, in which it was buried—is a modern project.[96] Contemporary politics call not only for the recovery of what has been suppressed or ignored but also for establishing the principle that sexual categories, norms, and experiences themselves have a history. Against the reification of "traditional values," for example, historians can document how sodomy was "invented";[97] against a static assertion of "traditional marriage," historians can even pose the question "whether forms of premodern same-sex couplings constituted 'marriage.' "[98] The present book would exist neither in the listing of the publisher nor in the mind of the author were it not for the urgency of issues surrounding lesbian, gay, bisexual, and transgender voices, relationships, and rights that are relevant to our particular time and place.

Yet the timeliness of the subject need not distort the investigation of the past. On the contrary. First, reflections on modern experience have been integral to the formulation of productive historical questions. For example, Adrienne Rich's personal and political concept of "compulsory heterosexuality" was at once an expression of U.S. feminism at a particular historical moment (1980) and—albeit in a way different from Michel Foucault's *History of Sexuality* (1978–1984)—an exhortation to investigate the power dynamics of sexual regimes.[99] Second, once such questions are posed the historical sources are often found to authorize the serious investigation of what might otherwise have been mistakenly construed as anachronistic and thus unanswerable questions. Pietro d'Abano reassures us that, as early as 1300, it was possible to imagine a history of sexuality. He commented that in Aristotle's time anal penetration was more common, whereas intercrural intercourse is "what many do today."[100]

This generic authorization for the historical study of sexuality is, however, an insufficient basis for the investigation of ideas about passive men and sodomites and male homosexual behavior from the specific perspective of the history of natural philosophy in late medieval Europe. Also required are both a line of approach and an account of the empirical evidence in its context. Formulating an approach (loosely speaking, a "method") has been greatly facilitated by decades of scholarship and debate on the history of homosexuality. Authors' choices have sometimes been manifested and often been schematized as a series of dichotomies: essentialism vs. social construction; familiarity vs. alterity; continuity vs. discontinuity; acts vs. identities; homosexual vs. lesbian/gay/bi/trans vs. queer; and (most recently) pride vs. shame.[101] Although still relevant (and occasionally still animated), discussions of these axes of analysis have produced a degree of theoretical complexity and profited from a proliferation of substantive research, to the point where it is tempting to declare a cease-fire, if not a synthesis. In her recent manifesto on lesbian historiography, Valerie Traub, after reviewing some of the debates, proposes a pragmatic strategy for accommodating the diversity and malleability of elements relevant to the history of homosexuality. It offers a way of avoiding both the nihilism of declaring all modern categories inadmissible, and the unwieldiness of declaring everything to be potentially queer. She proposes a preliminary list of recurrent themes, some but not all of which will be present in any given time or place, culture or society.[102] Among them are several with which the present book intersects, ranging from the relevance of the genitals, through the dynamics of predisposition and habit, to the significance of age. At the same time the checklist includes a variety of questions—from those about clothing to those about homosociality—about which this book has nothing to offer.

Indeed, there are many things that this book is not about. Perhaps precisely because *Problemata* IV.26 featured male anatomy so prominently and had so little to do with persons and relationships, none of the commentators made any allusion to anything resembling lesbian pleasures or acts, much less anything resembling lesbian relationships or identities. Other medical and natural philosophical sources contained scattered elements that, when taken together, might have constituted an identifiable female type, but such a convergence did not occur until the early modern period.[103] Thus, for example, the Arabic authority Avicenna made mention of a female disorder involving "added flesh" in the genital region, producing "a thing that is like a penis." In such cases, "It sometimes happens to her that she does with women [something] like what is done to them when there is intercourse with them."[104] And one thirteenth-century Bolognese medical writer mentioned that women with a particular kind of uterine growth that could take a form similar to a penis might engage in "intercourse" with women, but he did not elaborate.[105] But neither he nor other medieval medical and natural philosophical writers regarded the phenomena and causes of irregular desires in women as in any way noteworthy, much less comparable to those in men.[106] This lack of sustained attention does not mean that women were free to enjoy themselves with each other. There were prosecutions of female sodomites as early as the late thirteenth century. But like the texts that mention female anatomical irregularities in passing, the similarly masculine legal and judicial sources that have been uncovered seem concerned with men acting "as men do with women" or in a passive, womanly role. For example, Carol Lansing argues that the Bologna court was not interested in investigating female "sodomites," but that once faced with a case, it imposed a very heavy fine. The sentence was, however, less harsh than those received by some convicted men.[107]

Insofar as it is possible to map medieval anatomy onto modern, it seems that the form of the clitoris was recognized in the Middle Ages, but its function was not linked to arousal.[108] In contrast, by the end of the sixteenth century, the clitoris, with its homoerotic possibilities, had emerged not only as the anatomical site of female desire but also as a destabilizing element in the cultural and political economy of sex. Although the change is due in part to the recovery of certain ancient texts in the Renaissance, it is also related to specific historical tensions surrounding gender relations in early modern Europe.[109] Similarly, medical and natural philosophical writings in the Middle Ages (with very few exceptions produced by men) occasionally discussed masculine women and hermaphrodites. But until the late sixteenth century the erotic and sexual potential embedded in those categories was only occasionally alluded to and hardly considered.[110] Nevertheless, lurking behind the standard, sexually passive medieval woman is a sexually voracious one for whom sexual pleasure does not require a man.

Recent scholarship has done much to counteract the invisibility of medieval women's social, emotional, and erotic bonds with other women, and has both refined and broadened the approaches and categories that can be brought to bear on a wide range of archival, legal, spiritual, and literary sources.[111] In the same way, research on (mostly male) homosexuality and on sexuality in general now encompasses various dimensions of social, cultural, and religious history not treated here.[112] But although this study is not directly concerned with the history of subjects like the family or canon law, those domains and others constitute a context for the analysis of the *Problemata* literature. For example, though neither literary nor Judaic studies are central to its concerns, the material treated here intersects with vernacular literature when Evrart de Conty makes reference to *The Romance of the Rose* and with the history of Jews, who are implicated in the language of menstrual blood. In some instances, the evidence from the sources treated here differs from that derived from other genres. For example, although amid a great deal of "rubbing" there is only one mention of anal penetration in the texts examined here, the depiction of sodomites in medieval visual representations of the Last Judgment routinely shows them skewered through the anus to the mouth.[113]

Also missing from center stage in this drama of Aristotelian scholarship are the lives and feelings and speech of the men whom these texts purport to explain. Only by recourse to other types of sources, especially judicial, can they be discerned, and then mainly through the lens of official language.[114] But they too form part of the context of this book at specific points. Pietro d'Abano's reference to "what many do today" and George of Trebizond's story about a man named Poggio are exceptional in a sea of formal academic talk.[115] But the arguments and opinions, reactions and textual gestures of authors, copyists, and readers examined here represent an aspect of the world in which real men who derived pleasure from sex with men were living. True, the discourses of medicine and science in the late Middle Ages had nothing like the cultural and social power they wield today, and they were the purview of a small group of elite men. Nevertheless—now congruent with, now divergent from the discourses of other social sites—they still participated in the medieval dynamics of sexuality and gender.

The central subject of this book, then, is what can be learned about the history of male homosexual desires and pleasures from their appearance in medieval natural philosophy, and, conversely, what can be learned about the history of medieval science from its treatment of those desires and pleasures. The project involves the historical environments introduced above: the textual context of Aristotelian *Problemata* and the social context of those whose engagement with the subject matter has come down to us in medieval manuscripts. More concretely, it is about the way in which medieval scholars brought their knowledge of the natural world and the tools of scholastic investigation to bear upon a question that had been posed in Greek antiquity and passed on to them as one of many scientific conundrums.

"Aristotle Asks Two Questions": The Subject of *Problemata* IV.26

What question did medieval readers think was being posed by Aristotle in *Problemata* IV.26? In one sense, the rest of this book is an attempt to find out. The commentaries and the margins of the manuscripts mention sodomites and shame; anuses, penises, thighs, and vaginas; habits and birth defects; bodily fluids and planetary conjunctions; boys, adolescents, and men; eunuchs and women. In keeping with standard scholastic expository practices, which called for dividing an authoritative text into lemmas, or segments, and unpacking them one by one, a reader's first task was to explicate the question itself. The opening words of each section were often underlined or otherwise marked off, and these tag lines served in turn to identify the passages. In the remainder of this chapter and throughout the following two chapters, an English translation of each segment of Bartholomeus's Latin rendition precedes a discussion of the medieval commentaries on it. The underlined Latin tag lines can be used to pinpoint the corresponding portions of commentaries in the manuscripts, editions, and the Appendix of the present volume.[116]

As a general practice, late medieval commentators undertook one or more of a series of increasingly ambitious tasks. They secured the text that was the object of their project, often by comparing two or more manuscripts; they presented a close reading of the work, explaining key or difficult terms and interpreting obscure passages; they sorted out apparent contradictions, responding to possible objections (often referred to as "doubts"); and they interpolated independent passages (often referred to as "digressions") that expanded on a concept or connected the subject matter to related material. The remainder of this chapter and Chapters 1 and 2 of this book mirror those practices, providing a commentary on medieval commentaries on a text that was itself a medieval Latin translation of a Greek text. Just as Bartholomeus de Messina attempted to give his readers a literal translation of the work attributed to Aristotle, so I have attempted to provide an unexpansive rendition of his text. As a result, some passages that appear in English at the heads of the sections below are cryptic, recalling Jean de Jandun's complaint about the Latin. Like Bartholomeus, of course, I cannot extirpate my assumptions or avoid misunderstanding. And just as the medieval expositors, named and anonymous, both built upon the interpretations of others and inserted their independent judgment, so I have been guided but not limited by their understanding in my interpretation of their views. This replication of the scholastic method presents challenges to the modern reader. At the same time, it is the necessary basis for a serious exploration of the broader themes of gender, moral responsibility, and the struggle between science and silence that permeate medieval treatments of *Problemata* IV.26.

> [Propter quid quidam] Why do some people with whom intercourse is had experience enjoyment. And indeed some also acting, some not?[117]

In the case of the *Problemata*, an expositor's first task was to explicate the question with which each chapter opened. The text of Bartholomeus's Latin translation of *problema* IV.26 inquires about the cause of pleasure in a passive role in intercourse, dividing those who experience it into two groups: the ones who also take the active role and the ones who do not. However respectful of authority a scholar might be, the act of explication was also one of interpretation, and when Pietro d'Abano laid out his understanding of the question posed, he reshaped the ancient text and introduced some of the structures and themes that were to govern the rest of his commentary on this question, as well as influence those who came after him. First, he both highlighted and subverted the concepts of "active" and "passive" participants in sex acts. Second, his comments on this first passage hinted at some ambiguities surrounding which acts he and others imagined were associated with the pleasures at the heart of the *problema*.

Pietro observed that Aristotle was in fact asking two questions. To clarify the first, he started by expanding Bartholomeus's cramped opening line, which refers to those "with whom intercourse is had," substituting the less awkward "with whom the act of intercourse is carried out," and adding "such as women and those equivalent to them." The action word, *peragitur*, carries the connotation of thrusting through or piercing, and thus may suggest a distinction between the penetrator and the penetrated, although, as he will later make clear, penetration is not a defining feature of the acts he has in mind. Bartholomeus's Latin word for "some people" (*quidam*) can grammatically be read as referring to men alone or to both men and women, but nothing in the Aristotelian text itself suggests that women are intended. Indeed, in what follows there, men are clearly the subject. Pietro was not alone in his thinking, however. Between the lines of a fourteenth-century Italian copy of the Aristotelian text itself, above the words "those with whom" someone wrote, "namely passively, in the manner of women"; and, above the words "some acting," he wrote, "as males."[118] The introduction of women (and the contrast with real men) served to clarify what it meant to be acted upon, and it is the first of many feminizing references made in the *Problemata* texts and manuscripts. Whereas others, such as the less expansive Prague commentator, limited the subject of the *problema* to men who like to rub and or be rubbed by other men,[119] Pietro's reframing of the question brought homosexual acts into association with heterosexual acts. Jean de Jandun's edition of Pietro's work appears to strengthen that association by rephrasing the comparison in terms of two types of couples: "a man and a woman or those equivalent to them."[120] And in his final restatement of what he has identified as this first of Aristotle's two questions, Pietro expands the field of inquiry even further, claiming that the Philosopher is asking why anyone at all "enjoys intercourse, whether by reason of acting or being acted upon."[121]

This move on Pietro's part had two effects. One, presumably intended, was to set up his explanatory strategy, the first element of which was to establish the universal physiological principles of sexual pleasure that would serve, in turn, as a context for the special case presented in *problema* IV.26. That treatment of men's passive pleasures as a subset of all pleasures plays an important role in his naturalization of the phenomena in question. The other effect of Pietro's extension of the subject to all sexual enjoyment, "whether by reason of acting or being acted upon," was presumably unintended. It opened the door to possible confusions or even evasions. As a later scholar pointed out, just a few chapters earlier the Aristotelian text (and Pietro) had already taken up the question of why intercourse is especially pleasurable.[122] So there seemed to be no call for treating it once again, especially since, in other cases, Pietro made a practice of skipping questions he regarded as repetitious.[123] But for some readers the question had come to be about the pleasures of those "by whom" intercourse is had, even though Bartholomeus's translation had specified it was about those "with whom" it is had.[124] The move away from the specific subject of men passive in coitus with other men either caused or permitted Walter Burley to swamp the subject of IV.26 by importing substantial material from the earlier *problema* that asked why intercourse was the source of the greatest pleasure.[125] In short, by stepping back to ask about sexual pleasure in general, Pietro created the basis for explaining passive men in terms of ordinary natural processes and even for allowing them to disappear.

Related to how broadly or narrowly the subject of the *problema* should be construed was the question of what acts might be involved in producing the pleasures. In his first version of the commentary Pietro added that some people engage in the act "separately" (or perhaps "differently," *seorsum*).[126] He eliminated the phrase in his second version, perhaps because the Aristotelian text said "coitus" and made no reference to masturbation; perhaps because this particular combination of acting and being acted upon had no place in his subsequent explanations; or perhaps for other reasons. Jean, who was reading the first version, understood Pietro to be saying that "some do it in another way, alone, as will be evident near the end of the discussion"—referring to a passage late in the commentary that mentions in passing that "the sodomitical act" can be committed "with the hand."[127] The manuscripts' inclusion and omission of this phrase present a challenge to modern scholars seeking to determine the specific meanings that medieval intellectuals attached to the word "sodomy." In any case, especially after Jean's interventions, all sexual and grammatical options were in play: active, passive, and reflexive; men, women, and anyone comparable to either. Evrart de Conty summarized the inclusiveness developed by his predecessors, saying that the first of Aristotle's two questions was about the pleasure experienced in intercourse, "however it is done."

Insofar as the original *problema* posed two questions, they would seem to have been (1) why some men like to be acted upon; and (2) why some of those men also like to act. But Pietro carved out a different "first" question about the causes of sexual desires and pleasures of any kind and only then turned to the specific pleasures of sexually passive men. Most other commentators accepted (or at least did not reject) this basic division. Nevertheless the distinction that dominated most late medieval *Problemata*-related texts was the one between anatomical and psychological causes. And that dyad in turn prompted heated debates among readers who accepted the challenge of the Aristotelian text about what is "natural." For the task was, by definition, to supply a natural explanation for a natural phenomenon, the possible answers for which "battle in the mind."

Chapter 1

Moved by Nature

Commenting on a chapter about marriage and "venereal matters" in an astrological work, George of Trebizond noted the planetary conditions which could produce a man born to "filthy things [*immunda*]." In conjunction with Venus, the cold and dry influences of Saturn might cause the obstruction of his seminal pores, that is, the anatomical deformities described in the *Problemata*. The result, he explained, would be "certain monstrous things contrary to universal nature" as well as a "particular nature" of a feminine sort.[1] This particular nature, as Trebizond explained, itself has more general natural causes: "It is necessary that, by nature, the semen is driven out toward the genitalia. When it finds no channels and no ways through, it is drawn off to nearby passages."[2] The coexistence of these two senses of "nature" produced paradoxes and tensions surrounding the phenomenon of men who were susceptible to anal sexual stimulation. The project of explaining this set of deviations from the ordinary course of things thus presented a serious challenge. The efforts of Pietro d'Abano and others to meet that challenge to the discipline of natural philosophy is this chapter's central subject. What could it have meant to say, as one reader did in summarizing *Problemata* IV.26, that "some men commit the sodomitical sin [*peccatum*] moved by nature"?[3]

At one level, the two manifestations of nature are at odds with each other. Universal nature, sometimes referred to as "unqualified nature [*natura simpliciter*]" is a force and embodiment of an ordered existence. It constitutes the teleological principle that subsumes the purpose, the final cause, of everything in the physical world. And it is, not incidentally, consonant with medieval Christian notions of divine Creation and Providence. In contrast, a particular nature, sometimes referred to as a "qualified nature [*natura secundum quod*]," may stray from the path defined by universal nature. That is, it may, at the very least, fail to meet the imperatives of teleology, and might even appear to defy them. But the two ways of looking at nature were not regarded as entirely contradictory in the Middle Ages. First, the particular natures of individual objects usually did succeed in realizing the prescribed goals of universal nature, if not to perfection, at least adequately. Second, objects and events that were sometimes referred to as "against nature" were not usually beyond natural explanation. No supernatural intervention was required to account for most deviations from the norm—especially those that occurred with some frequency.[4] The medieval scientific notion of a natural world governed through final causes was not equivalent to the modern notion of a natural world governed through natural laws: individual imperfections and even failures were expected. As portrayed in the literature of the period, such as the French *Romance of the Rose* and Chaucer's English work, the allegorical figure Nature, who rules Creation as God's viceroy, was not always able to keep things—especially animate things—in line. She had particular difficulty regulating desire.[5]

What is more, medieval scholars wished to defend the teleological order against encroachments by the threatening philosophical notions of randomness or "chance," incompatible with both Aristotelian principles and Christian theology. For that reason, they sometimes went so far as to argue that irregularities were meant to happen. In commentaries and questions on Aristotle's *Physics*, the text that defined "nature" for them, they discussed "whether monsters, which are called errors (*peccata*) of nature, are intended by nature."[6] Jean de Jandun and Walter Burley, both figures in the history of the *Problemata*, wrote on that subject. Defending the fundamental axiom that nature always acts toward some higher end, they nevertheless both concluded that, in a weak and contingent sense, nature could be said to "intend" the occurrence of a specific individual with an "inappropriate disposition [*dispositio disconveniens*]." Thus, bending over backward to contain randomness itself within the ambit of a purposeful natural order, Burley declared that "if 'monster' is taken to be the whole aggregate" of individual substance and inappropriate disposition, as he believed it should be, "then I say that qualified nature intends such a monster, and that which it intends occurs by chance."[7] The *Problemata*, free as it was from the dominion of final causes and filled as it was with questions about curiosities, implicitly raised doubts about the sway of nature in its most commanding, universal form.

Such explicit distinctions between "universal nature" and "particular nature," "unqualified nature" and "qualified nature," do not appear in late medieval discussions of *problema* IV.26. Nevertheless, the senses and behaviors of "nature" were at the heart of attempts to understand the men "with whom intercourse is had." Taking his cue from Bartholomeus de Messina, whose translation uses nature-related terms ten times within this *problema*, Pietro d'Abano's commentary contains more than fifty such words. In most instances these are words like "nature," "natural," "naturally," and even "most natural"; in a dozen cases they are words related to birth—*nativitas* (birth) and *natus* (which I have usually translated as "constituted"). Along with anatomical terms ("penis" occurs twenty times and "anus" sixteen), the vocabulary of nature dominates the *Problemata*-related texts that follow Pietro's commentary. Within the sections of the commentaries that discuss the irregular bodies of men with irregular desires, the notion of monstrosity reinforces the theme of nature. It both stands in opposition to nature in its "universal" guise and, at the same time, represents an example of "particular" or "qualified" nature. George of Trebizond held that the natural and regular motions of the heavenly bodies caused "monstrous things contrary to universal nature." This ambiguity is embodied in the concept of a "monstrous nature" that figures in Pietro's exposition written half a century earlier—a nature "against nature."[8]

As the first, the most extensive, and the most influential medieval commentary on the *Problemata*, Pietro d'Abano's work offers the best introduction to how medieval readers understood the answers to the question posed by Book IV, *problema* 26. For that reason, this chapter and the next are built around a detailed account of Pietro's reading of Aristotle's brief discussion of deformities and habits. They constitute an exposition of his exposition. They explain his interpretation of the ancient text as he had received it in Latin, and introduce the basic natural philosophical concepts and sources to which he had recourse. In addition, the explication takes account of views of Walter Burley, Evrart de Conty, and a number of anonymous authors and readers that sometimes varied from Pietro's.

As a close reading of medieval texts, what follows mirrors some of the laborious practices of scholasticism from which Renaissance humanists so proudly liberated modern inquiry. But it was precisely those practices that enabled—indeed, in some sense, required—intellectuals of the late Middle Ages to speak in medical and philosophical terms about men who had sex with men. This chapter pieces together a concrete picture of superfluities pressing to get out, what happens when ejaculation through the penis is impossible, and which men are able to emit what substances. Yet it involves more than a summary of medieval arguments. It reveals the rhetorical practices and explanatory strategies that rendered the condition of men who are passive in coitus familiar and ordinary, conforming to the usual ebb and flow of bodily fluids and responding to the universal psychological and physiological mechanisms of desire. And it suggests that those processes and principles simultaneously created the very standards that some men fail to meet. In particular, it shows how late medieval scholars struggled to fit the pleasures of "those with whom intercourse is had" into the natural world and to subsume them under natural science.

The format of Pietro's work conforms to standard Latin scholarly methods, themselves indebted to the earlier models of biblical exegesis on the one hand and Arabic philosophical commentaries on the other. Conservative insofar as it was closely tied to the words of the text, the method nevertheless afforded scholars ample opportunity to redefine the issues being addressed, impose meanings on the content in question, introduce new material into the discussion, and in general place their personal stamps on a subject. Pietro, intellectually aggressive in all his works, took advantage of each of these avenues to assert and develop his own perspectives. His freedom to do so was, of course, enhanced by the very character of the compilation of unresolved *problemata* with which he was dealing.

Why Those Having Intercourse Enjoy It

In the previous chapter I introduced the first lemma of *Problemata* IV.26, which posed the question itself and the ways that various readers understood it. Having proposed his own view of it, Pietro proceeded to answer the first part of the *problema* as he had construed it—Why is sex a source of enjoyment? Like his reconfiguration of the question itself, his answer has the effect of subsuming the explanation of some men's particular desires under an explanation of the universal nature of sexual pleasures. And thus, in a sense, he makes their peculiar pleasures themselves a special case of a natural process. In order to explain the physiological predicament faced by the men in question, Aristotle had stated certain axioms of anatomy and physiology concerning the ways superfluous materials are drawn off and excreted from the body. At home with the topics of bodily parts and fluids, Pietro took the opportunity to display and deploy his deep familiarity with natural philosophy and theoretical medicine. Like his expansion of the *problema* itself to include active and passive heterosexual pleasures, this incorporation of standard terms and explanations associated with familiar bodily functions serves at once to normalize and to obscure the men whose pleasures gave rise to the *problema* in the first place. Indeed, when Burley summarized Pietro's treatment of IV.26, he allocated four times as much space to the twelve general causes of coital pleasure and to a comparison of men's and women's enjoyment as he devoted to his few comments about "effeminate men." The ratio is reversed in Pietro's own commentary, and Burley no doubt had other reasons to shape the *problema* as he did. Nevertheless, his move was facilitated by Pietro's virtuoso treatment of the anatomy and physiology of desire in the early sections of the exposition.

Indeed, Pietro not only frames the inclination to "be acted upon" within the context of sexual dynamics more generally, he places all sexual desire and pleasure within the context of a more general physiological pattern, namely the buildup and release of bodily substances. He identifies three passages in the Aristotelian text as the response to the first of the two questions posed by the *problema*—why people enjoy sex. In the first (at the lemma "Either because") he stated the axiom that each superfluity produced by the body has an innate and proper place in which it collects and from which (with the involvement of effort and a certain "windiness") it is expelled. In the second (at "Like urine"), he gives the examples of urine, feces, tears, mucus, and blood. And finally (at "Surely similarly"), he makes the connection with the collection of generative seed in the testicles and penis. Neither subjective experiences nor persons, much less relationships between persons, were the subject here for medieval scholars.

> [Aut quia] Either because for each superfluity there is a place in which [the part] is appropriate according to nature that it is separated out. And the effort having been made, the departing windiness makes [it] inflate and separate out through it.
>
> [Ut urina] Like urine into the bladder, the superfluity of food into the intestines, and tears into the eye, mucus into the nose, and blood into the veins too.[9]

In explaining what he construes as the cause of pleasure, Pietro follows closely the mechanisms itemized in Aristotle's text, adding physiological elaboration. For example, he reminds his readers that food undergoes several stages of refinement ("digestion"), for each of which there is a supporting receptacle by which waste products ("superfluities") are drawn off, contained, and eventually expelled. Aristotle had said here that each excess was segregated "according to nature" and later that its receiving organ was "innately suited" to it; Pietro added that the attraction and excretion occur "naturally," adumbrating his concern with the "nature" of the men he will later be inquiring about. The dynamics of these processes commanded Pietro's attention and he explains that the effort of which the text speaks is closely associated with motion, both of which help a windiness (a subtle, spiritous substance) to effect the collection, thickening, and expulsion of each substance. It is this propulsive windiness that causes the discharge of superfluities to be accompanied by swelling, he explains, citing the authority of Avicenna.[10] Pietro adds both detail and technical anatomical terminology to Aristotle's list of examples of superfluities and their receptacles: mucus is generated in specific areas of the brain and evacuated into the nostrils; blood is evacuated into the veins, and so forth. Scholastics used enumerations of this sort to display intellectual virtuosity, but this particular list also served to domesticate the subject matter. What could be more familiar, more ordinary than urination or sneezing?

Pietro's appeal to specific doctrines and vocabulary of Galenic medicine may have had a similar effect. Even scholars with no specialized medical knowledge were familiar with the theory of natural faculties—active powers inherent in organs that attract, retain, and expel the substances appropriate to them. Less familiar to Pietro's northern readers in the university arts faculties, but nonetheless legitimizing, was his appeal to Arabic medical expertise.[11] Taken together, the science and the rhetoric of Pietro's treatment provided a learned and dignified introduction to *Problemata* IV.26. Standard physiological knowledge conveyed in neutral philosophical and medical terms placed the men who were its subject within an unexceptional and even respectable environment.

The same tone of scholastic sobriety is sustained in the excursus on blood that follows. Clearly demarcated by phrases like "It should be noted that" and "Someone might doubt," digressions of this kind were an important dimension of the commentary form, often intended to respond to objections that might occur to listeners or readers. One might ask, says Pietro, why blood, a vital fluid, is mentioned in the same breath with the disposal of waste products like urine? Not all readers were bothered by this apparent problem, as is attested by the Erfurt commentator's addition of milk—another useful fluid—to the list of superfluities.[12] And the issue was, after all, tangential: blood was just one in a list of examples, ranging from feces to tears, of substances that collect in the body. But the digression is significant on two levels. First, precisely because it was standard scholastic practice, it formed part of the apparatus of normalization that characterizes the early sections of Pietro's commentary. Its expected, routine character is probably the reason that the Aachen and Bavarian commentators included it in their short summaries of Pietro's work. Second, the passage introduces for the first time the ambiguities of "nature" that will permeate the rest of Pietro's analysis. And, for some readers at least, the use of a superlative even evoked the notion that there was a continuum of naturalness, from less to more. Blood, they were given to understand, with its natural place in the veins, was "the most natural [*naturalissimus*] of humors."[13] Yet Aristotle appeared to be treating it as a waste product that needs to be expelled. Pietro's solution is to distinguish the positive sense of "blood" from "blood" that is "mixed with many superfluities of other humors and is not natural."[14] This explanation makes clear, however, that the unnatural blood itself is a regular product of the (natural) processes of digestion. The digression thus foreshadows the transition between the unexceptional physiology of superfluities to the less natural natures at the heart of the *problema*. It is here too that Pietro first introduced the word "anus," as one of the sites at which the less than fully natural "melancholic" blood is manifested.

> [Similiter utique] Surely similarly to these, and the generative material in the testicles and penis.[15]

Before confronting directly the subject of anal sexual stimulation, however, Pietro added an element to his elaboration of the normal workings of the body. It involved what commentators on Aristotle's *Physics* called "universal nature." The text of the *Problemata* simply incorporated semen into the list of superfluities routinely produced and the testicles and penis into the list of appropriate receptacles. Pietro explicates, distinguishing between the testicles, which are intermediaries that contribute to the forcefulness of the seed's exit, and the penis, though which it is emitted. But he also takes the opportunity to introduce and highlight the subject of pleasure. The text he is commenting on subsumes the buildup of semen under the general movement of superfluities; his explication similarly subsumes the pleasure of sexual release under the general experience of expelling them. "Thus, therefore," he explains, "the cause of pleasure and joy is due to expelling what is superfluous. And it is expelled at the place in which it is naturally constituted to be expelled, so that, in this, the harmony of our constitution is preserved."[16] The effect of this expansion is to add a layer of normalization to the sexual pleasure. Indeed, the notion that the evacuation of seminal material (like that of all superfluities) contributes to "harmony"—to physiological equilibrium—acknowledges one of the final causes of the sex act, namely, health. At the same time, Pietro's phrasing, which includes the words *naturaliter* and *natus*, recalls his earlier references to appropriate receptacles and hence prepares the contrast with irregular pleasures at inappropriate sites.

Pietro places pleasure in the foreground even more emphatically in what follows, another formally introduced excursus. This one concerns why "there is the greatest pleasure in coitus compared to other evacuations."[17] The digression implicitly harkens back to his treatment of an earlier *problema*, "Why is it most pleasurable to have intercourse?" (IV.15). There Pietro had included answers that linked sex to various higher causes. At IV.26, he starts by outlining some of the more proximate, physical reasons the pleasure associated with intercourse is so great. The animal spirit (which is an essential component of semen) and the matter associated with it pass though the most sensitive parts of the body, stimulating them. Most of the digression, however, proceeds in a more teleological (and conventional) mode. In contrast to what he has just said about the role of expelling seed in preserving the harmony of one's constitution, he asserts here that this evacuation (unlike others) serves not to preserve the individual but to perpetuate the species. Furthermore, the pleasure associated with it is especially important, because animals might otherwise shun the act of intercourse out of disgust. With a quotation from Aristotle's *Ethics* to the effect that pleasure results from the attainment of something good, Pietro further underscores the orderly arrangement of things that is implicit in sexual pleasure.

Since neither the *Problemata* nor Pietro's commentary on it prominently features final causes, their introduction here, marked off by a formal digression is significant. Its meaning is, however, ambiguous. Since the digression stands at the transition from the consideration of sexual pleasure in general to the explanation of peculiar sexual pleasures, it points to an implicit divide between purposeful, universal nature on the one hand and contingent, individual natures on the other. Yet the words "nature," "natural," and so forth, so visible in the previous passages of Pietro's exposition, do not appear in this one. Omitted in a group of manuscripts that tended to streamline his text, this elaboration on the causes of pleasure was ignored by some of the authors who followed Pietro most closely.[18] Although they differ in other respects, those works have in common that none emphasizes the effects of intercourse beyond the individual and thus none entertains the question of how that pleasure relates to some higher good. For other scholars, however, the matter was crucial. In Evrart de Conty's commentary procreation is always in the foreground and reinstates the notion of the "arrangement of nature" that Pietro had dropped.[19]

Far more than Evrart's arguments and rhetoric, Pietro's methodical treatment of ordinary natural processes, from the buildup of superfluities to the experience of pleasure, gave his readers at least two ways of measuring passive men against the normal patterns of nature. But the two perspectives were not entirely commensurable. On the one hand, with the early analogy to women and the later insistence on reproduction, he had set up a natural world to which such men could not properly belong. On the other hand, with his broad view of accumulating and expelling superfluities and its role in the "harmony" of the individual, he had set up a world in which they could participate.

Such ambiguities persist throughout Pietro's exposition, sometimes manifested as balance or restraint, sometimes as vacillation and confusion. Yet his treatment of this *problema* and others commanded the respect of many later readers, in part due to his mastery of standard scholastic methods and unimpeachable authoritative sources. In the sections of his text discussed up to this point, Pietro has established a pattern of citations; he has made the first of five references to Aristotle's *Nicomachean Ethics* on the interplay of the natural and the good; he has cited Aristotle and Avicenna on the mechanisms by which superfluities are expelled; and he has called upon Galen to support the commonplace that there is more pleasure emitting than in receiving. The invocation of so many authorities has not only lent weight to his specific assertions and broadened the base of his discourse but also allowed him to maintain a sort of rhetorical equilibrium in the face of so volatile a subject. As he addresses the specific matter at hand, Pietro will make clear that homosexual desires, acts, and pleasures are to be scorned, but also that they are to be explained by the concepts and methods that he has illustrated in these opening sections. The subject to which he turns involves breaches in the order of nature in its broadest and most exalted sense, at the same time that it involves the natural processes that cause the regular occurrence of such breaches.

So far, Pietro has been dealing with what he had designated as the first of Aristotle's two questions; now he arrives at what he construes as the second—the reasons for the pleasure experienced by "those with whom intercourse is had," in the words of Bartholomeus's Latin translation. Thus, if Pietro has opened his explication of this *problema* in a way that postpones this central question and normalizes its problematic subjects, he does not, in the end, shirk his duty as a commentator. More than three-quarters of his exposition of IV.26 is devoted to men whose innate bodily conditions and ingrained habits instill in them capacities for enjoyment that, if they do conform to the basic outlines of generalized sexual pleasure, nevertheless deviate significantly from the norm in their particulars. These proportions roughly mirror those of the classical text, but following the original ratios was far from compulsory: Walter Burley reduced the desires of "effeminate men" almost to invisibility; the Prague commentator scarcely spoke of anything else, and others chose options somewhere in between.

The structure of Pietro's remaining exposition is complicated, due in part to the rambling nature of the original *problema* itself and in part to his attempt to pursue each of the causal explanations suggested by it. First of all, unlike many of the phenomena addressed in the *Problemata*, the suggested answers to this one are not phrased in a manner that lays out alternative explanations: "Is it because of this or is it because of that?" Rather, in this case, the ancient text starts out as though there is only one line of speculation (anatomy), with the second possibility (habit) raised only toward the end. In addition, the question posed by Aristotle is about two distinct groups: those who enjoy the passive role and those who enjoy both the active and the passive.

Partly for these reasons, the taxonomy of the answers that commentators and readers supplied to *problema* IV.26 was variable and often complex, if not confused. Insofar as the "first" question was understood to be about the cause of sexual pleasure in general, the answer was fairly straightforward. It results from the release of a physiological superfluity. But, both as Aristotle had stated it and as medieval readers construed it, the "second" question—about the reasons some men experienced sexual pleasures in irregular ways—involved a cascade of subquestions. Was there a place in the taxonomy of this disorder for always active men, not mentioned in the Aristotelian text itself? In any case, there were two other types of men to be accounted for: the always-passive and the sometimes-passive–sometimes-active. In addition, the two types of causes proposed in the original text—innate anatomical flaws and habits acquired in youth—might but did not necessarily match those two groups of men with irregular pleasures. And upon further examination at least some medieval scholars discerned branches within each of the causal pathways themselves. Of those men with anatomical defects, some had blockages in their seminal passageways that were complete, others that were partial. And among those in whom experiences had developed into a bad habit, some readers made distinctions according to the age at which the experiences occurred. As a result, the organizing principles of the *Problemata*-related texts are far from clear. The variables overlap and intertwine.

Pietro's extended exposition is the most attentive to the structure of the *problema* but does not entirely eliminate its ambiguities. He announces the organization, the *divisio*, of Aristotle's question as he understands it.[20] He sees the "second" question, which inquires about the sources of irregular pleasures, as having two parts—one about men in whom the vessels to the penis are completely obstructed and the other about those in whom they are somewhat open at the anus. He goes on to say that the answer to this double question is itself divided into two parts: innate causes and habits. It might be possible to envision two groups, one with complete obstruction, formed at birth, and always passive, the other with partial obstruction, shaped by habit, and only sometimes passive, but the opinions of the commentators do not map out in that way.

The choices readers made about how to divide and organize the text reflected their interests and positions. The Erfurt *Problemata* mentions birth and habit but does not distinguish between the always passive men and the sometimes passive men. Walter Burley places his emphasis on what Pietro had treated as a general introduction. And for Evrart de Conty, the main division of the text is the line between what is in conformity with nature and what is not, in which context the distinction between two causes does not serve his purposes. The details of organization will become substantively relevant at several points. Here the significance of authors' attempts to discern the argument of the *problema* is that, like the uncertainties about what question Aristotle was asking, these struggles with the details of organization contribute to the picture of earnest scholars attempting to understand a phenomenon of nature not easily explained—a *problema*. They carried on, lemma by lemma, constructing a natural philosophy of distorted pleasures. The remainder of this chapter explores their explanations of innate anatomical and physiological causes; the following chapter continues at Pietro's (somewhat misleading) transition to psychological causes, treating the problem of those who are sometimes passive as well as the processes and consequences of habit formation.

Those Whose Pores Are Not Naturally Constituted

Blinded by Nature

> [Quibus utique] Surely in these people the pores are not constituted according to nature. And if because those that are in the penis are blocked, as occurs in eunuchs and effeminates, or alternatively the humidity flows into the anus and goes out in that way.[21]

The analogy with the ordinary evacuation of various superfluities established the general type of bodily function to which the emission of men's semen belongs. The Aristotelian text then moved to suggest possible answers to the core question: Why do some men enjoy being acted upon and some of those enjoy acting as well? If there had been any ambiguity about the inclusion of women in the central subject matter of the *problema*, it is eliminated here by the explicit discussion of the male anatomy. In general what happens, according to Aristotle, is that the pores—the hollow spaces and passages for the sperm—are disposed in a manner that is "not according to nature." But the irregularity can take two different forms. Either a man's vessels simply come to a dead end at the base of the penis or else some moisture can collect and be evacuated through the anus. According to this taxonomy, in one group of men—those associated with eunuchs and effeminates—there is no outlet for the superfluities in question; in the other group, a way out exists, but at the wrong anatomical site.

No such distinctions disturbed the views of some commentators. The very brief Erfurt text takes the question to be entirely about "effeminate men such as eunuchs" and says that in them "moisture is expelled at the muscles and veins of the anus."[22] That is, it conflates the two types. Even Evrart's more elaborate treatment discusses simply "effeminate and imperfect" men in whom the proper pathways are blocked. For Evart, therefore, the matter that needs to be purged is violently restrained, and these men seek to expel it another way. Thus, for many (perhaps most) readers, subtle somatic differences had no significance, and the ascription of effeminacy applied to all the men discussed in *problema* IV.26. For Pietro, however, the distinctions were not only present in the ancient text at this point but also necessary in order to sort out what followed. They had implications for the ascription of femininity and for his subsequent treatment of the mechanics of sexual desire, pleasure, and frustration.

The central feature of both forms of the innate condition is that the proper route for the seminal fluid is cut off. The reason is that some material gets in the way, a situation comparable to blindness. The Greek and Latin texts had used terms (ἀποτυφλωθῆναι and *excecantur*) that carry both a primary sense of blindness and a more general sense of obstruction. In the mind of some medieval scribes, the broader sense prevailed, and they read "cut off" (*exsecantur*) rather than "blinded."[23] And, as the author of the Erfurt commentary explained, "When those places are cut off [*preciduntur*] or closed, such [substances] will be expelled to the nearby empty spaces by the force of nature."[24] One reader even offered the gloss "Surgeons cut," suggesting that the association with eunuchs was foremost in his mind.[25] But Pietro chose to capitalize on and explicate the root meaning, and in so doing he gave greater specificity to the blockage: it is like blindness caused by a thick humor obstructing the optic nerve.

Blindness figured in medieval discussions of moral responsibility, but at this point Pietro's particular account of it has another effect: it weakens the hold of anatomy on those whose irregularities are innate. Whereas Aristotle appears to have been speaking of simple structural defects, Pietro's elaboration of the analogy with blindness suggests a physiological component: a viscous substance. There is also a physiological dimension to his reading of "eunuchs and effeminates." As he had explained elsewhere, the term "eunuch" can apply to the consequences not only of artificial castration but also of comparable natural conditions, only some of them related to the configuration of the genitalia: "A eunuch is an imperfect man, womanly, soft, unindustrious, of bad mores, unable or barely able to engage in coitus."[26] And, again going beyond the simple anatomical phrasing of the original text of *problema* IV.26, he compares the men with complete blockages to "eunuchs who lack testicles or who have small and weak ones."[27] (At least two fourteenth-century Italian readers of the *Problemata*, however, had no inclination to interpret the term so vaguely and offered "deformed [*occasionatus*]" as a gloss on "eunuch.")[28] Unpacking the "feminine" embedded in "effeminates," Pietro developed the physiological dimension in a similar direction: such men have constitutions so cool and moist that their natures and indeed their behaviors are those of women. This softening of the anatomical explanation affected the ways he and others understood the processes of habit formation and, more particularly, the possibility that a habit might effect a physical change. For Pietro, the comparison to eunuchs and women apparently did not apply to those men in whom the seminal passageway is "blinded" to the penis but in whom it remains "somewhat open toward the base of the penis and the anus." The manuscripts of his commentary exhibit a little confusion about the specifics of the blockage, but agree on its effect: "In such men, indeed, the spermatic 'moisture runs into the anus' and is dispersed from there. Thus moved and rendered subtle through rubbing together [*confricationem*], it eventually comes forth to the outside."[29]

In this section, then, Pietro has introduced some of the men who will be the subject of the rest of the *problema*, and has described two possible underlying anatomical configurations—complete and incomplete blockage—adding a physiological dimension. He has not fully elaborated on their divergences from the norm; he has not begun to address the distinction between those who are simply passive and those who both act and are acted upon; and he has not raised the subject of habits. Yet he has opened up several major themes of his commentary on *problema* IV.26. First, and most concretely, he has established the anatomical and physiological parameters of his analysis, including deflected seminal fluid as the basis for anal sensitivity. In this connection he has mentioned a behavioral dimension of the condition, rubbing, which will make further appearances in his and others' commentaries. Second, he has set in motion a cluster of associations with eunuchs, effeminates, and women. Finally, he has introduced what will be a continuing tension between, on the one hand, the characterization of the men as "not natural" (here in their bodies; later in their desires) and, on the other hand, the application of natural philosophy and medical theory to the project of explaining their natural causes.

The apparent paradox of undertaking a natural explanation of phenomena deemed unnatural produced visible unease among readers of the *Problemata*. Things occur in nature and by the operation of natural processes that nevertheless breach the order and purpose of nature, deviating from its normal course. The challenge resided not only in the ambiguity of the term "natural" but also and more profoundly in the generally regular but not strictly compliant behavior of the physical world itself. The fourteenth-century scribe who made a copy of Bartholomeus's Latin text reflected this confusion. Rather than "not according to nature" (*non secundum naturam*), he characterized these men's pores as "truly according to nature" (*vero secundum naturam*).[30] At one level, this reading is simply an error, permitted by medieval practices of letter formation and abbreviation. The substitution made sense in context, however, since the questions posed in Book IV and throughout most of the *Problemata* were all about natural occurrences. And *problema* 26 in particular was about something that, although "not according to nature" was indeed "truly according to nature." For that reason, the person who, not long after the production of the manuscript, went through it correcting textual errors let this one stand.

"The Opposite of What Should Happen in Natural Intercourse"

> [Signum autem] And [there is] a sign: the drawing together in coitus of that place and the loosening of those that are around the anus. If, therefore, someone has too much intercourse, in these it accumulates in that place. Wherefore when this desire may well up he desires rubbing where it is collected.[31]

In Pietro's schema, the two groups of men described under the previous heading have in common some sort of blockage of their seminal pathways. No doubt because he himself is committed to a complex taxonomy and because the text on which he is commenting provides no markers that would help the reader see the order that he himself discerns, he takes special pains at this point to delineate his approach to the crucial question of causation. He reiterates the division between anatomy and habit, and points to the sections of Bartholomeus's text that (in his view) pertain to the two types of malformed vessels. In one, the natural superfluity has no release; in the other it has a displaced outlet at the base of the penis and around the anus. Both require an explanation.[32] But for the time being Pietro alludes to but does not pursue the origins of what he construes as birth defects, referring the reader to a passage in Aristotle's *Generation of Animals* about reproduction gone awry.[33] Nor does he introduce astrological causes. Both of these lines of analysis will come into play later. Rather, he here embarks on a detailed exploration of the mechanisms by which fluids and pressures, emissions and accumulations produce sexual desires and pleasures, in both their acceptable and their questionable manifestations.

Having just introduced the idea that the deflection of seminal moisture can lead to an anal emission, Aristotle offered evidence that such a deflection of fluids is plausible. In particular, he called attention to one "sign" of a connection between the area around the anus and the usual pathway of semen. This relationship is, as one commentator put it, "experientially confirmed."[34] According to Aristotle, when people (implicitly men) have intercourse (implicitly active), their buttocks contract and then weaken.[35] For that reason, a person who engages in excessive sex will want to be rubbed there when he feels desire. Perhaps his original point was that (as everyone was presumed to know) the buttocks are always involved and thus, if they lose the power to impel the semen to the penis, they themselves may become a site of accumulation and desire. Led by Pietro, however, the commentators took the text to set up a contrast between proper and improper intercourse and featured the juxtaposition of natural and unnatural natures. His version was inevitably complex, and not all medieval (or modern) readers were likely to be interested in the details. In the margin of one manuscript, someone has construed the passage in simpler terms: "Note that someone seeks to be subjected in the anus."[36]

For Pietro, the evidence or "sign" cited by Aristotle is not just an illustration of the way in which the buttocks and anal area are connected in all forms of coitus but rather a point of clear contrast between properly and improperly constituted men. As he noticed, the contraction of the buttocks was an aspect of an earlier *problema* (IV.2), which asks why the eyes as well as the buttocks are adversely affected by excessive sexual activity even though the eyes are far from the action. In his commentary on that earlier question, he had pointed out that all the muscles and vessels "around the anus" (a phrase common to IV.2 and 26), as well as the testicles, are exerted in and affected by intercourse. (He does not include the penis.) As evidence, he had cited the observation that "the sperm is not emitted unless the aforesaid parts located around the anus are fully helping together."[37] His commentary at IV.2 thus painted a picture of crisscrossing vessels and closely collaborating parts in any kind of intercourse. The men described in IV.26 might be taken as a special case of this general anatomical and physiological pattern, but Pietro chose to emphasize difference rather than similarity. As he explains, whereas in normal men the contraction of the buttocks in intercourse, "as though squeezed by a hand,"[38] pushes the semen out through the open penis, in this group of men, the constriction and narrowing occurs not in the buttocks but around the upper area of the penis. At the same time, there is an opening of parts not around the penis but around the anus, so the fluid is forced out in the wrong place. "Thus what properly happens [to them] is the opposite of what is supposed to occur in natural intercourse."[39] The Bavarian commentator, after quoting that last sentence of Pietro's, summarized the consequence. In natural coitus, "the spermatic material would be pressed out from the anus into the penis, while here [in unnatural coitus] it would be pressed out from the penis into the anus."[40]

If the physical arrangement of expelling this superfluity is contrary to the dictates of nature in the men described in IV.26, what they experience is proper for them, at least in the weak sense that their particular inborn traits constitute their nature. Pietro may thus be suggesting that their sexual responses are suitable, since they "occur properly [*recte*]" in a way contrary to natural intercourse. An alternative interpretation of this phrase is possible: he may be saying that these responses are "exactly [*recte*] opposite" to natural ones. But the more paradoxical reading is confirmed by related usage later,[41] and would echo what Pietro has announced as the basic puzzle addressed in this passage, namely, the existence of men whose spermatic moisture is not emitted naturally (*naturaliter*), because of the way they are from birth (*a nativitate*). The double sense of "nature" surfaces even more clearly when he paraphrases the rest of the Aristotelian passage. If a man of this sort desires intercourse, he "wants to expel the spermatic material by rubbing and strong motion at the inborn place at which it is collected in accordance with his monstrous nature."[42] The anatomical locus of desire is *nata*, that is, "inborn" or "shaped by nature." As he had done early in the commentary, Pietro stressed the link between *nata* and *natura* by stating unequivocally that this desire does indeed correspond to the nature of a certain group of individuals. The specific phrase "monstrous nature," however, leaves no daylight between the empirical assertion that nature produces such men and the metaphysical conviction that such productions violate the order and purpose of nature in the larger sense. Later in the commentary, Pietro repeats and expounds upon the notion of monstrosity, giving it a more technical meaning in the context of natural philosophy. Here it resonates rather as a dogmatic and perhaps even moral declaration.

One consequence of the connection between penis and buttocks, as Aristotle explained in IV.26, is that in those who engage in excessive intercourse (with ejaculation through the penis), some substance collects in the rear area and, when subsequent arousal occurs, they desire to be rubbed there. But, where the ancient text reads, "If anyone has too much coitus," Pietro moved gradually away from an emphasis on the act itself. In his first version he had added overabundant semen as a cause: "If someone has too much coitus and its material." In his second version the semen comes to the fore: "too much material of coitus or perhaps [its] use.[43] Pietro's commentary on this passage makes no further allusion to behavior-induced appetites, thus keeping the focus on irregularities at birth. But although he bypasses this opportunity to exploit medieval associations among various forms of sexual debauchery, including uxoriousness, homosexual appetites, and effeminacy, sexual excess will soon enter his depiction of those with innate defects. And it will resurface in some of his later assertions about the effects of habit.

Desire Arises from Food and the Mind

> [Desiderium] And desire comes about both from foods and from the intellect. For when it will be moved by whatever thing, windiness accumulates and such superfluity flows where it is appropriate, whether it be subtle or windy.[44]

When he arrives at the next topic enunciated by Aristotle—the causes of desire—Pietro repeats the pattern he established in connection with the movement of superfluities. That is, he juxtaposes familiar and regular mechanisms (supported by references to authoritative texts) with the particular and irregular mechanisms that apply in the case of *problema* IV.26. In this way, he sustains the antiphony between the two senses of "natural." Aristotle's text provides the occasion by identifying food and the mind as the general causes of desire in relation to the flow of physiological substances and the collection of excess material. Pietro takes the opportunity not only to enumerate the dietary stimuli and explicate the psychological processes but also to elaborate at great length on abstract Aristotelian theories of motion in animate creatures. That is, he explores some of the underlying principles related to the ways that appetite can lead to action. Such a display of learning was entirely uncalled for. Thus, although Pietro included it in both the first and the second version of his commentary, the manuscripts of Jean de Jandun's redaction omit this digression altogether, as do the commentaries dependent on it.[45] Aside from enhancing Pietro's reputation as the consummate scholar, the extravagant excursus reinforces the ambiguous status of those born with inappropriate parts. Their bodies are, in large measure, explained by the ordinary processes of nature and ordinary principles of natural philosophy; at the same time, they are contrasted with ordinary bodies.

How, then, does sexual desire work? According to Aristotle, whether as a result of specific foods or of the mind a certain windiness arises and the resulting superfluity (whether subtle or windy) flows to the place appropriately constituted. The "windiness" here is a standard element in ancient and medieval medicine that ranges in its associations from flatulence to spiritousness. One of its functions is to convey motion and it thus plays a role in driving many physiological processes. It was particularly implicated in male sexual functions, notably erection and ejaculation. In the present context, it serves to connect the experiences of men with little or no fluid emission to the ordinary rhythms of accumulating and releasing superfluities. Thus, once again, the classical text invites the medieval reader to ponder the relationship between the usual and the unusual in the men of *problema* IV.26.

In the preceding section on the collection of semen around the fundament, Pietro had resisted the implication that the text was speaking of men in general; in this section, he accedes to it. He treats the lemma as a digression on desire, akin to the digression on superfluities and their receptacles. In this case, the subject is the sequence of physiological events, which Pietro takes as an opportunity to elaborate on the psychology of appetite and the physics of motion in animate beings. These he applies to sex and to the specific desires and emissions of men with partially blocked vessels. Although he does not fail to allude to their peculiar "nature," his commentary on this passage places them in compliance with the fundamental principles of Aristotle's natural philosophy. Summarizing the *Problemata*'s assertion that desire can arise from food or the mind or (he adds) from both, he quickly glosses the commonplace that arousal is related to diet, mentioning, for example, the effects of specific stimulants such as turnips and new wine. These have the effect of "inflating" (or, as a couple of manuscripts would have it, "inflaming"), presumably by contributing to the windiness and spiritousness that cause erections.[46] Other commentators exhibit more interest in the role of foods. The Aachen text adds beans to the list of those that produce windiness. And Evrart de Conty, whose other major work was a long and elaborate commentary on an allegory of love, used this occasion to remind his French audience that the classical poet Ovid had warned those who wanted to avoid "foolish love" to avoid certain foods.[47]

Pietro, however, discerned nothing in the subject of diet that would contribute to an understanding of any particular kind of sexual arousal. So he soon turned to the term "intellect," a key element in the Aristotelian account of how appetite, which arises in the soul, produces corporeal actions. In medieval scholastic terminology, *intellectus* often referred to the highest mental faculty, itself associated with the defining human characteristic, reason. But, given the subject matter and level of explanations entertained by the *Problemata*, Pietro equates it rather with certain higher faculties of what Aristotle called the "sensitive soul," shared by the more perfect species, including dogs, horses, and humans. He quotes specifically from Aristotle's treatises *On the Soul* and *On the Motion of Animals*, and refers the reader to Avicenna's commentary on the former—a formidable density of respected authorities.[48] These same sources undergird his later explanation of habit formation, but here his interest is in the general causes that set animals in motion. A beetle can, of course, be washed away in a flood, and a cat can be tossed out the door by its impatient owner. But an essential aspect of beetles and cats is that they are capable of moving themselves, and that the motions they initiate are purposeful. Although there were many ways to understand the internal forces that motivate animals—in terms of reason, imagination, choice, will, and desire—Pietro followed Aristotle in asserting that they all boil down to appetite and intellect, which operate in concert.[49] By "appetite" he means something more general than hunger and thirst; it is an object-oriented quest or desire. By "intellect" he means what he elsewhere calls "practical intelligence," the mental functions that perceive and remember; distinguish past, present, and future; evoke images of objects and experiences; and make the rudimentary judgments and decisions that direct actions. He pictures a chain of behavioral causation, also borrowed from Aristotelian theories of the soul, according to which sensation triggers imagination (located in the practical intellect), imagination (or "fantasy") triggers appetite, appetite triggers passions, and passions move the body.[50] The "passions," which are emotions embedded in physiological states, thus mediate between the psychological and the physical.[51]

Pietro continues to sustain his cool philosophical distance from particular behaviors in this section, but he is delivering a causal sequence that will become more volatile when he arrives at the subject of habit formation. He goes on to address the external causes of animals' movements, that is, the objects of desire at which intellect and appetite together take aim. What makes something desirable? What incites the intellect or appetite? What draws a creature toward it? His starting point is "the good." The passage in Aristotle's *Motion of Animals* to which he refers asserts that appetite always has a specific target which is a specific good.[52] Using vocabulary that also evokes Aristotelian physics and metaphysics,[53] Pietro explains that the ultimate source of motion is the eternal and divine Good, itself immovable. Needless to say, the notion that the ultimate Good was the source and goal of animate motion was adaptable to Christian theology: the creature desires its Creator. But at a more worldly level (the level of the questions posed in the *Problemata*), the external sources of motion, that is, the objects of animals' appetite and practical intellect, are mundane derivatives of that higher good. These are more local or contingent goods, suited to particular circumstances, for a creature never flees what it apprehends as beneficial nor seeks what it apprehends as harmful. In the context of *problema* IV.26, the object of desire is pleasure itself (not another person), which is often an appropriate goal for the calculations of the practical intellect, as in the case of the desire for food, which is mediated by the pleasures of taste. Students and domesticators of Aristotle's psychology and ethics knew that, especially without the help of higher faculties, animals (and negligent humans) might mistakenly pursue an apparent good rather than a real one. And as they move through the text of the *problema* that involves men desiring anal stimulation, the commentators increasingly confront the moral dimensions of desire, the subject of Chapter 4. But although he alludes to "apparent good," Pietro does not elaborate on the distinction. He limits his discussion here to explicating general mechanisms. Indeed, it may well have been its generality that led Jean de Jandun to eliminate the whole discussion of psychological processes. Hence Pietro's versions continue to present the men who desire to be acted upon within the framework of normal forms and functions, whereas Jean's, by omitting the digression, continues to stick to the point, declining to veil those men in abstract and theoretical drapery.

Jean did follow Pietro where his analysis of this passage introduced concrete details about the effects of defective anatomy, though Evrart regarded such information as irrelevant and other commentators were content just to mention food and intellect. When desire is triggered by one or both of those usual causes, motion and "rubbing together" excite the windy spirit.[54] Both rubbing and windy spirit are standard features of sexual arousal discussed throughout Book IV of the *Problemata*. And, once again, the substance induced is headed for the place provided by nature. But, of course, something is amiss: the spermatic material collects around the base of the penis and from there flows out into the "defective [*orbatum*] place intended for it according to this nature" or, as some readers received the text, "the place intended for it according to this defective nature."[55] In other words, medieval scholars could at least imagine locating the innate flaw either in a specific part of the body or in the individual's essence. Given their apparent lack of enthusiasm for the philosophical niceties that Pietro took such care to lay out, the distinction would probably not have caught their attention. Whether it modified an anatomical site or more general characteristic, the word "*orbatus*" would have been evocative for them, since it appeared in Aristotelian discussions of females, where it also has the sense of "deprived," "imperfect."

Blocked Pores, Limited Emissions, and Excessive Desires

A Subtle or Windy Material

> [Hoc exeunte] This going out, something like swellings in children and in these people who [are] of age sometimes relax with nothing moist going out, sometimes when the moisture is extinguished. And if he should experience neither of these, he has desire until one of them occurs.[56]

Pietro has continued to sustain the balance between subsuming the defective persons under the sheltering principles of orderly nature and banishing them from its philosophical protection. As he brought his exposition of the passage on desire to a close, he temporarily restored their compliance with the normal accumulation and expulsion of superfluities established at the beginning of the *problema*. The gathering spirit joins and propels spermatic moisture to emission—albeit in the wrong place. But this process posed a question that may have arisen as much from reflecting upon men's sexual experiences (or at least lore about them) as upon the logic of natural philosophy: What exactly is the result of anal stimulation? For those whose blockage is not complete, Pietro provided a range of possibilities that depend on just how constricted the individual's vessels are: "And what comes out would be either a subtle watery moisture, as in those men whose passages may be broader, or it would be just windy [moisture] that comes out rarified by rubbing in those men whose passages are extremely narrow."[57] That is, a seminal (or semen-like) substance that exists internally in all men may become more or less attenuated or evaporated by the friction, depending upon how much of it can reach the outside. Air being "moist" in the medieval system of qualities, the material called "windy moisture" would not necessarily be wet like a liquid.

With the sources of desire firmly established at both physical and psychological levels and with the pressure of superfluities illustrated in the ordinary mechanics of intercourse, Pietro turned his attention to the immediate purpose of "coitus" in a sense broad enough to incorporate *problema* IV.26. First continuing his examination of those with some outlet at the anus, then proceeding to those with none, Pietro extended his exploration of the moist and spiritous materials he had just introduced and of the ways in which they do or do not escape from the body. In his efforts to make sense of the substance excreted, how it gets out, and the feelings that result, he develops a string of analogies. Men and women, boys and girls, the sick and the insane are all implicated in his account of the pursuit of sexual satisfaction. And in the course of these elaborations, the causes and manifestations of imperfect sexual organs, impulses, and emissions multiply. One result is to blur the distinctions between those with partially and wholly blocked passages, suggesting more of a continuum. A second result is to supplement the circumscribed anatomical, physiological, and psychological contexts already established with a context populated by people—some normal, some pathological. Other commentators appear to have been less interested in such particulars, distinctions, and comparisons. Evrart de Conty and Walter Burley ignored them entirely, and the authors of the Aachen and Prague texts gave them short shrift, but some readers exhibited their interest in one or more points.

The text of Bartholomeus, which does not deal with degrees of blockage, says that the men in question experience certain swellings, as do boys and grown men. Such people can be satisfied with the most meager excretions, "with nothing moist going out." Pietro first builds on Aristotle's comparison to boys. He glosses their "swellings" as "expansions." For the Erfurt and Bavarian commentators these are simply erections, but Pietro does not see them as the real thing. Rather, they are like the pressures created in the mouth and armpits that give rise to yawning and stretching. These are caused by a windy or spiritous substance that produces titillation as it rushes through the sensitive parts of the body, causing feelings of attraction and "the goose bumps of a sensual itch [*pruritus oripilationes*]." Similarly, the moisture (*humiditas*, *humidum*) produced by boys is not what Pietro elsewhere calls "spermatic" or "seminal," even though semen has a similar spiritous component. He and his readers may have had in mind a particular kind of moisture—either a preliminary lubrication or a premature ejaculation regarded as ineffectual.

One candidate for such a fluid would have been *alguadi*. According to Avicenna, the penis had three vessels, each dedicated to the evacuation of a different substance, of which the first two were urine and sperm. As Pietro had explained in his commentary on an earlier *problema* (citing the Arabic authority), "The third is the passage through which passes the subtle moisture called *alguedi* [*sic*], which is mainly emitted when someone, touching his wife in copulation, becomes weak."[58] There is circumstantial evidence that Pietro and others had that substance in mind. The Italian scribe who produced one early fifteenth-century copy of Pietro's commentary referred to the chapter in Avicenna about "those who are accustomed to men lying on them" as *alguagi*.[59] Even more suggestive is the association of *alguagi* with youth. In his enumeration of life stages, Avicenna identified an age called *alguadi*, "when sperm comes out of him, which is until they grow beards."[60] (Before that comes the period when one gets all one's teeth but before experiencing emissions; after it comes the period when one stops growing.) Taken together, these texts weigh heavily in favor of interpreting the subtle or windy anal emissions of at least some men as (or as comparable to) *alguadi*. But their exact nature and quantity remain indeterminate. In one context Avicenna distinguishes the substance from sperm (as well as urine); in another he calls it sperm. The Erfurt *Problemata* speaks of "imperceptible exhalations"—an echo of the spiritous or windy substance.[61] As Pietro pictures the process, the *humidum* mentioned by Aristotle causes titillation, "on account of the windy material, heated by friction, passing across over their sensitive organs."[62] For another scholar, however, the process was much simpler. Having read that the pathways are closed around the penis and open to the anus, he wrote in the margin, "Sperm can come out through the anus by means of friction."[63]

More certain than the precise quality of the emissions was the underlying connection between men with partially blocked seminal passageways and adolescent boys. As the author of the Bavarian text succinctly put it, "The pleasure that happens to them in such an act [is] just as in boys who do not have the spermatic material for it, but only something windy causing the erection of the penis."[64] Adolescents will surface again at the discussions of habit formation. Here they serve as a model of sexually incomplete males. And in that context Pietro minimizes the importance of emitting full-fledged semen, emphasizing that boys may nevertheless experience pleasure. To make the point, he incorporates a lengthy quotation from Aristotle's *Generation of Animals*: " 'Pleasure occurs in intercourse not only on account of the emission of sperm but also on account of the emission of spirit, from which spirit, fixed and condensed, sperm is made. And this is evident in these boys, who cannot yet produce sperm, when they have been close to the time in which they should produce sperm. And it is similarly evident in sterile men. For in all these, pleasure occurs from the release of spirit,' " and, Pietro adds, rubbing.[65] The passage makes explicit that the boys (*pueri*) in question are at the point of puberty. It postulates a lack of semen, so the only component of the seminal fluid involved here is what the Erfurt commentator had referred to as "imperceptible exhalations" that contribute to an erection.[66]

Pietro's quotation from Aristotle's work on reproduction also lent force to his extended comparisons of the men with blocked vessels to other types of people. For they are not only like boys approaching adolescence but also like "sterile men," and, according to Pietro, like men in various other states of physical imperfection. He points to a number of circumstances in which full and fully effective semen may be temporarily or permanently lacking in mature men. Too much copulation can lead to an "inflammation of the body" causing men to ejaculate "no corporeal [*corpulento*] moisture"; certain hot illnesses, such as "hectic fever" can have similar effects. Indeed, the natural "complexion" of some mature men is so hot and dry that their moisture is converted into "windy spirits."[67]

In keeping with the general tenor of the *Problemata* and Pietro's commentary, all these causes of imperfect ejaculation are natural in one sense or another. All share a physical basis, in particular heat that evaporates moisture, turning it into windy spirit within the body. Yet nature intervenes in a different way in each case. With excessive sexual activity, the causal chain begins with a voluntary behavior which, we have learned, can itself have physiological or psychological roots. With illness, the state is similarly temporary and reversible, but usually arises independently of an individual's intentions (although poor regimen may also be responsible). Complexion, in contrast, is innate—the steady, if not entirely permanent, constitution of an individual when in a state of health, even though the specific ratios of humors or qualities may fail to conform to the ideal. In that respect, the concept of complexion embodies a paradox similar to the one at the heart of IV.26: such men's healthy condition is unhealthy. As when he speaks of men's "unnatural nature," Pietro is explicit about this group's physiological imbalance, saying that "their complexion [is] by nature excessively hot and dry."[68] In an aside, fortified by references to three ancient authorities, he points out that men who are too hot to produce semen when young may grow out of their condition.[69] In all of these cases, the natural operations of heat in the body give rise to an unnatural state of affairs, and it is for this reason that Pietro gathers them together here.

The various conditions expatiated upon by Pietro—debauchery, illness, immaturity—have no implications concerning the location at which the male sexual substance exits. The author of the Erfurt *Problemata* was alone in his conclusion that the common denominator was anal emissions: "Just as in these [boys]' arousal and pleasure occurs in the anus."[70] Nor do these phenomena bear upon the gender of the sexual partner, or indeed whether there is a sexual partner (masturbation being a possibility mentioned in the commentaries on IV.26). Rather Pietro is concerned with the traits that distinguish this group from their completely blocked brothers: the existence and expulsion of some substance that can produce relief. Indeed, although the Aristotelian text says simply that boys and their counterparts are calmed (*quiescunt*), Pietro holds that either a boy or man can get real pleasure from such irregular emissions—"he desires to go and to be emptied of the subtle or windy material contained in this way, since, once it is emptied, he can be calmed and experience joy [*gaudere*]."[71]

Later passages in the *problema* address the intensity of desire and satisfaction, but Pietro has already introduced clues to the level of libido involved in men who desire to be acted on. Here, in addition, he mentions sparrows. The condition under discussion may occur when a person's hot, dry complexion results in "resolving the moisture into windy spirits, just as takes place in sparrows."[72] Each animal species has its characteristic physiological profile, often measured along the axes of hot-cold and moist-dry. Thus, at one level, the point is just that there exist other creatures in whom, by nature, the ratio of moist to dry is low. But sparrows had a special status in medieval sexual lore—one that reached well beyond the circles of academic natural philosophers.[73] In his work on natural history, Aristotle had included sparrows in the group of animals in which the female does not squat down during intercourse. In them, because of the mechanics of copulation, the act is quick.[74] Interpreting this information, the Dominican scholar Albertus Magnus (ca. 1200–1280) explained that this "is apparent in sparrows who are very fast in intercourse and they repeat the act frequently, one time right after another." Later he referred again to their sexual and reproductive proficiency, claiming that "they copulate perhaps twenty times in an hour." Albertus introduced the same sort of physiological explanation that Pietro was to invoke: "The males only live a short time because of copulation and motions, which dry up the vital humor in them." (Consequently, their meat is hot and dry, capable both of acting as an aphrodisiac and of inducing constipation.)[75] These are ordinary, not deformed or deviant sparrows; their copulation is heterosexual and reproductive. But excessive sexual appetite can, as Pietro has explained, spill over into—indeed create—the physical conditions for diverted desires.

In the course of discussing the series of passages dealing with the innate partial blockage that permits some substance to seep out, Pietro has elaborated the general mechanisms of desire and release, of which the men in question experience a modified but recognizable version. He has introduced both a set of traits and a set of issues attached to the men of *problema* IV.26. Some characteristics he attributes to them directly. Thus, for example, he describes such a man's unnatural nature and pursuit of friction: "When the appetite for intercourse takes over, he seeks by rubbing and strong motion to expel the spermatic moisture this way, at the place where it is collected by nature, in accordance with his monstrous nature."[76] Much of what we learn, however, he conveys by a series of associations with those who are oversexed, sick, young, immature, and sterile, as well as with lascivious birds. Less directly marked have been questions concerning the nature of nature—its status as a norm, its variety as manifested in different species, its idiosyncrasy as manifested in different individuals, its malleability in the course of a lifetime. These traits and themes recur and take on added dimensions in the course of the subsequent segments of the commentary, expressed especially in terms of femininity and monstrosity.

"Their Monstrous Nature"

> [Qui autem] And those who are effeminate by nature have been built in such a way that it is not (or a little) separated out there where it is expelled in those holding in accordance with nature. And into that place since they have [been] constituted outside nature. For existing males are arranged this way.[77]

Decisions about where and why to divide the text of the Bartholomeus translation affected the classification of men who enjoyed the passive role in intercourse. Pietro characterized one group as distinct not only in its internal anatomy (fully blocked, with the seminal substance cut off from any release at all) but also in its effeminate behavioral traits and physiognomy. He did so in spite of the fact that the ancient work did not discuss degrees of obstruction and ascribed effeminacy to those who give off nothing or almost nothing (*modica*), which would seem to cover all those affected from birth, whether wholly or partially blocked. Faced with this ambiguity, Pietro himself was not entirely consistent in sustaining the differentiation, leaving open the question of whether all or only some of those marked from birth were in any sense feminine. Certainly his parsing of the population of innately defective men was of interest neither to Walter Burley and Evrart de Conty, who cared little about the anatomical aspects of the question, nor to the author of the Prague text, for whom anatomical and physiological matters were simpler. There was only one "innate" category, men whose seminal passages were closed through the penis and opened to the anus, and whose emissions were semen or, at least, semen-like: "And in these men the spermatic material flows toward the fundament and they emit sperm in intercourse. And these want to be rubbed by the placement of a penis at their fundaments."[78]

For Pietro and for those readers who followed him closely, however, setting apart the men with no outlet at all sometimes seemed to be essential for attributing sexual frustration and drawing lines of similarity with other types of people. In his brief exposition of this passage, he highlighted two themes that he had introduced previously: their association with effeminates and eunuchs; and the unnatural nature of their natures. Whereas earlier he had reasonably interpreted Aristotle to say that these men were merely similar to eunuchs and effeminates, here Bartholomeus's translation refers to men "who are effeminate by nature." Accordingly, Pietro shifts from analogy to identity. "These men," he says, "from their basic component parts, are 'effeminates' or eunuchs."[79] The "or" is ambiguous, since it is often used in scholastic explications to mean "that is." Thus "eunuchs" may either denote a separate group or be a synonym for "effeminates."[80] Commenting on other chapters of the *Problemata*, Pietro used an expanded concept of eunuchs that included those with small, weak testicles, which may justify their inclusion here. But for most readers, the rhetorical effect of their presence at this point in the text would have been to emphasize the sense of mutilation and the consequences of anatomy without outlet.

Distortion and deformity are the defining features of such men, even if some slight emission does occur. For, according to Aristotle, it would occur in a place that is not "according to nature [*secundum naturam*]" but rather "outside nature [*extra naturam*]." In an expository passage of just over a hundred words Pietro used the word "nature" and its derivatives five times, combining it with three instances of "monstrous" and its derivatives. The effect is to hammer home the painful paradox of the unnatural natures. Since they are effeminates (and, for some readers, eunuchs) "naturally" built with their passages blinded, they can expel nothing or almost nothing "at the place in which it is 'expelled' in those men who naturally have those passages in keeping with nature."[81] This formulation expresses a stark contrast between those who naturally diverge from nature and those who naturally comply with it. The former are the products of nature not "properly operating." With the normal pathway blocked, the alternate route is in play: something can come out "at the place [that is] in keeping with their monstrous nature."[82]

Lest a reader—medieval or modern—suppose that Pietro is speaking of monstrosity in a loose sense, he provides a reference to the full, technical explanation of "monsters" in the *Generations of Animals*.[83] There Aristotle discusses many ways in which the reproductive process can go awry, producing offspring with irregularities. His examples include various frequently occurring defects such as polydactyly. These were not the kind of monsters whose very humanity was open to question, much less those of a singular, prodigious, or supernatural kind.[84] Commonplace flaws were widely understood to arise from such natural (and ordinary) accidents as weak seed or superfluous matter. But, by the fourteenth century, few intellectuals would fail to think of the astrological causes involved. Thus Evrart de Conty mentions not Aristotle but Ptolemy in the context of inborn flaws. Because Pietro, a far more accomplished astrologer, brings the subject up at the end of his commentary on *problema* IV.26, a fuller discussion is located below. But the significance of Evrart's reference here to "the misfortunes of the heavens" is that it confirms the difficulties posed by what he calls "the evil disposition that nature has given them from the beginning of their gestation."[85] Nature can and often does produce outcomes that are not only defective but even bad.

The logic of monstrosity does little to clarify Pietro's taxonomy of failed males. On the one hand, he has taken pains to specify that not all those defective at birth, only those with no possibility of release, suffer the frustration that gives rise to insatiability. On the other hand, the natural history of monsters would seem to apply to those with partially as well as fully blocked vessels. Since he was the only commentator to employ the language of monstrosity in the context of *Problemata* IV.26, and since there are no marginal comments in the manuscripts that record other medieval opinions, the question of who counts as monstrous remains unresolved. Circumstantial evidence, however, weighs in favor of including all those innately incapable of ejaculation through the penis. The deflection of seminal fluid toward the anus is frequently called "unnatural," without reference to its subsequent fate. Certainly, for those like the authors of the Erfurt and Prague commentaries, who did not trouble themselves with the fine points, what applied to some applied to all.

In What Manner Such Men Should Be Treated

> [Quare necesse est] Wherefore it is necessary that that place of them be cut. Indeed this cutting either produces complete destruction or deterioration. That, therefore, is not, for a woman would surely come about.[86]

One of the most controversial issues associated with *Problemata* IV.26 was one not raised by the Greek work itself: Can this condition be treated? In one sense, it resulted from the difficulties of translating the cryptic and rambling text into Latin. In another sense, it emerged from the convergence of medieval factors as diverse as the history of eunuchs, the scope of medicine, the juxtaposition of respected authorities, and the weight of social and moral sanctions. What seems to have started in antiquity as an abstract question about how to understand the relationship between defective men and proper ones turned into a question about genital surgery. The author of the Aachen *Problemata*, who usually stayed close to the wording in Jean de Jandun's redaction of Pietro's exposition, was sufficiently struck (perhaps alarmed) by the thought of such an intervention that he posed the possibility largely in his own words: "And then [Aristotle] continues about these men—about a cure of those who have this kind of obstructed pores and passages. [And he says that] they should not be operated on or cut with any bronze or any other sharp instrument, since possibly a complete excision and obstruction would occur; or they would face death or a perversion of the mind; or they would become completely like women. Thus they would become effeminate and acquire the behavior [*mores*] of women."[87]

The Greek original had conveyed the idea that, although such men are male, they are deformed. In this situation, what they are (male)—their essence or definition—can be either completely destroyed or distorted. Presumably because it is absurd, Aristotle ruled out the former: if a male's maleness as a whole were annihilated, he would become the opposite, a female. So the other alternative must apply: an otherwise male individual will simply have a degraded, imperfect maleness.[88] This was the way in which the two fifteenth-century translators, George of Trebizond and Theodore Gaza, understood the original text.[89] The term that describes the men's deformed state is πεπηρῶσθαι, which can suggest "mutilated."[90] Perhaps for that reason, in the thirteenth century Bartholomeus, whose translation Pietro used, had eunuchs in mind. In any case, his Latin gave rise to a very different set of issues. The Greek had offered, "This area [of the body] is necessarily mutilated." It was almost certainly alluding to the philosophical notion of "necessity," by which the realization of a natural and proper end is disrupted by nonteleological accidents. This is the type of mechanical or material causation Aristotle assigns to the formation of "monstrosities" in the section of *Generation of Animals* to which Pietro referred. As understood by Pietro and others, however, Bartholomeus's translation says, "It is necessary that this area of these men be cut."[91] That is, for most Latin readers, "necessity" referred not to the forces that produced the defect but rather to the conduct of a surgeon. Hence they believed Aristotle was implicitly posing the question, "If innate anatomical structures are at the root of the problem, why not correct them?" Bartholomeus names the same two possible consequences that appeared in the Greek, "destruction [*corruptionem*]" and "deterioration [*perversionem*]." He concludes, somewhat opaquely, "that, therefore, is not" (or, perhaps, "therefore the former is not"), "for a woman would result." One thoughtful reader supplied a gloss at this point that captured the original philosophical sense that the complete destruction of a man's maleness would be absurd. At "Therefore it is not," he added "possible";[92] most, however, simply considered surgery a bad idea.

Pietro, as a proficient Aristotelian, seems to have grasped the more abstract philosophical issue, but he nevertheless felt called upon to deal with the implication that surgery was at issue, since the verb *secare* means "to divide," "to operate on," and even "to castrate." Just what this hypothetical surgery might have involved is not clear. Some readers may have had castration in mind, but an attempt to open the seminal pathways would have made more sense. The testicles are not implicated among the causes of the condition, nor was their state generally thought to be a sign of homosexual inclinations.[93] But, having taken on the question of cutting, Pietro reads the phrase "it is not" to mean that surgery "should not be done."[94] That is, the issue raised by this passage had become whether there was a surgical "cure" for those in whom the semen was deflected away from the penis. And the answer was "no." What in the Greek had been alternative logical possibilities here became alternative medical outcomes. The destruction of an abstract type, the male, is transformed by Pietro's interpretation of Bartholomeus's translation into "the 'corruption' of pores of all kinds and very often death." Apparently the best surgical outcome that could be hoped for would be " 'perversion' and change into something worse," namely, a woman—hardly a goal to be pursued. Pietro hastens to add that the latter result should not be understood in a literal sense. Rather the process, which he likens to the effects of treating wounds and scars by lancing, involves adverse effects on the relevant pores. In particular, "The parts around the testicles would be destroyed [*corrumpentur*] in such a way that such a man would be rendered a eunuch and effeminate, when previously he was not like that."[95] In other words, although he would not actually be a woman, he would acquire "womanly actions and behavior"—a paradoxical risk for a group already tarred as "effeminate." For medieval readers, the connection with women was sealed by the next lemma of the Aristotelian text.

> [<u>Necesse est igitur</u>] It is therefore necessary that [some]
> of the seminal excretion must therefore be diverted and

> moved elsewhere, because of which [they are] insatiable, like women. For a modest amount of moisture does not manage to get out and is cooled off quickly.[96]

At this point, according to Pietro, "Aristotle teaches in what manner such men ought to be treated [*curari*]." In fact, neither the Greek nor Bartholomeus's translation gives any indication of this intention, indicating simply that, because the seminal fluid cannot follow the ordinary path, being "diverted [*perverti*]," it must follow some other route—it must go "elsewhere [*alibi*]." Aristotle is merely reiterating the idea that it seeks an alternative outlet. As the author of the Bavarian text pointed out, "As long as neither windy spirit nor watery material is emitted from [those with anatomical blockages], they do not rest, but desire to be rubbed until one of these things occurs."[97] Unsatisfied appetite gives rise to perpetual desire. Worse still, according to the author of the Aachen text, "However much more they are rubbed, so much more [superfluous moisture] is attracted . . . [but] almost nothing is emitted and thus the more they are extended in their madness [*furia*]."[98]

Pietro does not decline to comment on the theme of insatiability. Nor does he pass over the reference to the trace of moisture that evaporates. (The phenomenon reminds him of a more ordinary process and, in one of his intertextual gestures that invoke the familiar, he makes reference to an earlier *problema* that alludes to sweat passing through small channels and evaporating.)[99] Once the idea of medical intervention had been broached, however, it was easy to focus on the apparent suggestion that the physician (rather than the relevant natural forces) must effect the redirection of fluids to an alternative outlet. Thus, instead of seeing "elsewhere" as a reference to the anus, that is, the "unnatural place" to which they will flow, Pietro interprets the word more broadly. By the proper medical interventions, he suggests, the trouble-causing substance can be dispersed, rather than channeled. With the surgical option ruled out, he (falsely) attributes alternative medical advice to Aristotle: "Therefore he teaches here how such men should be treated, saying that the spermatic moisture needing to be purged that is in them must be diverted from that part and moved to another part. This is done with desiccating foods and medicines diverting the sperm and its material to another area, such as to urines and sweats, etc."[100] For Pietro, who wrote a treatise on poisons, the physiological mechanisms involved would have been obvious. First, the accumulated fluids that were creating a pressure for release would be reduced, using drying substances, either ingested or externally applied. Then, perhaps with the aid of a different set of foods and materia medica, an appropriate form of excretion would be induced. For most of Pietro's readers as well, this sort of process would have been familiar from the common conceptualization and treatment of other disorders resulting from various superfluities. Indeed, Walter Burley, though he greatly compressed the contents of Pietro's commentary, inserted specific suggestions about the types of drugs to be employed. In short, Pietro rejected the putative suggestion of a surgical remedy but embraced a medical approach consisting of drugs and diet. This remedy would not correct the underlying anatomical cause, but it would alleviate the symptoms produced by the pressure of trapped, superfluous seminal moisture.

Once he had introduced the subject of a cure, Pietro was almost required to address the views of Avicenna, whose massive *Canon of Medicine* was the foremost and most comprehensive work on theoretical and practical medicine available around 1300, when he was writing his exposition of the *Problemata*. Not only in that work but also in his *Conciliator* and elsewhere, Pietro demonstrated both a detailed knowledge of and a critical skepticism about the views of this towering medical authority. In connection with this *problema* he had already had recourse to Avicenna to support his account of the mechanisms of expulsion, in the same way that he had cited works of Aristotle to support other points. Now he referred for the first time to the passage in the *Canon* whose subject most closely approximates that of *problema* IV.26. Preceded by a chapter on men who defecate when they ejaculate and followed by one on hermaphrodites, the chapter entitled "De halubnathi" in the Latin translation concerns "the illness [*egritudo*] that happens to a man who becomes accustomed to men lying on top of him."[101]

Pietro believed the authority of Aristotle supported his conviction that the condition could be "treated" or "cured" (*curari*) following medical principles. He proposed specific types of medical regimen and drugs that would relieve the pressure of unreleased superfluities; and he explained how the treatment would work by drying them up, diverting them, and excreting them in urine or sweat. Nevertheless, he called on Avicenna, citing chapter and verse (or, in this case, the Arabic "fen") to support his argument: "Whence Avicenna in the third book, twentieth fen of the *Canon*, speaking about this kind of disposition, and more about the former [kind, i.e., innate] says: 'The people who want to cure [*curare*] them are stupid, for the origin of their illness is mental, not natural. But if a cure does apply to them, then it is that which breaks their desire through sadness, hunger, staying awake, being imprisoned, and being beaten.'"[102] Pietro was appealing for support, and in one sense Avicenna's prescription was comparable to Pietro's own: neither proposes to cure the underlying condition, that is, to correct the anatomical blockage; both propose to resolve the problems that arise from it. But according to this version of his text, Avicenna had categorically rejected the idea that the condition has any somatic basis at all. His incarceration and corporal punishment are not in the same category as Pietro's approach ("foods and medicines"). Avicenna's goal is psychological, formulated in terms of breaking desire (*frangit desiderium*); Pietro's goal is physiological, formulated in terms of moving the superfluity to other parts of the body (*movere ad aliam partem*). Avicenna's list is decidedly punitive. Even if fasting did sometimes figure in a medical regimen, incarceration and blows did not. Such measures underscore his insistence that the cause of the condition is not natural and thus not really susceptible to medical management, properly speaking.

In addition to this internal logic of Avicenna's approach, Pietro's extensive knowledge of the *Canon* would suggest that he understood (and rejected) the assertion that the condition was psychological or spiritual in origin. And his medical strategy of dispersing the seminal moisture fits with that picture. Furthermore, Evrart de Conty, who tracked Pietro's analysis closely even when he disagreed with it, sided with Avicenna, whom he understood to be postulating psychological causes. Yet there is a significant textual problem associated with the passage from Avicenna and Pietro's use of it. In the Arabic and in the original Latin translation as reflected in some medieval manuscripts, Avicenna's position is that the root of the disorder is "mental, not natural [*meditativum non naturale*]." But most extant copies of Pietro's first and second versions and Jean de Jandun's redaction read Avicenna's explanation as, "unnatural medicine [*medicamentum non naturale*]," giving rise to a decidedly different meaning regarding the condition's cause.[103] The variant also occurs in some early editions of the *Canon* itself and may thus have been present in some manuscripts, perhaps including one Pietro had access to.[104] As *meditativum* appears in two manuscripts of Pietro's second version, it is possible that Pietro himself made a correction. Or, in those cases and in that of a single manuscript of Jean's version, learned scribes may have been the ones to insist on "mental." In either case at some point readers would have been faced not only with the general claim that the cause was "not natural" but also with the implication that some kind of misguided or nefarious human action was to blame—not simply the bad thoughts of the individual himself. How could some potion, some "unnatural medicament," be involved? Sorcery is one possibility. At the end of this chapter of the *Canon*, the translator may have tacked on an interpolation.

According to "some people," the Latin version says, the condition has its origin in a "wicked craft [*arte maligna*]."[105] The indefatigable fifteenth-century French commentator Jacques Despars, writing on this chapter of the *Canon* takes note of this apparent reference to magic in Avicenna,[106] but none of the Latin commentators on the *Problemata* mentions it. If Pietro was thinking along those lines, he must have imagined that whatever poison or spell was involved operated upon the individual's father, mother, or embryo, because he continued to insist that the condition was congenital in some men.

Conclusion: The Problem of Nature

Lacking a manuscript that can be attributed to Pietro himself, we cannot know exactly how he understood Avicenna. We do know that some of his readers encountered one explanation and other readers the other. But whether taken in terms of mental failure or bad drugs, Avicenna's position that this disposition cannot be a naturally caused innate condition points up a cultural struggle at the heart of medieval encounters with *problema* IV.26. In particular, the Aristotelian assertion that there were two distinct groups of men, the constitutional and the habitual, made it difficult to ignore or elide those with inclinations arising from accidents of birth. Yet, among Christian intellectuals in the late Middle Ages, the pursuit of pleasure through anal stimulation was difficult to treat as a mere analogue to other common "monstrosities" such as polydactyly or even congenital blindness. Both within formal philosophy and in the context of immediate personal responses, authors and readers could not escape their concerns about the psychological and moral resonances of this particular irregularity—and often did not wish to. Translating the standard version of Avicenna's remark into French, Evrart de Conty insisted that "'the origin of their illness is not natural', thus it is engendered by meditation and strong thought." Furthermore, he concluded (citing Ovid for good measure) that therefore no medical remedy is possible. "For no medicine can change that insane thought."[107] Evrart's position is that where nature is not the cause, medicine is not the remedy.

Pietro would not have disagreed. But he had been arguing the converse: where nature is the cause, medicine is the remedy. He pursued a systematic exposition of this chapter in Aristotle, as he had of the hundreds of chapters that preceded it and the hundreds that followed it in the *Problemata*. In doing so, he made the case as fully as his considerable intellectual tools permitted that some men who enjoyed the passive role in sex with other men were, at least in some sense, natural, even if, in another sense, the way their bodies worked was "the opposite of what is supposed to happen in natural intercourse." Just as blood, that "most natural" of fluids, could sometimes be so poorly and impurely composed as to require expulsion from the body, so human males, ordinarily the most perfect of living creatures, sometimes fell short. There was no doubt that they were, as Aristotle had said, "against nature," in the sense of its teleological plan, yet Pietro repeatedly referred to their own particular nature that inhered in them from birth.

Medieval readers of the *Problemata*, even those who took the trouble to summarize or comment on it, were not always interested in the exhaustive details and intricate scholastic niceties that characterized Pietro's exposition. For that reason, his views on the text, though authoritative in the sense that they were original, comprehensive, and influential, cannot be taken as typical. Other scholars, including some who were relying on his work, had their own views about *problema* IV.26, often leaving Pietro's subtle distinctions in shreds. For one, the subject was "the sodomitic sin."[108] For some the emphasis was on "defective" structure; for others the idea of rubbing was central.[109] One concluded simply that "semen comes out their ass," and another discerned an additional cause: "This sometimes happens when so much spermatic material abounds that . . . he cannot purge [it] enough through coitus."[110] Nevertheless, many readers show clear signs of having recognized the ascription of natural causes, even when, like Evrart, some proceeded to reject them. Whether he agreed with Pietro or not, the reader who wrote in the margin, "Some men commit the sodomitical sin moved by nature," certainly got the point about causation.

In other words, scholars whose intellectual inclinations had led them to pick up the work in the first place grasped the general concept that such men were a natural phenomenon. Indeed, that was the premise of the *Problemata* itself. No matter how puzzling or unpleasant, whether exotic or mundane, a *problema* was about something that existed or happened and that begged to be explained. And, in that specific textual context, most readers expected the explanation to come from natural philosophy. Pietro d'Abano put into practice the principle that Walter Burley had articulated by opening his prologue to the work as a whole with a quotation from Virgil: "Happy is he who can know the causes of things"; and by adding a special preface to the series of questions about sex in order to remind the reader that "nothing natural is shameful."[111] With less fanfare, Pietro's explication of *problema* IV.26 constituted an implicit defense of the academic project in which he was engaged, an assertion of the ample jurisdiction of natural philosophy. As will become clear, neither he nor Burley was completely successful at banishing shame. Furthermore, the discipline of natural philosophy had limits, and, not surprisingly, this subject matter tested them. Nevertheless, the early portions of the *problema*—with their insistence on a condition from birth and their mechanical account of internal pores and passages in relation to external body parts—offered a generous opportunity for the application of scientific principles drawn from medicine, psychology, and even physics. At the same time, interpretations of those passages contributed to a sketch of the men who were born to enjoy the passive position in sex with other men.

Although some saw fit to skip Pietro's excursus on the mechanisms of appetite, and although others, reading Aristotle directly, took note of different details, the evidence of many manuscripts suggests some areas of convergence concerning the men whose bodies and desires were being explained. The accumulation of a superfluity blocked from an appropriate exit caused pressure at the base of the penis and around the anus. Exactly what, if anything, got emitted was also not always clear, nor was the distinction between the fully blocked and the partly blocked vessels. But a lively interest in the question of emissions with seminal, spiritous, and/or watery components figured both in the commentaries and in the marginalia. Associations with eunuchs, adolescent boys, and women commanded attention, providing hints about these men's traits beyond the particulars of their anatomical and physiological irregularities. These included imperfection, desire, and even madness. What emerges, then, is an identifiable, permanent category of men who share a specific set of somatic traits centered on a small region of the body. There is no cure for them, but their desires can (and presumably should) be managed with appropriate treatment. With the combination of susceptibility to anal stimulation and effeminacy medieval scholars focused more on the "condition" than on the men in whom it inhered, and, insofar as their sexual partners were represented at all, only their penises appear in the texts.

When Pietro d'Abano and others turned to the passages of *Problemata* IV.26 that deal with men who enjoy both the active and the passive roles and with those whose desires have developed as habits through the repeated experience of pleasure, a somewhat fuller picture can be discerned. A few even show some signs, however slight, of having life histories. Because it introduces an alternative to physical deformity as the cause of men's irregular desires, and thus abandons the specific innate and permanent mechanisms that grounded the congenitally obstructed group in nature, the remaining portion of the *problema* also reopens the challenge posed by Avicenna and echoed by Evrart. If the cause is not an innate defect (or an illness or a surgical operation), how could it be natural? And if it is not natural, how could it be susceptible to explanations drawn from natural philosophy? Thus, as commentators and readers moved to deal with the formation of sexual habits, they faced an apparent contrast between nature and habit and an implicit challenge to the disciplinary resources of natural philosophy.

Chapter 2

Habit Is a Kind of Nature

Thus far, Pietro has consistently upheld the physical answers to the questions posed by *problema* IV.26. The men who take pleasure in the passive role in coitus are constituted in particular ways that give rise to specific desires and pleasures. Even when he represented the men's bodies as monstrous, he ascribed the effects to natural processes grounded in the bodies' material configuration. Of course, the resultant natures are not as they should be. Pietro contrasted them to the norm—to what ought to happen according to nature. Yet they are natural in the sense that they do occur with some regularity, and, contrary to Avicenna's assertion, their causes are natural, not psychological. The extent to which Pietro was committed to this explanation based on a congenital condition (which was, after all, one of two answers proposed by the Aristotelian text) is suggested by the fact that he treated it with great seriousness and at great length. Furthermore, his approach was consistent with the naturalistic positions he took not only elsewhere in the *Problemata* commentary but also in his other works. Indeed, the same sense of natural processes persists even in his treatment of the second line of causation offered by the ancient text: "In some, this condition comes about from a habit."[1] Pietro and others treated this second group very differently, both from the point of view of the relevant philosophical principles and in terms of the moral issues they perceived. Nevertheless Pietro laid out for subsequent readers two ways in which this second line of causation converged with the first. For, first of all, habit formation itself is susceptible to natural philosophical analysis and, second, habits themselves operate as a sort of nature.

The specific causes and consequences of the habits that incline some men to irregular sexual behavior are the subject of the final three segments, or lemmas, of *Problemata* IV.26. The anonymous author of the brief Aachen exposition summarizes them: "For those who get used to doing such things around the time of their youth [*iuventute*], habit is thereafter nature in them, so that they almost always rub those parts on account of the pleasure or want to be rubbed by others. And these in particular are sodomites [*sodomite*]. And what they do in their youth is in them in their memory. And since they are soft [*molles*] then, sexual indulgence [*luxuria*] happens often to them then or later."[2] Thus, in their interpretation of these final sections of the *problema*, medieval readers convey their understanding of the formation of sexual habits and their relation to nature; of who was doing what to whom (and which of them qualified as sodomites); and of what role age played in the sexual economy. It is also in these final sections that they suggest just how abominable these individuals' inclinations and acts were. In contrast to their treatments of the earlier parts of the *problema*, readers' assumptions and opinions on these subjects diverge considerably. Even those authors who followed Pietro's commentary (as did the one just quoted) manifest a significant degree of disagreement or confusion.

In part, but only in part, the lack of clarity arises from the Aristotelian text, from Pietro's exposition, and from the tension between them. Pietro, downplaying the division between the always-passive and the sometimes-passive in favor of the distinction between the two causal pathways, announces the transition to the subject of habit at the point where the original text is making the transition to the subject of men who enjoy both the active and the passive roles: "[Aristotle] treats those in whom the aforesaid things occur on account of a twisted and repulsive habit. And such men are sodomites." He admits that Aristotle does not in fact address habit at this point but rather "first repeats what was said about the second element of the previous part of the text [about those] in whom moisture flows to the anus."[3] Nevertheless, Pietro's premature introduction of habit makes it all the more difficult for those who read his work to sort out the relationship (or lack of relationship) between those with double desires and those whose desires and pleasures resulted from experience rather than birth. In addition, it clouds what might otherwise have been a clear ascription of the word "sodomite." But the fine points of Pietro's organization cannot fully explain these ambiguities. They seem rather to reflect broader uncertainties among medieval intellectuals that affected both the science and the stigmatization.

The final passages of the *Problemata* and of Pietro's commentary thus introduce complications of two sorts. First, the purely passive individuals of the earlier parts are joined by individuals with mixed appetites; second, the anatomical and physiological explanations are joined by psychological and astrological ones. The present chapter opens with an attempt to understand those men who enjoy both acting and being acted upon, centered on the lemma Et quibus. It then turns to the twin subjects of the next two lemmas (Quibusdam autem and Propter hoc): How do habits become second nature? What are the processes by which they are formed? The medieval explications of those issues make possible in turn the following sections of the chapter on the way authors and readers pictured the acts and the participants in them. The chapter concludes with a subject absent from the ancient text but introduced by Pietro: the possibility that the heavens influenced the sexual dispositions of men. The astrological perspective draws together many of the questions and categories that late medieval readers considered in connection with this *problema*. It conveyed the idea of natural causation but lacked the concreteness suggested by anatomical structures; it implicated both the physiological and the psychological; and it raised issues familiar to medieval readers about the relationship between natural constraints and moral responsibilities.

Halubnathi and the Sources of Passive Desire

> [Et quibus] And those in whom [it goes] into the anus they desire to be acted upon; and in whomever into each, they [desire] to act and be acted upon. And into whichever one, they desire that more.[4]

From the perspective of Bartholomeus de Messina's original Latin version, the text reaches a major dividing line here between the two groups referred to at the opening of the *problema,* those with one source of pleasure and those with two. And because of Pietro's apparent haste to introduce the subject of habit, this passage forms a bridge between the two causal explanations that structure his commentary. It did not function the same way for other readers,[5] and even Pietro himself makes clear that the cause of some men's dual desires is anatomical, not the result of experiences. He explains quite simply (and in keeping with his source) that those whose spermatic moisture is entirely directed toward the anus desire only to be acted upon and "to be rubbed vigorously around the anus, so that the moisture may be expelled." Those whose fluid goes partly to the anus and partly through the penis, on the other hand, seek stimulation in both places: "They desire to rub with their penis and, with the penises of others, to be rubbed in the anus by them."[6] Men in the latter group may tend more toward one pleasure or the other, depending on where the "stimulative moisture" predominantly flows.

If dealing with the distinction between always passive men and those doubly inclined is less central to his interests than elaborating on the relation between birth and habit, Pietro nonetheless makes good use of the occasion provided by this passage. In a digression (introduced with the usual marker, "It should be noted that") he brings the reader back to the dialogue between the two key texts that dealt with male homosexual pleasure, the *Problemata* and the *Canon of Medicine*.[7] (Such juxtapositions of authoritative texts constituted the organizing principle of Pietro's compendium of conflicts between natural philosophy and medical theory, the *Conciliator*.) Earlier, when speaking of the possibility of a cure, he appeared to have been citing Avicenna's view that the condition pertained to the mental or spiritual realm (*meditativum*), not the medical. While agreeing with Avicenna that there is no cure properly speaking, he advocated pharmacological treatment rather than punitive treatments that aimed at the spirit. Now, once again, he confronts the fundamental difference between his own (Aristotelian) view of causation and Avicenna's.

Pietro read the translation of this portion of the *Canon* carefully—he cited it five times and used the term of its title, "halubnathi," three times. Furthermore, when it suited him, he mustered elements from it. But Avicenna had insisted on the unimpeded (if irregularly induced) emission of semen along the ordinary route through the penis, a position incompatible with Pietro's ideas about blockage and deflection. Avicenna described the error of "certain people" who held that in those who take pleasure in being subjected the nerve responsible for genital sensation is split, the sturdier part running to the glans, the finer part running to the anus. According to Avicenna these unnamed authors held that such a man requires "vehement rubbing in order to have sensation." When someone ejaculates on him (literally, "when it is sprayed over this man"), his sexual desire is aroused. Pietro observes that Avicenna appears to be discussing the same phenomenon as Aristotle and to be paraphrasing and attacking Aristotle's reasoning.[8] He expresses outrage that Avicenna has so emphatically rejected the ancient philosopher's view. Here again—and more explicitly—Pietro insists that these men's inclinations are embedded in their very constitutions, pointing out that Avicenna is quite willing to see other, comparable defects as the proper subjects of medical concern: "Avicenna, who seems opposed to the opinion of the Philosopher, does not introduce a reason. I do not see what would keep such a disorder and monstrosity from coming about from birth, since immediately afterward Avicenna mentions hermaphrodites, who appear more monstrous with respect to nature."[9] This adamant declaration (underscored by the unusual interjection of the first person singular) is at once the conclusion of Pietro's argument about the physical causes of homosexual desire and the point of departure for his transition to psychological causes.

In this same digression Pietro introduces two new terms that color the meanings associated with the men of Book IV, *problema* 26. The first is "halubnathi," which, as he explains, is the term Avicenna applies to the "condition [*passio*]" of a man "who is accustomed to having men lie on top of him"; the second is the word "sodomites," with which Pietro announces his transition to those whose condition is based on experience and habit. "Halubnathi" is a transliteration of the Arabic "al ubna," introduced to Western readers by Gerard of Cremona, who translated the *Canon* into Latin in the second half of the twelfth century. Gerard, whose mode of translation was word for word, frequently let the Arabic stand when he could find no Latin match, which was probably the case here.[10] The word "sodomy" had already been coined in the context of same-sex transgressions but may have been unknown to Gerard and, in any case, belonged to a discursive domain distinct from the medical. As Mark Jordan has remarked with reference to another twelfth-century author, "Nature does not have the right word for this sin."[11] That is, no proper Latin philosophical vocabulary existed. And more than a century later, although Pietro and most other commentators had a much larger body of medical and natural philosophical literature at their disposal, they could not find a better term than "halubnathi." Or they preferred not to. The exception was a Parisian reader who refused to beat around the bush and wrote in the margin at this point in Pietro's commentary, "On *alubnati* according to Avicenna, and it is the sodomitical act."[12]

Scribes copying Pietro's commentary (the only one to incorporate the Arabic term) were apparently unfamiliar with it and had trouble reproducing it correctly.[13] That evidence suggests that the passage in Avicenna was not a particular site of academic curiosity or fun. By the time the *Canon* appeared in print at the end of the fifteenth century, the chapter has come to be entitled "On *aluminati*," which may be the result of an attempt to settle on a more Latin-sounding word but may also be related to yet another term transcribed from Arabic that conveys the sense of "having been cut."[14] On the one hand, the general uncertainty about the word underscores its foreign nature, and the specific association with Arabic and thus with Islam may have added a layer of aversion.[15] Although, in no small part because of Gerard's translations, Arabic terms had become common in Latin medical texts, Western Christians had early come to associate male homosexual practices with Muslim elites. On the other hand, the specific forms that the uncertainty took suggests the possibility that late medieval readers vacillated among various ways of identifying the group of men being discussed—in terms of their behavioral subjection, their physiological insufficiency, or their comparability with eunuchs. Neither the compressed and cryptic final passages of the *problema* nor the commentaries upon it offer a secure consensus. But they do introduce an additional, psychological dimension into the mix of possibilities.

Back to Nature

> [Quibusdam autem] And in some this condition comes about from a habit. For surely whatever things to do [?], enjoying and emitting semen occurs to them in that way. And thus they desire to do the things that happened to them, and to a great extent habit becomes nature in a way.[16]

Having attacked Avicenna for denying the possibility of a congenital cause, Pietro now turns to the psychodynamics of habit formation. For the Aristotelian text has finally arrived at the group of men whose condition derives from experience, not birth. Together, this passage and the next introduce the terms in which the processes of habituation are to be explained, namely, pleasure and memory. And they both suggest that, in a sense, habit becomes nature. Perhaps because the second point was more interesting to him or perhaps because it was more controversial, Pietro treated it first and gave it fuller attention. The fate of the habit thus precedes its origins in his commentary.

Aristotle had declared (and Pietro would later demonstrate) that the inclinations under discussion may arise from experience as well as from birth. But, having insisted on this second causal path, Pietro accepts the textual invitation to minimize the distance between the two. His initial strategy is to remind his readers that the notion of habit becoming nature was a familiar one.[17] They will, he points out, find it in Aristotle's short treatise on memory (where habit is said to displace nature),[18] as well as further along in the very work they are currently studying. Indeed, at *Problemata*, Book XXVIII, *problema* I, the question is why a habitually intemperate person suffers when he adopts a temperate regimen. That is, it deals with a bad habit that has come to inhere so materially in a person's nature that contravening it is not merely difficult but harmful: what would be healthy for most people would be dangerous for this individual. In addition, drawing on medical authority, he refers to a Hippocratic work, *Airs, Waters, Places*, which, he recalls, describes the practice of certain women in Asia who use their hands to elongate the heads of their infants by pulling at them over and over.[19] In both cases, Pietro takes the position that repetition can create a new nature in the most concrete sense; at the very least, habituation creates a stable, deeply rooted state. The principle was indeed widely accepted: in the mid-fifteenth century George of Trebizond, elaborating on the same subject, would observe:"In all things habit is like another nature, especially, indeed, in sexual matters."[20]

The invocation of recognized authorities did not, in Pietro's mind, exempt him from presenting rational arguments for a position that, to some extent, undermines the clear distinction between men whose deflected desires are based on innate structures and those whose desires result from experience. If, for him, the central distinction in the *problema* was between nature and habit, then whether and how habit can be reduced to nature is a critical issue. Using conventional scholastic rhetoric, he challenges himself by raising a "doubt." Certainly, one might object, it does not appear that a thing can form a habit that overrides its basic nature. You can throw a rock up in the air a thousand times, he says (referring to a passage in Aristotle's *Ethics*), and it will never form the habit of moving upward against its heavy nature.[21] Pietro's response to this challenge takes a form common in the works of Aristotle and his medieval followers: he makes a distinction. We speak of how things tend to act—the "motion and inclination" of natural entities—in two ways. The first, he says, applies to an inanimate object. What it does is determined by its material composition, instilled by whatever first produced it. In particular, whatever element is dominant in the mix of elements from which a thing is constituted will govern its motions and cannot be overridden by a habit. The rock, earthy and therefore heavy from the time it came into being, is of this kind, and so its return to earth is inevitable as long as the rock exists. The other sense in which a natural motion or inclination may be understood pertains to animate beings and this is the one, Pietro says, that Aristotle intends in *problema* IV.26. In such a case, an object may "follow a higher form that has resulted from a mixture of the elements or that is coming from the outside," that is, from something other than its original physical composition.[22] The extent to which an object might have properties that transcend the ratios of its combined elements was a frequent topic of discussion among medieval natural philosophers. Here, however, Pietro's reference to the *Ethics* suggests he is making an analogy with the way in which individuals may arrive at a state or condition (*habitus*) in which they were not born. Becoming a virtuous (or evil) man requires the same thing as becoming a good (or bad) house builder, musician, or physician: practice.[23] Pietro's implicit argument is that as one becomes a just man by engaging in just acts, one becomes a man susceptible and inclined to homosexual pleasures by engaging in them.

Pietro's introduction of the term "inclination" is, on one level, inconsequential.[24] Although scholastic natural philosophers did not usually speak of inanimate objects as having inclinations, the word could refer to the rock's inherent tendency to move downward, even when it was not actually falling. But here, rather than stressing the distinction between the physical necessity embedded in a natural object's essence and a person's disposition (presumably less rigid), Pietro seems to blur the line between those whose behavior (like that of stones) is the result of their fundamental constitutions and those whose behavior is influenced by something beyond their material composition. The coupling of "motion and inclination" fits well with the project Pietro had before him. Motion had associations with nature, inclination with habit. What is more, nature as well as habit can, in Pietro's view, result from the operation of both internal and external causes.

Two circumstances, however, suggest an unresolved problem embedded in his exposition. The first is that although Pietro alludes to Aristotle's *Ethics* in this passage, Aristotle himself insists on a sharp distinction between motion and inclination. One of the distinguishing features of virtues and vices is that they are not compelled, and therefore cannot arise from our inherent nature in the way that, say, emotional responses do. "With respect to passions," he says, "we are said to be moved; with respect to virtues and vices, not moved, but disposed in some way."[25] The second sign that Pietro's conflation was problematic is that the word "inclination" is absent in Jean de Jandun's redaction, though it appeared in Pietro's first version.[26] The omission of a single substantive word is rare among the *Problemata*-related manuscripts, and it is possible that it resulted from a lapse in scribal attention. But it is at least as likely that Jean perceived difficulties analogous to those associated with the vocabulary of astrology.

Against this backdrop, the final sections of Pietro's exposition of *problema* IV.26 center on the origins and characteristics of sodomitical habits. For the commentator and his readers that subject intensified questions about the general relationships among nature, habit, and moral responsibility that are the subject of Chapter 4 and about the specific role of astrological influences that fueled some of the troubles treated in Chapter 5.

Habit Forming

> [Propter hoc] Because of this surely whoever, not before puberty but around puberty, have become accustomed to being subjected, because of that, what happens to them in experience, memory. And at the same time as memory, pleasure. And because of habit they desire to be acted upon just like those born suited. And surely often habit becomes just like in those born.[27]

Pietro's choice to place in the foreground the transformation of habit into nature (rather than the distinction between always and sometimes passive) was consistent with his perspective on the *problema* as a whole, particularly his explorations of the scope of the natural. As he turned to the less contentious subject of how habits form, he invoked the role of pleasure, featured in the first lemma on habit (Quibusdam autem), and the role of memory, featured in the second (Propter hoc). Just as he had dutifully elaborated the standard view of the buildup and excretion of superfluities, providing an ordinary framework for an extraordinary group of men, here he offered a routine Aristotelian account of habit formation, cushioning the causes of "sodomitical acts" in the comfortable environment of familiar natural philosophy.

In concise terms, the Aristotelian text sketches the steps by which some individuals come to desire the passive role, just like those born suited for it. They become accustomed to being "subjected"; they experience pleasure; they remember the pleasure; and thus they desire it. Bartholomeus consistently translated the general Greek term for sexual pleasures (ἀφροδίσια) by the concrete act of intercourse (*coitus*) and the active form of the related verb as "to copulate" (*coire*). Here he rendered the middle (neither active nor passive) form of the Greek verb with the grammatically passive "to be subjected." What some people become accustomed to is, literally, "to be placed underneath" (*supponi*).[28] Passivity (linguistic and behavioral) is thus imbedded in the Latin received by medieval scholars.

Pietro availed himself of this passage to stress once again the role of nature in the dynamics of desire and pleasure, alluding to principles of psychology in his brief exposition. These involve a pattern of pleasure, memory, experience, and desire akin to what, in modern terms, might be called reinforcement. The process starts with pleasure. As he had said in connection with the prior lemma, "those who 'emit semen' when they are engaging in coitus" are "moved to this act on account of pleasure," and they enjoy " 'whatever' [pleasure] happens to them." Apparently, this generalized appetite leaves some (or perhaps all) men susceptible, "and 'thus' it happens to those to whom such things are done." Having experienced "this wicked act, they seek to engage [in it] to the point where they may experience enjoyment and receive pleasure."[29] An individual's susceptibility depended in part on his age, and, in the second lemma concerning habit, Pietro explains that adolescents get used to being " 'subjected' and rubbed around the anus." Because of the "specific sensations and delights," the practice is retained as a memory. As he has before, Pietro calls upon the authority of a canonical Aristotelian text, in this case the *Posterior Analytics*. He refers to a passage concerning the more pedestrian subject of how states of knowledge are acquired, in order to elucidate the manner in which men may acquire the habits that make them comparable to their counterparts with monstrous seminal pathways. Sensation (in Pietro's description, pleasure) produces a memory, and multiple memories resulting from multiple sensations get consolidated as experience, which becomes settled in the psyche or "soul."[30]

This account echoes the one about motion and inclination that Pietro had just adapted from the *Ethics*, and he uses it as a bridge back to the close similarity (almost identity) between habit and nature. The text he is commenting on repeats that a habit becomes just like what is in those who are born with such desires. And, as though Pietro had not mustered sufficient authority earlier with his reference to Aristotle's treatise on memory, his summary of Hippocrates's anecdote about mothers shaping their children's skulls, and the allusion he has just made to the way experience inheres in the soul, he calls up yet another classical ally, Cicero, whose work on rhetoric he quotes: "Habit, from long duration, acquires the force of nature."[31] The result, says Pietro, is that the habit comes to be "just as though it inhered in boys [*nati*] from their birth [*nativitate*] by nature [*natura*]."[32] A reader of one manuscript underscored the striking character of that assertion by copying it out in the margin.[33] The high density of words containing the shared linguistic root of "birth" and "nature," the last of which Pietro added in his second version, is remarkable, given the emphasis he has placed on the distinction between innate and habitual inclinations. It is all the more remarkable in the context of the invective that infuses his language as he speaks of the men themselves in whom such habits have formed and become second nature.

"Abominable Habit"

[Propter hoc] Because of this—continued.

Starting with Pietro's premature introduction of habit and running through the end of the *problema*, although the language of the Latin Aristotle remains fairly restrained, the rhetoric of Pietro and the other commentators becomes markedly more derogatory, sometimes reaching an extravagant pitch. To the extent that the commentary on these final lemmas involves interactions between persons (as distinguished from the more disembodied pathways followed by superfluities), they bring out the moral resonances of *problema* IV.26. When Pietro introduces "habit" he asserts that it "ought rather to be called a source of corruption [*coruptela*]." In the last sentences of the *problema*, Bartholomeus used the words *luxuria* (corporeal indulgence) and *mollis* (literally, soft), but it is Pietro, going beyond ascriptions of effeminacy and monstrosity, who employs terms like "shameful" (*turpis*) and "filthy" (*obscenum*).

Such reactions on the part of Pietro, the other commentators, and their readers underscore how, in contrast to the relatively straightforward phenomenon of the innately, anatomically alway passive men, those whose causes were more elusive, whose bodies were unmarked, and whose desires were unconfined gave rise to a frightening array of sexual possibilities. Those with blocked pores sought to be rubbed around the anus, but the person doing the rubbing was hardly alluded to; these other groups seemed to require readers of the *Problemata* to imagine actual individuals (or at least their body parts) interacting. The resulting images undoubtedly contributed to the impulse to express disapproval and discomfort. They also provide us with evidence about how the medieval intellectuals pictured the sex acts and actors.

If the use of Avicenna's word "halubnathi" had been evasive, the appearance of the word "sodomite" was the opposite: it placed in relief what was shocking about the subject matter. Although one annotator conflated the two terms, most seem to have understood that "halubnathi" referred to a condition rather than a person. Pietro's way of dividing and organizing the text introduced an additional source of ambiguity by mentioning "sodomites" just before embarking on the explication of those who both act and are acted upon: Are sodomites characterized by a proliferation of appetites? Pietro was clearly applying the label to those, whether susceptible to one pleasure or more, "to whom the aforementioned things happen on account of a perverse and abominable habit. Men of this kind are sodomites [*sodomite*]."[34] Taking their cue from Pietro, scribes who made indices or wrote tag words in the margins of his commentaries made use of "sodomite" and related terms, connecting the content of an authoritative ancient work, the *Problemata*, with a set of norms firmly located within the context of Christian Europe. It was in this textual environment, too, where habit is distinguished from innate condition, that Pietro uttered his most vehement expressions of outrage, first in terms of morals, then in terms of laws. These, he holds, are the men whom Aristotle declared to be "beyond the boundaries of vice,"[35] and it is in order to root them out that laws are imposed.

Although Pietro provides an outline of the mechanisms by which a habit is formed, the exact content of the habit is not immediately clear. At first it seems as though it might be the same as the state of the men with blocked pores, for he says that "the 'condition [*passio*]' that was in the previously mentioned men from their birth" comes about in other men as the result of a habit. The term *passio* contained in Bartholomeus's text (like the Greek that it translated—πάθος) refers, in an abstract sense, to the result of an action (hence its relation to "passive"), but it often means "emotion" or "suffering." In this last sense it can refer to an illness. Pietro had earlier suggested "disposition [*dispositio*]" as a synonym, and that sense accords with his approach in what follows: what is being formed is the inclination to pursue particular sources of pleasure.[36] But he is seeking to understand more than a disposition or desire. For him the explanation upon which he is about to embark accounts not only for the "condition" specified in the Aristotelian text but also for the resultant behavior—the "wicked practice." In this way and others the final segments of the exposition sustain the shift in emphasis from the anatomical or physiological to the psychological and from the material to the social. But what exactly was this "perverse and abominable habit"?

Rubbing and Being Rubbed

[Propter hoc] Because of this—continued.

The transition from the relatively neutral language of the earlier parts of the *problema* to the invective of the final sections reflects the importance of the distinction medieval readers made or sensed between a condition that was materially compelled and a comparable condition that was not. It also colors and is colored by the parallel introduction of the men with two ways of achieving sexual pleasure and the men whose desires are shaped by habit. And finally it may reflect the fact that, since explaining habit formation involved contemplating acts and experiences, the latter part of the *problema* almost compelled readers to picture who was doing what to whom. Indeed, the language of concrete interaction intensifies at the point where Pietro (after making the premature introduction of habit) takes up the situation of those whose spermatic moisture is anatomically divided between paths to the penis and the anus. Whereas in the simpler case those whose seed runs only to the anus want to be rubbed, in this case those affected want to rub others and to be rubbed by them. Even though one glossator concluded simply that the resultant desire was "to be rubbed,"[37] Pietro and others held that the preference would be for one role or the other, "according to whether the stimulative moisture is configured [*nata*] to flow into the penis or into the anus."[38]

Some readers, influenced by Pietro's early mention of habitual sodomites, took this final, two-path and double pleasure resulting from anatomical irregularity as congruent with what resulted from the second line of causation, habituation. Thus the author of the Aachen *Problemata* followed Pietro's words closely, but applied them differently. Whereas Pietro attributed individuals' preferences to the location to which more "stimulative moisture" flowed, this scholar attributed them to what stimulation they had previously experienced: "Here [Aristotle] shows who is subjected to such rubbing through a distorted [*pravam*] habit. He says that there are certain people in whom such moisture is around the anus and certain people in whom one part is around the anus and the other part around the penis. Now, those in whom that moisture is around the anus desire to be rubbed and be acted upon, but [those] in whom part is around the anus and part around the penis desire to act and be acted upon, to rub and be rubbed. And this more or less, according to which they are more or less accustomed and stimulated."[39] The Erfurt epitome of the *problema*, less closely tied to Pietro's work, entertains no such complexity, stating only that "it sometimes comes about from a habit that the superfluity may collect in the anus." In this case, the result of habit formation would be simply the physiological equivalent of an anatomically based desire for anal stimulation.[40] George of Trebizond seems even to have had some sort of physical realignment in mind. Those who are both active and passive arrive at their condition from habitual actions, which stretch the pores at the anus, and they are to be distinguished from those rendered "soft" by innate obstructions.[41] In short, readers of the *problema* disagreed about two issues that emerged from the distinction between innate and acquired desires. First, some saw the results of habit as merely analogous or similar to nature; others took habit to effect actual material changes in the body. Second, while most attributed dual desires to habitual sodomites, others continued to be concerned only with the passive inclinations elaborated in the first part of the *problema* and emphasized by Avicenna.

Pietro's own understanding of those propelled by habit involved both acting and being acted upon, but the states and processes in play are not easy to sort out, in spite of his willingness to be blunt. After presenting his abstract account of the way habits form and become ingrained, with its citations of classical works on logic and rhetoric, he introduced a brief digression in which he spoke in the most explicit terms about the acts involved. In his first version it read in full: "It should be noted that some practice this abominable [*nephandum*] sodomitical act by rubbing the penis with the hand, and others by rubbing between the thighs of boys [*puerorum*], which many do today; others, though, by making friction around the anus and putting the penis in it as it is put into a vagina [*vulvam*]. And Aristotle's discussion seems to be more about such men."[42] The passage contains a wealth of information about the way Pietro understood and viewed male homosexual behavior. Having announced that the term "sodomite" referred to those guided by a habit, he portrays the "sodomitical act" in active terms; the person to whom the thighs and the anus/pseudo-vagina belong is not the subject of the verbs here. And one scribe preferred the even more explicit term "act of the sodomites" to "sodomitical act."[43] This perspective is consistent with an earlier passage, in which Pietro had attached condemnatory language to the active partner in contrast to the one with fully blocked pores, saying that the latter "would readily bend over [*condescendat*] to those who engage in this by an extremely shameful transgression."[44] Although his phrasing at these two points does not authorize the conclusion that those with ingrained bad habits invariably take the active role, it makes clear that he does not regard passivity as the defining feature of this group of men as a whole. Given that the *problema* and the commentaries center on explaining the susceptibility to perianal stimulation (whether exclusive or occasional) with particular emphasis on being acted upon, the shift of focus to the active "sodomite" introduces new questions, not all of which get answered. Is enjoyment of the active role acquired by experience? By experience of the passive? Are the active habit and the passive habit the same or distinct? If passivity is at the heart of the *problema*, why would the main stigmatizing term appear to be ascribed to those who act?

Pietro offered some clarification when he asserted that the specific actions covered by the *problema* and the category "sodomitical" may vary from person to person and from time to time. In this context, sodomy may be protean, but it is hardly vague. To be sure, the inclusion of "with the hand" was worth noting, for example by a reader who commented, "They do it with the hand rubbing the penis."[45] The nonreflexive wording rules out an interpretation of mutual masturbation, so the action described would seem to fall outside the active-passive framework. And in his later version, Pietro eliminated the reference to the hand, apparently judging it to be out of place. But since Jean de Jandun based his redaction on the earlier form, the hand makes its appearance in most of the extant manuscripts.[46] So too does a related phrase that Pietro likewise eliminated from his revision: at the very beginning of the commentary, where he explicated the central questions of the *problema*, along with the always-passive and the sometimes-passive, he had originally added that some people "do it separately, as will appear at the end of the discussion." Perhaps Pietro was giving some thought to the taxonomy of sexual sins that preoccupied some contemporary theologians when he made his changes; more likely he considered that the acts and pleasures of masturbators were not relevant in this context—whatever their social and moral status might be.[47]

Whether or not masturbation was included, Pietro's excursus on sodomitical acts places no particular emphasis on penetration. In the context of those affected by congenital causes, the Prague commentator had specified that "they want to be rubbed by the insertion of a penis into their anuses."[48] But in Pietro's account, "between the thighs" and "into the anus" differ historically but not substantively.[49] The enumeration of the acts not only emphasizes the active individual but also implies that the site for these sodomites' own pleasure is the penis. Only the final phrases of the *problema* bring Pietro back to consideration of passivity. There, the text he is commenting on states clearly that those who experience and remember the pleasures "desire to be acted upon because of the habit, just like those born suited [for it]."[50] And Pietro makes explicit that for them there is "great pleasure and the emission of seed at the anus."[51] In this context, the man who does the acting, along with his sexually sensitive penis and its emission of seed, has faded into the background, as has the term "sodomite."

Since Pietro had clearly asserted that those who acquire their desires from experience are the ones called "sodomites," we are left without an explanation of how they become active, escaping the passivity that is the very source of their habituation. How do those who become accustomed to the pleasures of being acted upon come to enjoy acting upon others? One possible clue to Pietro's thinking on this point is contained in his earlier comments on Aristotle's discussion of habit. The passage is difficult, and Pietro struggled to interpret its specific sense, though its gist is clearly that pleasurable experiences lead men to seek them again. Different manuscripts of Bartholomeus de Messina's Latin translation divide the phrases differently, betraying a similar lack of certainty on the part of copyists. A plausible (but by no means satisfactory) sense is, "Whatever happens makes them be glad and emit semen. Thus they to whom these things occurred desire to do [them]."[52] A certain obscurity may derive from Bartholomeus's Greek source.[53] The humanist translators appear to have detected explanations for both active and passive behavior in this account of habit formation. George of Trebizond says the men in question enjoy what they do and thus desire to do it.[54] He does not clearly specify which role they take and which they then seek to repeat, but a century and a half later a critic took him to task for suggesting that the result might be "doing" in an active sense.[55] Theodorus Gaza speaks more explicitly: whether they are the ones who do the bearing or the ones who act, they are capable of emitting semen, and desire to behave on the basis of their experience.[56] Both express in the grammatically active voice what the men involved end up desiring—they want "to do" or "to act."[57] Acknowledging these difficulties, Pietro had offered two different paraphrases:

> They are moved to this act on account of pleasure, since "whatever [pleasure] occurs makes" those men who "emit generative material" when they are engaging in coitus properly "experience enjoyment" and pleases [them]. It happens "thus" to those to whom such things are done. On account of this wicked act, they seek to engage [in it] to the point where they may experience enjoyment and receive pleasure. Or, understood in another way, [it is] because "whatever", etc.,

> that is, their entire effort is nothing other than to "emit generative material", that is, the seminal moisture, so that, from this, it "may happen" that they "experience enjoyment" and receive pleasure.[58]

One plausible interpretation of Pietro's readings is that the men in question, who have no impediment to ejaculation, learn to enjoy indiscriminately any behavior that satisfies them. Readers exhibited some confusion about who was rubbing and being rubbed where and with what.[59] Those that have anything to say on the subject do seem to be in agreement that habitual practitioners have multiple potential sources of pleasure, although they may prefer one to another. The Aachen commentator does not specify which organs are involved, but asserts that those in whom "habit is nature . . . always rub those parts on account of the pleasure or want to be rubbed by others—and those [*isti*] especially are sodomites."[60] Whether he is agreeing with Pietro on the active posture of sodomites depends on whether he intends "those" to refer to one group, the other, or both.[61]

Like some of Pietro's readers, the more independent Prague commentator joined the causal options (birth and habit) to the behavioral options (always passive; sometimes passive and sometimes active). But his clear language nevertheless leaves the processes of habit formation and its results unclear: "And there are also some who, through seeing and habit from [their] youth, want only to rub, since they remember the corrupted (*perverta*) pleasure. But there are others who want to be rubbed and to rub who have spermatic material [going] toward the virile member. And in them there is no malformation (?), but the spermatic passages are straight and open. And these same men sometimes have intercourse with women. And perhaps in them the spermatic material is divided, so that one part flows toward the rectum and the other toward the penis."[62] Here physical and psychological causes, rubbing and being rubbed, homosexual and heterosexual acts, deflected and undeflected semen, and penile and anal involvement all figure among the permutations of male sexual experience. Furthermore, unlike the original *problema*, which asks solely about the only-passive and the sometimes-passive, this text includes consideration of the only-active. It asserts that habit, with the concomitant memory of pleasure, causes the desire to rub, but does not explain how this comes about. It is possible that "sight [*visum*]," or watching, plays a role. An earlier version of the work may have read "practice [*usum*]," but the chapter in Avicenna's *Canon* mentions that a man of this kind "desires to see two people having intercourse."[63] Whether the creation of the original author or of the person who made the copy, the text as it stands in the only extant manuscript reflects one way that the process could be imagined.

According to this view, men with no anatomical irregularities and thus able to emit semen through the penis may develop habits from certain unspecified but "corrupt" experiences that they remember with pleasure. Some of them become rubbers as a result, whether because their original enjoyment came from acting upon others or from being acted upon and (by unexplained psychological processes) forming the desire to do what was done to them. Others are habituated to the pleasures of being rubbed and in addition—whether by a similar, unexplained process or because it is the default desire of men—are inclined to rub others. Finally, "these same men [*ipsi*]" can and do have presumably ordinary intercourse with women. Sandwiched as it is between a sentence about double desires and one about divided spermatic pathways, the assertion that the same men have intercourse with women would seem to apply only to those seeking both to rub and to be rubbed. If so, the passage leaves open the possibility that the active men with good anatomy and bad habits, that is, the equivalent of Pietro's "sodomites," may not have sex with women even though they could. In the final sentence of this passage, the Prague commentator adds to the already crowded array of figurations the suggestion that those with habit-based dispositions may have their spermatic material flow to both destinations. Pietro too had asserted the possibility of two outlets (against Avicenna's position), but he did so in the context of the "monstrous nature" of men with anatomical defects rather than bad habits; the Prague commentator postulated two outlets in men with "no malformation."

Taken together, the Prague text, Pietro's exposition, and the comments of other readers—sometimes internally inconsistent, sometimes in disagreement with each other—give the impression that late medieval natural philosophers had difficulty producing a clear picture of what all these men were doing and why. Whatever their personal knowledge of the world was and however familiar they were with doctrinal rhetoric or judicial practices, their attempts to clarify the Aristotelian text and to characterize the states and experiences of the people it purported to discuss leave us anything but a coherent picture. Like many of the differences that emerge in readers' views about the moral dimensions of the phenomena, their disagreements about who does what to whom do not take the form of scholastic debates. Rather they appear to be individually improvised. The lack of clarity on these points undoubtedly has something to do with the nature of the *Problemata* itself—its miscellaneous content, its focus on phenomena not easily reduced to higher philosophical principles, its distance from the academic core curriculum—as well as with the cramped Latin translation. But it is probably also related to the delicacy of this particular subject matter and to the confusions and silences about it within other realms of medieval discourse.

As a result, even those readers who followed Pietro d'Abano quite closely were left to scramble as they elaborated on how habits occurred, in what sense they became (like) nature, how the nature/habit distinction was related to the type and number of desires experienced by an individual, who did what to whom, and who should be called a sodomite. But for at least some readers, a few of these questions were made clearer by the final variable introduced in the last passages of the *problema*: the age of the participants.

Mollis: A Tender Age

(Propter hoc) Because of this—continued.

Whether they thought there were specific conditions, such as divided spermatic vessels, that made some individuals susceptible to the formation of a "filthy habit" or whether they left open the possibility that any male might develop deviant desires, medieval authors were generally in agreement about the importance of age. The age at which the experiences occurred and the memories formed was a significant factor in determining an individual's potential to develop an abominable second nature. The text of the *Problemata* itself was specific about the age at which sexual desires are inculcated: "not before puberty but around puberty."[64] Bartholomeus's Latin made it clear that the result was a man with the desire to be acted upon as though he had been born that way. But one reader thought the passage referred to a process centered on a time before puberty and explained where active sodomites came from. He expressed this view in the index he prepared for a manuscript of Pietro's commentary. In the margin of the manuscript he was indexing, this scribe wrote, "Those who are sodomized before puberty, etc." and then, "They are the ones who later become sodomites."[65] In the index itself he confirms his understanding of the passage: "Those sodomized [*sodomitati*] before puberty and the time at which they could emit sperm are those who are later made into sodomites [*sodomite*] on account of memory, etc., since habit turns into nature."[66]

Although it is no more enlightening than are other interventions concerning the means by which memories of passive pleasure produce active desires, and although it places the founding experience "before puberty," this account seems to correspond to a model familiar to historians of premodern male same-sex relations. Court documents from late fifteenth- and early sixteenth-century Florence, for example, underscore the distinction between the active and the passive partners, and record that, in the overwhelming number of cases, the active were older than the passive (who were usually adolescents). Furthermore, adult men were expected to have outgrown passivity, and those who did not were subjected to special stigmatization.[67] That is, actual cases seem to have corresponded to an idealized model according to which the transition to full manhood was to be a transition to full and exclusive activity. Taken together, the index maker's summary and the evidence available from reports of behavior produce a diachronic view of the relationship between the (passive) sodomized and the (active) sodomite.

Whether or not Pietro d'Abano himself had imagined a developmental transition from acted upon to acting, for him it is the older, active partner with no anatomical excuse whose behavior is particularly transgressive. When he first introduces the "wicked practice," he speaks of "those who do these unspeakable things [*nephanda*]": "The law cries out for the eradication of these men: 'Let the laws rise up, let judgments be armed; let them be slain by the avenging sword.' Whence Avicenna [says] in the *Canon of Medicine*, book 3, fen 20: 'Intercourse with young children [*infantibus*] is [considered] filthy among most peoples and legally prohibited.' "[68] The first of the citations is from Roman law.[69] It is an exhortation that appeared in connection with sexual transgressions in a variety of medieval sources, and the fact that Pietro's quotation is inexact suggests that he was familiar with it from his cultural milieu, not from reading the law.[70] Together with the line from Avicenna and a reference to Aristotle's *Nicomachean Ethics*, both of which he had read carefully, the avenging sword forms a triad of invective—law, medicine, philosophy—of a sort completely absent from Pietro's treatment of the behavior of the congenitally passive. It also establishes, if through the voice of Avicenna, Pietro's sensitivity to the dynamics of age in relation to the subject of *problema* IV.26.

When Pietro reached the mention of age within the Aristotelian text itself, he set aside extravagant rhetoric temporarily and returned to the task of scholastic exposition. Governed by the apparently clear phrase "not before but around puberty," he left behind Avicenna's mention of "young children." As a prelude to his explanation of habit formation, he specified his understanding of puberty in a manner consistent with his earlier comparison of men with blocked pores to boys: "[Aristotle] shows how those beginning to practice it at the time of puberty relate to this act, mentioning thus 'whoever, not before puberty' and the time when they are capable of emitting sperm, 'but around' it, has become accustomed to being put on the bottom [*suponi*] and being rubbed around the anus. For that reason such a memory is retained for them in a habit."[71] That is, the process does not take place before a boy is able to emit semen. Thus Pietro gives no indication that these male adolescents are even temporarily similar in constitution to the men with blocked seminal pathways. On the contrary, if they do not form the habit before they can emit seed, then their seminal fluid *can* get out. Even on this point, however, a hint of confusion creeps in. In one manuscript, readers would find this critical moment of susceptibility as "not before puberty and in the time in which they *cannot* omit sperm."[72] And Jean de Jandun's version also diverges subtly from Pietro's.[73] All clearly convey the sense that the youths involved are at a moment of transition, but the differences foreshadow a more significant opacity in connection with the passage that follows.

> [Si autem] And if licentiousness emerging should occur and soft, and each of these happens quickly.[74]

As the *problema* comes to a close, the taxonomy of males inclined to act upon and to be acted upon by others becomes even harder to discern. The final sentence of *problema* IV.26 seems to add simply that seminal emissions will take place quickly in those who are soft (μαλακός) and lustful (*λάγνος*). Modern and humanist translators are in general agreement about its meaning. It does not mention age, but treats a special or extreme case: such things will happen more easily or quickly to someone with such weaknesses.[75] Bartholomeus's translation, however, is difficult to parse. It might be understood to say, "If, however, arising desire should occur and [he were] soft, each of these things happens to him quickly." Perhaps because of the difficulty of Bartholomeus's syntax, perhaps because of his addition of the word *existens*, which may mean "coming forth" as well as "existing," or perhaps because the element of age differences seemed important to them, Pietro and other readers understood the subject of the Latin text differently. They took it to be about the relevance of age in the process of habituation, rather than about what type of individual was particularly susceptible.

By importing the phrase "before puberty" into his exposition of this final passage, Pietro expanded the significance of age. He had evoked it earlier, quoting Avicenna's declaration that intercourse with boys is reviled among nations. At that point he was similarly introducing the element of age into a textual environment from which it had been absent—Avicenna's views on *halubnathi*. To medieval readers with philosophical and medical educations, the connection between passivity and age was reasonable but not imperative.[76] Pietro's decision to emphasize it was highlighted in some manuscripts of his work and was to persist in later treatments of the text. In his view: "[Aristotle] shows how those who are subjected [*supponuntur*] 'before puberty' relate to this. He says that if this kind of shameful licentiousness [*luxuria*] should happen at the time when the nature of those who are subjected is 'soft' and weak, as when they are not [yet] able to emit sperm, all of the things just mentioned happen 'quickly' to them—enormous pleasure and emission of seed at the anus, etc. And the things that have become habits are turned into nature—and more than in those who do not begin 'before puberty.' "[77] An annotator wrote in the margin "sodomized before puberty," further underscoring the point.[78] By the late Middle Ages, *luxuria* (often translated as "lust") was firmly anchored to notions of sexual excess,[79] and so ascribing it to boys would have been unusual. More important, even though the Aristotelian text had insisted (and Pietro had acknowledged) that such habits are not formed "before" puberty, only "around" that age, here the commentator and scribe alike seem to propose that younger boys are especially susceptible. Not only do they form such a habit more quickly, that habit is even more likely to become their nature.

The logic operating at this point centers on the word "soft [*mollis*]," which had two fields of meaning that intersected in Pietro's commentary. First, it had associations with childhood. Arguing in his *Conciliator* that youths [*iuvenes*] have more temperate complexions than children [*pueri*], he pointed out that children are "excessively soft and smooth."[80] In that context, medieval medical and philosophical traditions offered ample insight into the physiological and psychological malleability of the young. For example, they are by nature particularly impressionable, and have better memories than adults.[81] Thus not only their ability to learn good or bad behavior but also the robust persistence of what they learn figured in late medieval educational theory. As Christine de Pizan (1364–ca. 1430) remarked, children should be exposed to devotional literature and good conduct, not foolishness, "because the principles and teaching that the child retains in its earliest youth it usually has recourse to all its life."[82]

But *mollis* had a separate field of meanings that applied to adults, especially to adult men, many, though not all, explicitly sexual.[83] For example, in

Book VII of Aristotle's *Nicomachean Ethics*, which Pietro quotes several times, the term is absent from the discussions of children but appears in connection with general incontinence and emotional weakness.[84] The kings of the Scythians are said to be soft by nature, while others may be soft because of a sickness.[85] In this connection, the word had sexual connotations ranging from men given to excessive indulgence with women through men who persistently assumed the passive role in intercourse with other men.[86] Although it may sometimes refer to men who, in their weakness, have sex with both men and women, that association does not surface in the context of *Problemata* IV.26, where the commentators pass over the opportunity to use the word in connection with the men who enjoy pleasure at two anatomical sites or those who have sex with both male and female partners. George of Trebizond used the word in his Latin translation, where Bartholomeus de Messina had used *effeminati*. Later, in his note on this passage, he appears to suggest that (as might be the case with eunuchs), softness was a result of sexual passivity, rather than a precondition for it. "Thus the humor is diverted to the anus. . . . Hence they are made *molles* by nature. They are the ones who passionately desire to be acted upon, nor can they in any way act."[87] Hence, in many contexts, this sense of "softness" is ascribed to adult men rather than to the very young.

These patterns of usage make Pietro's interpretation of the final sentences of the *problema* all the more original, if not strained. That is, in addition to his apparent abandonment of Aristotle's insistence that the desires are inculcated "around puberty" and "not before," his reading of "soft" as a mark of extreme youth would seem to run at a tangent (if not counter) to the usually adult associations between softness and sexuality. Some readers were nevertheless willing to follow Pietro in stressing special effects upon the very young. One manuscript of the Bavarian *Problemata* seems to express the same relationship between softness and susceptibility: "And here Aristotle says that this huge pleasure from the emission of seed, and emission of seed at the anus . . . happens quickly to boys [*pueris*] while they are delicate and soft, more than in those who do not begin before puberty."[88]

In contemplating the very young, this author introduced a source of pleasure that is absent or at least invisible in other discussions of *problema* IV.26. Most commentators emphasized an individual's desire to be rubbed and entertained the question of his own ability to enjoy the release of some form of accumulated superfluity. But in this case the child is, by definition, incapable of emitting seed, so the pleasure imagined by the commentator must be the result of the presumably adult sodomite's ejaculation.

Other students of the text, however, appear only to have been struggling to understand the age of susceptibility. Some may have been puzzled by the apparent contradiction between the initial insistence that the subjects be around puberty and Pietro's subsequent inclusion of those who were not. They may even have had more general questions about the role of age in the production of this particular habit. Whatever the cause, the manuscripts reveal a certain dissonance. Another copy of the Bavarian text says that the soft boys (who are manifestly prepubescent) are more susceptible "than those who have *not* yet begun puberty."[89] It is difficult to make sense of this apparent slip, and it would have little significance were it not for a similar confusion manifested in the manuscripts of Pietro's commentary itself. A number of them contrasted the swiftness of the process in these soft young boys with what happens in "those who begin 'before puberty'" (rather than "'not before puberty'").[90]

Some late medieval readers, particularly those less tied to Pietro's exposition, found ways around or through such difficulties. The Prague commentator declines to split hairs, and refers vaguely to habits from "youth [*iuventute*]" that result in a desire to rub.[91] George of Trebizond provided a striking solution that retained the logic of the Aristotelian text with respect to the proper age of habit formation, while entertaining the possibility that preadolescent experiences might also shape future pleasures and behaviors. It was in this context that he explained how, in contrast to the pleasure of being acted upon around puberty, being acted upon before puberty produces a physical stretching or breaking of the passages in the anus. "And from this cause, rather than from memory, they are brought to this shamefulness. Indeed those [who are] somewhat older who subject themselves to men desire to be acted upon just by the memory of this sort of pleasure."[92]

Not only did scholars disagree about the significance of age, individual readers, including Pietro d'Abano, were sometimes unable to articulate consistent positions. Since the state of the congenitally blocked individuals was fixed, the logic of their condition left no place for consideration of the age of those engaging in sex acts, and neither Pietro nor other readers attempted to create one. Nor have they left evidence that they regarded age difference as a regular aspect of homosexual relations once the habit had been established. But, largely under the influence of Pietro, they expanded its role beyond what was obvious in the Aristotelian text by opening up the possibility (seized upon by one annotator) that the sodomized would later become sodomites, who were the active partners in this context. And they extended the importance of age by locating exceptional susceptibility in boys younger than the *Problemata* itself would deem possible.

In its particulars, which strained the meaning of his Aristotelian source, Pietro's discussion of the age of susceptibility provided a tenuous ending to his exposition of *problema* IV.26. Its more general point, however, closed the circle formed by birth and habit. As Aristotle had put it, the habitual becomes just like the congenital, or, in Pietro's stronger formulation, "things [to which one has become] accustomed are converted into nature."[93] Thus, for readers of the *Problemata,* the particular malleability of children suggested yet one more way in which an inclination acquired by habit approximated one acquired at birth: in both cases the disposition existed from childhood. For here Pietro reiterates the effective equivalence of the two causes that he has taken as the main subject of *problema* IV.26: both innate properties and reinforced experiences produce some sort of "nature." He likewise brings together the disparate authorities on the subject, implying that their views reinforce his own. Avicenna, for whom the condition was "mental not natural," becomes an unwitting ally once Pietro has dismantled the distinction between those two alternatives.

Pietro added a postscript to his exposition of *problema* IV.26, without the slightest prompting from the text itself: he cites two astrological works by the foremost authority in the field, the second-century Greek astronomer Ptolemy. His inclusion of these references, though unaccompanied by any quotations or elaboration, draws together significant elements of his argument. The stars influence us from the time of our birth or even earlier. They often leave a physical imprint upon our natural physiological balance and even on our physiognomy. Yet much astrological literature is devoted to personality traits, inclinations, and behaviors. Thus, as Pietro concludes his commentary he implicitly restates the fungibility of the congenital and the behavioral, the anatomical and the psychological, simply by mentioning Book IV of Ptolemy's *Quadripartitum* and aphorism 80 of the pseudo-Ptolemaic *Centiloquium.*

In the Stars

In the fourteenth and fifteenth centuries, authors both sympathetic and unsympathetic to specific applications of astrology—from the medical to the political—inquired into what sort of force was exerted by the heavenly bodies. In those contexts many of them took care to distinguish astral influence from the determinism of "necessity." Such concerns entered into the arguments of

Evrart de Conty and his older contemporary Nicole Oresme, and it was sufficiently familiar to be deployed by a knowledgeable woman, Christine de Pizan: "In what is subject to the soul, namely deliberation of the will, the influences of the heavens have no sovereignty, notwithstanding that it may be true that the action of the heavens gives the individual a number of inclinations, such as to some people joyfulness, lust, or other natural motions. But this notwithstanding, the individual can rein them in by reason and resist with respect to deeds all such inclinations."[94] In different terms, Pietro himself made a related case, when he wrote about his set-to with the Dominican inquisitors of Paris and when he defended the scientific nature of astrology.[95]

Pietro was an expert and committed astrologer, if not as a practitioner at least as an intellectual. His work the *Lucidator* presents sophisticated arguments concerning the nature, principles, and problematic status of the science of astrology, as well as certain controversial questions of cosmology and mathematical astronomy. In addition, astrology makes frequent appearances in his main medical work, the *Conciliator*, and in his treatise on physiognomy.[96] In spite of the fact that the *Problemata* poses few questions that invite their consideration, astrological causes have a place in many chapters of Pietro's commentary.[97] Nevertheless, he made no direct appeal to that science in his exposition of *problema* IV.26 until its very last lines. Readers attuned to astrological language may have heard hints of celestial influence in his mention of the role of a "higher form" and his use of the word "inclination" in the course of his digression about the relation between natural motion and habit—and the stone thrown repeatedly upward. But whether the subject was uninteresting or uncomfortable, few of Pietro's readers took note of it. The Bavarian commentator copied his brief astrological discussion in its entirety; the Aachen, Erfurt, Vatican, and Prague commentators omitted it, as did Walter Burley.

For some scholars, however, the operations of the stars were relevant. George of Trebizond was probably inspired by Pietro when he cited one of the same astrological sources in his notes on *problema* IV.26, but he was original when, in his specifically astrological writing, he attributed the unnatural closure of men's pores to Saturn's sway over the "formative power" at work before an individual's birth.[98] The fifteenth-century physician Jacques Despars, on the other hand, explicitly rejected astral causation in his commentary on Avicenna's *halubnathi*, and Evrart de Conty implicitly took exception to blaming the stars.[99] The manuscripts of Jean de Jandun's redaction omit the word "inclination," but Jean himself discussed celestial influences in the production of monsters in his commentary on Aristotle's *Physics*.[100] The language of celestial influence was common currency in the late Middle Ages, but few scholars were familiar with the specific content and technical details of the texts Pietro referred to, and so scribes often confused the numerals (and thus the planetary positions) mentioned in his citations.[101]

Pietro himself is hardly expansive at this point. Following an allusion to Aristotle's *Ethics*, he says simply: "Indeed Ptolemy, in the eighth [eightieth] aphorism of the *Centiloquium* and in the fourth book of the *Quadripartitum*, indicates the cause [is] from the conjunction or aspect of bad [planets] with Venus in the sign of bad [constellations] and especially from a conjunction or aspect in the seventh and sixth house of Jupiter, far from these things."[102] He does not summarize, much less explicate, what these sources say about the predicted sexual inclinations and behaviors for men born under such stars. Yet for Pietro himself, for the astrologically informed members of his audience, and for modern readers, this bridge from the *Problemata* to the language and principles of celestial influence expands the meanings and evaluations associated with the subject of *problema* IV.26.

Although familiar with an impressive range of sources, Pietro relied especially on the two Ptolemaic works he cites here, along with the Latin translation of point-by-point commentaries attributed to the eleventh-century Arabic scholar 'Alī ibn Riḍwān with which they were usually copied.[103] Pietro held to a strong form of astrological causation, taking the position that many things are the necessary consequences of celestial influences acting upon susceptible terrestrial materials. When he addressed the matter directly, he took care to preserve some room for human free will by formulating a conditional or "soft" position on necessity.[104] Yet he often seems to operate at the edge of orthodoxy. Elsewhere in the *Problemata*, he appeals to astrological authority in his explanation of why people have widely divergent tastes in food. After having intimated that, although some people find it delicious, the very sight of someone consuming milk with bread in it makes him want to throw up, he refers the reader to Ptolemy's astrological aphorism on the subject. There one learns, as the Arabic commentator personally attests, that there is no accounting for taste—in food or sex—unless it be by the influence of the heavenly bodies: "Indeed, I saw a certain rich man, the lords of whose nativity were Mars and Saturn—and they were strong. He took pleasure in eating marine beasts that were salted and rotting, and he cooked them in the oven. He also swore that when he had intercourse with lovely and sweetly smelling people, his penis failed, and with a base and stinking person it was strengthened."[105] Pietro quotes only briefly from the text of the Pseudo-Ptolemy's *Centiloquium* in this *problema* (XXI.13), and he does not refer to this discussion of sexual predilections. Any specific association with *problema* IV.26 hundreds of chapters earlier would surely have been lost on the medieval reader and was probably not intended by Pietro, who worked for many years on the almost endless sequence of questions in the *Problemata*. Nevertheless, it illustrates a textual environment in which idiosyncratic or deviant human pleasures were understood to be imprinted from birth by the powers of the heavenly bodies.

The citations Pietro gives at the end of IV.26 confirm the sense conveyed by the body of his commentary that its subject is not simply men who assume the passive role in intercourse with other men. In general the men treated in the Ptolemaic passages to which Pietro refers are inclined to unclean and shameful intercourse, and in some cases they are able to resist their impulses. In particular, under the influence of Venus and Jupiter without the involvement of Saturn, they will be attracted only to women (sometimes especially low women and servants), only to men, or to both—depending on the relation of the planets to the sun. The age of the male partner may matter in some cases and not in others. In the Latin version favored by Pietro, the relevant passage of Ptolemy's *Quadripartitum* (in the chapter "On Marriages") reads: "If one of the two planets was out in the morning and the other in the evening, he will love the pleasure [*luxuriam*] of males and females, but will not enjoy one more than the other. And if both were morning [planets], he will enjoy pleasure, especially of women, a lot. And if they were in feminine [zodiacal] signs, he will allow the same thing to be done to him. And if both [planets] were out in the morning, he will only enjoy boys [*pueros*]. And if [they are] in masculine signs, he will enjoy lying with men of whatever age they are."[106] The Arabic commentary which was translated along with the ancient book, and to which Pietro referred frequently in his *Lucidator*, explains that a man born under feminine signs will "suffer to have done to him what he does to women," and that a man who likes men of any age does not like women.[107] A reader's picture might have been slightly different if the volume at hand contained the other widely circulated translation of the same work, with which Pietro may very well also have been familiar. It does not distinguish between men and boys but uses the more age-neutral term "males" [*masculi*].[108]

The other passage to which Pietro refers without elaboration comes from the pseudo-Ptolemaic compilation of astrological aphorisms, the *Centiloquium*. It states simply that the child born under certain configurations of Venus and Saturn "will be a child of unclean intercourse." In its Latin version, this work too circulated with a helpful commentary by the same Arabic author, so a reader would not have been left to wonder what kinds of behavior were involved: "[Venus] will stimulate the one born to prefer posterior defilement to anterior and [to take pleasure] in the stimulation of sperm with the hand. But if Saturn was in the East, he would enjoy rubbing together with boys [*pueris*], and he prefers black ones to white ones and blonds. . . . [If Mars is involved] he may have intercourse with several boys and they may have intercourse with him and with his wife. And it may happen that he would fornicate with his own children [*filiis*] and act with them against the law and other things that pertain to lewdness."[109] The *Centiloquium* was also the subject of a commentary by George of Trebizond, according to whom, "When [Venus] is in conjunction with [Saturn] it obstructs and closes the seminal passageways of men." That is, he understood the astral influence to have a physical rather than a moral effect in this case, just as he had favored a physical interpretation of the manner in which the *molles* came to be exclusively passive.[110]

A survey of the surviving manuscripts of the two astrological works cited by Pietro might yield some conclusions about medieval readers' interest in these passages and the range of their reactions to them. Trebizond, at least, ventured beyond the basics of scholastic commentary. As an example of the type of men with this physical defect, who "are in no way able to act," he cites the case of his enemy Poggio of Florence, whose name, he asserts, derives from *podex*, "asshole."[111] According to his account, included in an early version of the commentary and omitted in a later one, when Poggio was older than seventy, he took an adolescent wife and "did what he had never been able to do as a youth." Trebizond troubled himself neither with the facts of Poggio's life nor with fitting the change into his theories.[112] Like the science of physiognomy, the science of astrology—a source of insight about personal traits and behavioral patterns imprinted in individuals—was available for insults as well as explanations.

The passages Pietro refers to thus give some insight not only into his own tendency to resort to natural (and often specifically astrological) explanations but also into the diverse desires and behaviors he had in mind as he was commenting on *problema* IV.26. Taken together, the astrological texts predicted that a man with such a horoscope would have inordinate sexual appetites and perpetrate or be subjected to shameful acts. Such a man might enjoy sex with either women or men, as the Prague commentator had noted in connection with those with healthy pores who nevertheless like to be rubbed as well as to rub. And, depending on the details of the celestial configurations, the man described in the standard commentary on the *Centiloquium* might assume the passive role or the active role; he might prefer boys or enjoy sex with males of any age and perhaps in any role; he might have intercourse with boys and they with him (and with his wife); he might prefer "posterior" stimulation and enjoy masturbation.

This textual environment helps to clarify the meaning of "sodomy" in the medieval context of *problema* IV.26. Although only a few readers would have had direct acquaintance with the specific passages in Ptolemy, the widening circles of association reflected in Pietro's commentary were not limited to those particular works. At its narrowest, the question posed by the Aristotelian text concerned some men's susceptibility to anal sexual stimulation, but the *problema* itself incorporated eunuchs, effeminacy, and insatiability. Pietro elaborated on those, highlighted activity as well as passivity, introduced and then excised solitary masturbation, and was sufficiently equivocal about the age of the participants to admit the possibility of various pairings. The astrological citations he called up admitted heterosexual intercourse to the fringes of the subject, and added incest. Homosexual acts are thus embedded in a morass of problematic impulses, but male-male attractions and interactions are clearly the heart of the matter. Furthermore, their variety refuses reduction to a simple active-passive dichotomy, or even the tripartite division expressed at the beginning of *problema* IV.26 itself: men who are always passive, sometimes passive, and (implied but not mentioned) always active. What is at issue, in other words, is neither "sodomy" in its most diffuse sense of disordered sexual indulgence nor simple sexual inversion in which a man (by definition and nature active) is rendered passive—the hammer becoming the anvil, in the metaphor of Alain de Lille.[113] Rather it encompasses the erotic entanglement of males in a profusion of permutations. Understood in this way, the culminating invocation of Ptolemaic wisdom explains and authorizes much of the complexity of Pietro's commentary on *problema* IV.26.

The association with astrology had other implications, even for readers who lacked access to the particulars or technicalities. It conveyed the late medieval commonplace that celestial circumstances influenced objects in the terrestrial sphere. In other words, it alluded to the distant causes of some or all of the conditions Pietro had heretofore been explaining in terms of immediate ones, whether anatomical or psychological. His brief references to "conjunctions," "aspects," and "houses" offered enough detail to lend an aura of scientific precision. Astrology did not figure prominently in the exposition of the *Problemata* as a whole, precisely because the questions addressed there are framed and, for the most part explained, in terms of local conditions and contingencies. But astral forces were a widely accepted aspect of the late medieval world view in general and of late medieval medicine in particular. Placed at the end of *problema* IV.26, they reinforced the theme of natural processes that underlay Pietro's commentary on the work as a whole.

Received in this way, the final reference to nativities lent implicit weight to nature in the competition between birth and habit, yet it undermined the very terms of that dichotomy. Although Pietro made decisions about structure and interpretation that went beyond and even diverged from the text on which he was commenting, he had, to this point, appeared to accept its clear implication that "birth" pertained to the corporeal and "habit" to the psychological. Even though he linked the two by emphasizing that habit became a sort of nature and even though, as a consequence, the resulting desires and pleasures were hard to separate, the two causal pathways nevertheless remained distinct. But the medieval science of astrology operated in both domains: the stars reigning at one's birth could shape one's physical traits (e.g., healthy or unhealthy) and one's psychological traits (e.g., quick-witted or slow-witted). In the context of evaluating mates, which is the topic around which sexual matters come to the fore in astrological texts, Pietro's sources do mention anatomical features. For example, under certain conditions a child may be born a eunuch or lacking the hole in his penis.[114] And, more generally, an individual's predicted fertility was a topic of interest. But nativities were also concerned with future character traits, with social and economic status, and with how a person will behave. Pietro's audience would need no specialized background to expect astrological science to deal with the nature and strength of an individual's sexual appetites. These might in turn be grounded in an underlying physiological complexion. *Problemata* IV.30, for example, asks why melancholics are lustful, and a predominance of black bile was well known to be associated with sexual intensity.[115] Sexual proclivities were expressed as dispositions of temperament which, like physical features, were widely understood to originate at conception. They were not mechanical like blocked vessels, but they resembled the anatomical irregularities in that they inhered in an individual from birth. Thus they were not acquired from experiences, and yet, like habits, they inhabited a person's mind or soul.

Conclusion: Natural Habits and Habitual Natures

Pietro dismantled the dichotomy between nature and habit in two ways, first by emphasizing how acquired cycles of pleasure and desire get deeply embedded to the point of becoming indistinguishable from nature, then by referring his readers to celestial influences presiding over both body and behavioral traits from birth or before. Thus a medieval or modern reader hoping for answers to the Aristotelian question will not be entirely satisfied. "Why do some people with whom intercourse is had experience enjoyment. And indeed some also acting, some not?" On the source of pleasure, Pietro advanced anatomical, physiological, psychological, and astrological explanations. Not only did he not choose among them, but he also confused and conflated them. Together, however, they do constitute a general stance. The processes involved, even those of habit formation, are natural ones. Pietro's position is forcefully expressed in his rejection of Avicenna's assertion that the origin of the condition is "mental, not natural." Though the condition may have psychological causes, these are by no means independent of the natural processes with which physicians and natural philosophers are properly concerned. And thus to the secondary question of the *problema*—birth or habit?—he not only answered, "Both," he also added "Is there a difference?" Finally, on the division between the always-passive and the sometimes-active, or on the question of who does what to whom, Pietro's account is elusive: he and others who commented offered differing views of the roles and ages of participants in sexual contacts between males. These ambiguities of inclination and act are further heightened by the profusion of possible appetites that inhabit the astrological sources.

Like his insistence on material and efficient causes, rather than on the higher principles of teleology, Pietro's mapping of multiple causal pathways and varied behavioral results is consistent with the character of the Aristotelian compilation on which he was commenting. The phenomena presented in the questions were, by the definition of the genre itself, puzzling objects and events. They did not (or seemed not to) fit into the purposeful and harmonious system of the world as construed by Aristotelians, much less the providential one as construed by Christians. Yet they were natural, both in the sense that they were well recognized and in the sense that plausible explanations could be advanced using the standard tools and concepts of natural philosophy and medical theory. Pietro's final appeal to astrology reinforced the tractability of the questions about men with whom intercourse is had, as had his previous references to Aristotle, Avicenna, and other authorities. Yet whereas Pietro had quoted directly from those authors and sometimes explained their positions at length, he declined to do the same for his astrological sources. Perhaps the political and moral marginality of that contested science caused him to hesitate—not elsewhere in the work, but in this particularly inflammatory context. The language he applies to the *problema*, from the neutral "these men" through the label "sodomites" to the indignant "shameful," suggests a certain level of tension in the voice of this aggressive naturalist.

Such signs of disapprobation and disgust played themselves out at two levels in the literature surrounding the *Problemata*. From a philosophical perspective, the naturalization of the unnatural raised particular questions about moral responsibility, which Pietro underscored with multiple references to Aristotle's *Nicomachean Ethics*. These are the subject of Chapter 4. From a social perspective, Pietro's vocabulary of shame and transgression represents discomfort with the subject matter—discomfort that is profusely manifested in the positions expressed by Evrart de Conty and other commentators, as well as in the words and actions of scribes and readers as they have been preserved in *Problemata*-related manuscripts that circulated in the fourteenth and fifteen centuries. These are the subject of Chapter 5. But explanations of how physically healthy males can develop deviant desires and pleasures through experience also involved language that linked women to both anatomically monstrous men and men with unspeakable habits. Women provided the male readers with a rich array of resources for interpreting *Problemata* IV.26.

Chapter 3

"Just Like a Woman"

Passivity, Defect, and Insatiability

If commentators' references to "these men" and "such men" as well as "sodomites" suggest the existence of a category or type of person, they do little to illuminate what traits characterized its members, beyond the natural mechanisms invoked to explain their existence. But just as the Aristotelian text pointed in the direction of several possible causes, so it also suggested a cluster of associations that have the potential to enrich the picture of the men in question: it compared them in various ways to eunuchs, impotent and sterile men, boys, and adolescents. Though medieval readers did notice and even develop these signs of incompleteness and incapacity, the indication that the men in question were in some way like women especially commanded their attention. They entertained the notion that anatomical blockages of the seminal passages reduced men "to the nature and behavior of women"[1] and remarked on the similarity of their sexual appetite to that of adolescent girls. As a group, the *Problemata* texts enhance the portrait of the several sorts of men described in IV.26 by developing certain aspects of the female and the womanly to which they are, in various ways, tied.

Just Like a Woman

In some respects, these resonances helped to create a profile of the men whose astrological imprint, anatomical constitution, or long-standing habit inclined them to very specific forms of pleasure. But the temperamental and behavioral traits that clustered around these associations turn out to be in tension with each other or even contradictory. Thus the result of these medieval comparisons between women and the men of *problema* IV.26 is the sort of elusiveness familiar to modern scholars who have tried to pin down the medieval notion of "sodomy." Passive receptacles turn out to be both active and passive, and are often not receptacles; certain defects make girls and men unsuited to intercourse, which is why they so adamantly seek it—or something like it; men who are already "effeminate" face the paradoxical danger of becoming "womanly."

The attention given to the comparison of the men of IV.26 with women varies from one textual setting to another. Although it adds milk to the list of natural superfluities, the Erfurt *Problemata* omits any mention of women, even where the ancient original had explicitly set up the connection.[2] And a bowdlerized copy of the Prague text records nothing but the comparison between the ordinary manifestations of sexual pleasure in women and men—with no hint of the original role played by that point in the full version of the text. But in their views of causes and in their choice of language, other short renditions of the *problema* considerably strengthen and enrich the associations alluded to in Bartholomeus de Messina's translation of the Greek original. The fuller commentaries exhibit a similar range. Evrart de Conty's references to women in this context are oblique, except where he mentions how intercourse is properly accomplished.[3] Pietro d'Abano, on the other hand, not only addressed the original textual cues head on but added some comparisons of his own. For example, he offered a quotation from Aristotle's *History of Animals* indicating that girls around the age of fourteen require chaperones. And although he eliminated it from his second version, that vivid evocation of female adolescent desire circulated widely thanks to Jean de Jandun.[4] In spite of this variability in emphasis, the qualities that men with bad natures and bad habits shared with ordinary women reveal a great deal about what these men were thought to be like.

At least as much as those who composed or adapted commentaries, readers seem to have been interested in the analogies to women and particularly in the opportunity they created to reflect on the sexual appetites of young women. Passages relating to that subject were a favorite place for them to leave their marks on manuscripts of the *Problemata* and its commentaries with a "nota bene" or other signs of interest.[5] In an Italian manuscript of the Jean de Jandun redaction once owned by Christopher Columbus's son, one marginal note quotes scripture on the subject of lustful girls; another warns men about potentially adulterous women.[6] Linking men who engaged in homosexual acts to women produced opportunities for some of the prurient fun discussed in Chapter 5. But the comparisons also provided concrete images from which the *Problemata*'s readers built the personas of the men with blocked pores and bad habits. Furthermore, the currency of these associations was not limited to the universe of Aristotelian texts. In his examination of a group of legal cases from late fifteenth-century Florence, Michael Rocke noted that fully a third of the denunciations of men for the crime of sodomy used some sort of feminizing language, and of these more than half referred to the receptive partners *as* women.[7] Northern Italy around 1500 cannot stand for all the European environments in which *problema* IV.26 was read during the fourteenth and fifteenth centuries, but, as this chapter will show, similar ascriptions occurred in a variety of times, places, and contexts. They affected the way readers approached and understood the scholastic question about always passive men and sometimes passive men. And their approaches, in turn, resonate with more general attitudes and with aspects of medieval society and culture beyond the library of scholastic texts.

The academic strands, variegated in themselves, were entangled with many disparate threads arising from social, legal, religious, and cultural circumstances. These help to explain not only the compelling force of the ties between women and the men who engage in sexual acts with men but also some of the confusion that surrounded those associations. Indeed, discussing the legal case of a man who dressed as a woman and had sex with both men and women, Ruth Karras and David Boyd go so far as to argue that the central feature of his transgression was that other men had sex with him "as with a woman." In their view, the breach of gender roles was primary, the sodomitical acts secondary: the real issue in late fourteenth-century London was the defendant's perceived renunciation of a masculine physical presence and social role, rather than his choice of male sexual partners.[8] On the basis of a very different set of sources–literary, pastoral, and theological—Karma Lochrie comes to a similar conclusion.[9]

When the thirteenth-century French Dominican Gulielmus Peraldus found the "sin against nature" especially abominable, one reason was that a man "makes himself a woman."[10] Presumably men can become women by taking a subordinate position in sex, whether to a woman or to another man, thus assuming a gender-betraying passivity.[11] But Peraldus seems to have blamed both the giver and the receiver in an act that puts seed in the wrong vessel.[12] Is the "active" member also equivalent to a woman? Peraldus intended his work for the use of preachers and confessors, and thus he placed the ascription of womanliness (itself perhaps drawn from everyday discourse) into the growing stream of images and doctrines disseminated to late medieval Christians.

Like many of the men who shaped and read texts related to the *Problemata*, Peraldus wrote in Latin and expressed his concerns in general, often abstract terms. In other settings, distinct but overlapping discourses reinforce the web of associations. Court proceedings such as the case of the transvestite-sodomite are usually mediated by the reports of men more educated than those testifying, but they seem to reveal a common ascription of femininity to some men involved in sexual acts with other men. In the first half of the fourteenth century, Arnaud de Verniolle, a literate but not learned man of about thirty, answered charges before the Inquisition in southwestern France. According to testimony about a number of sexual liaisons with young men in his town, his bedmate "placed his penis between Arnaud's thighs and [moved] himself about as with a woman."[13] Some of his young partners seemed to distinguish among various acts and positions, one testifying that "although he had heard that sodomy is a graver sin than fornication, he didn't believe it was, unless a man lay on top of another like a woman or committed the sin through the rear."[14] This account alludes to three distinct sets of gender relations that may be enacted between men. The first concerns who moves and who is the object of the motion, resonating with a distinction between activity and passivity. The second has to do with the topographical hierarchy–who is on the bottom? And, unless "from the rear" is simply another form of topographical prohibition, the third alludes to the role of the receiving vessel. (Such reasoning may help explain why, according to Pietro d'Abano, men of his own time preferred rubbing between the thighs to anal penetration.) The young man's canon-law defense about what was and was not a grave sin suggests that the combination of subtle taxonomy and programmatic silence that characterized Latin discussions of sins against nature, rather than combating such transgressions, provided laymen with enough information and empty space to devise loopholes.

The presence of a strong theme of feminization in these contexts confirms that maintaining masculinity and gender boundaries was an important feature in the construction of this aspect of sodomy, although modern scholars have interpreted the connections in diverse ways. While Karras, Boyd, and Lochrie see the regulation of gender in the foreground of comparable cases, Rocke sees the persistent comparisons with women as an element in the reinforcement of social hierarchies, particularly as manifested between younger and older actors.[15] And, surveying a slightly different scene in the German-speaking lands, Helmut Puff sees in sodomy prosecutions of priests an attempt by municipal authorities to enforce uniform collective values and thus promote an integrated system of moral and political authority, often in competition with ecclesiastical interests and jurisdictions.[16] In common with each other and with the present book, these studies present evidence that the stigmatization of sex acts between men implicated far more than issues of object choice per se. The lines of stress that crossed medieval understandings of male-male sex acts varied from one dimension of culture to another. Both in the responses to the *Problemata* and in sermons, stories, and trial records, other axes of explanation and derogation came into play. Yet the impulse to patrol the line between feminine and masculine in the various settings and genres of medieval writing provides one rich, if unstable, ground for understanding the reactions of commentators, copyists, and readers to the subject of *problema* IV.26. The three themes that emerge from their confrontation with the text are womanly passivity, womanly imperfection, and womanly sexual insatiability. These drew substance and meaning both from the resources available in the intellectual milieu of philosophy and medicine and from the notions and conditions of the larger world in which late medieval physicians and natural philosophers circulated.

The Male and His Opposite in the *Problemata*

The impact of measuring men with blocked pores or bad habits against the defining characteristics of women depended on the clear distinction between the categories "male" and "female" and on the superiority of the male. Dimly reflected in the ancient text of *problema* IV.26 itself, the contrasts manifest themselves concretely in the medieval treatments of it. There both physical and behavioral differences set the terms of the parallels between deviant men and standard women. But the aspects of the feminine they invoked were not always consistent, even though the gender hierarchy remained constant.

At two points the language of the Aristotelian text offered this field of associations to its Latin audience. First, it calls upon the nature of women in the warning against surgery: "This place in them must be operated on. But this cutting certainly produces either complete destruction or deterioration. Therefore the former is not, for a woman would surely result."[17] As we saw in Chapter 1, there were complicated philosophical layers to Aristotle's point. Pietro the physician-philosopher took on both the medical and the theoretical issues, but he explained that his readers ought not to understand the imagined transformation literally. Rather the danger was that the individual would take on the "acts and behavior of women."[18] Most medieval readers, however, understood the horrifying result in concrete, not abstract terms, and they concluded, "This should not be done."[19] They saw this passage not as a hypothetical intended to elaborate the state of the men in question but rather as a reference to a possible (if too risky) surgical remedy. Given the repeated mention of eunuchs in *problema* IV.26 and the texts based on it, the womanliness in this passage was colored by association with surgically emasculated men, who were themselves like women. But the fearsome potential consequence here seems to have gone beyond the condition of a castrato. The Aachen summary warns against subjecting the men to the knife, even though it might remove the obstruction, because it might cause "death or degradation of the mind or they would be made completely like women."[20] All the *Problemata* texts conveyed the idea that "woman" is in some sense the opposite or privative form of "man"; some opened up a concrete, even anatomical, relationship between the men of *problema* IV.26 and the female body. And whether the surgery was a thought experiment or a dangerous procedure, the value hierarchy was explicit.

The second mention of women in the ancient text follows almost immediately upon the first. Indeed, its proximity may have encouraged the more literal, physical rendition of the previous one. Having argued that males of this description are characterized by an undermining (rather than a total destruction) of their masculine forms and functions, the Greek text turns to the nature of their imperfection, remarking not only on the displacement of their seminal emissions but also on the difficulty with which they are expelled, their small quantity, and their tendency to cool quickly. The effect is to make them "insatiable like women."[21] In a few brief phrases the Latin *Problemata* thus offered medieval readers the occasion to reflect on the womanliness of some men. Two instances of the word "effeminates" and one of the word "soft [*mollis*]" expand the semantic field, but it is the women themselves who dominate the late medieval texts that sought to explicate the natures of men with deflected and distorted desires.

Pietro d'Abano's elaboration of these meager textual cues drew on his broad acquaintance with natural philosophy and theoretical medicine in which female-male dichotomies were commonplace. His clearest statement of the distinction between men and women occurs in his explanation of the remark about surgery. After warning that cutting can result in death, he adds, "Sometimes it induces 'deterioration' and a transformation into something inferior in the parts of the pores, just as happens not infrequently with lancing sores."[22] Pietro has medicalized the consequence of the imagined intervention and framed the risk in terms of "deterioration [*perversionem*]".

The festering sores and the danger of death may have pushed aside temporarily the theoretical risk that a man might be feminized. But Pietro brings connotations of "deterioration" and "inferior" to bear upon the phrase "a woman would result." In addressing it, he lends emphasis to the contrast, if not quite the opposition, between the male and the female, and introduces the familiar value differential: "If this [surgery] were done, [he continues,] he who was a man at first would be changed into a woman and thus would be changed into something worse. Thus it is properly said that a 'deterioration' would result from it. . . . However, it is not to be understood that a man would be changed into a woman essentially but rather accidentally, acquiring womanly actions and conduct."[23] In mustering the scholastic distinction between the essential and the accidental, Pietro restored some of the philosophical sense of the comparison. He was asserting that the result of an operation gone awry would not really *be* a woman but rather would have some traits or properties that pertain to women. At two levels, then, Pietro maintained the strict separation between man and woman. First, he eliminated the possibility that a person could actually go from being a man to being a woman. Second, in discussing and ruling out the surgery, he drew a distinction between the original state of men with these anatomical irregularities and the effeminate state of survivors of the hypothetical surgery. In other words, if doctors and patients heeded his warning, there would be no men relegated to "womanly actions and conduct." Those with the innate condition described would apparently remain unlike women.

Yet if at one point Pietro reserves for these men some unidentified markers of masculinity, he denies them at another. In the course of elaborating a reference to eunuchs, he introduces a physiological dimension into the picture, "complexions": "The pathways that end at the penis are blinded, with obstructing material getting in the way (just as the eyes are blinded when viscous humor falls into the optic nerves), as one can see . . . in those whose complexions are chilled and moistened to such a degree that they are reduced to womanly nature and conduct."[24] That is, men with blocked vessels are like eunuchs who are like women, sharing women's cold and moist temperaments. Once again, we are left without any clues as to what aspects of "womanly" behavior are in play. The fact that Pietro has inserted the term "woman" independently of any prompting from the Aristotelian text, along with these new, physiological associations contributes to the case that Latin readers regarded the link between women and the men who desired anal stimulation as significant. But the specific lines by which they chose to connect the two groups were ad hoc and opportunistic. Here those who can emit no moisture are cold and moist. Yet at another point Pietro says their complexion is hot and dry, and George of Trebizond (who links the men with the planet Saturn) says their pores are cold and dry.[25]

The ambiguities produced by these passages are consequences of the force exercised by the comparison with women. To the extent that women are radically different from men, the correspondences between the two groups are at once impossible and powerful. The questions implicitly posed by the vacillation in Pietro's interpretations are, "Just how marginal, how unmanly, how unnatural are the men who deviate from the male norm in this invisible anatomical structure?" and "Just how potent, how insidious, how exportable are the (as yet unspecified) properties of women?" The dichotomous relationship between female and male and the difficulties that attend upon it formed the basis for much of what the readers of the *Problemata* had to say about the axes of passivity, deformity, and insatiability that join the men of *problema* IV.26 to women.

Associated with these elusive material connections are the more straightforward ways that specific roles are assigned to women and men in intercourse. Women are the ones with whom intercourse is had. They are, of course, sometimes the subjects of the verb *coire*, but, where sexual norms are placed under pressure, a properly supine woman can stave off the threatened disorder. Men have intercourse with women. The imperative force of this arrangement is apparent in Evrart de Conty's taxonomy of the men proposed by *problema* IV.26. The question clearly asks only about men who like to be acted upon and men who like both to be acted upon and to act. It does not ask about men who simply act. Perhaps, among those under scrutiny, there are no such men. In any case, Evrart introduces exclusively active men in order to highlight the contrast between proper and improper coitus: "Aristotle wishes to say that there are some who desire and want to accomplish this thing, to achieve the above-mentioned pleasure, in the company of women in the way nature and reason favors, and in no other way."[26] For him, at least, the sex of the partner is critical. But his insistence on proper pairing might also enhance the impression that one member of any copulating couple is always female in some sense.

The question addressed by IV.26, with its assertion that the natural internal destination of semen is the penis, clearly concerns men. Its phrasing, however, referring to "those with whom intercourse is had," created the occasion, if not the imperative, to picture women. Thus from the very outset medieval readers had reason to keep women in mind as they considered the men with bad pores or bad habits. Pietro's commentary clinched this association with its long exposition of the ordinary (hence heterosexual) mechanisms of pleasure, but more particularly with the way he restated at the outset the basic issues to be addressed: "Aristotle asks . . . why it is that those with whom the act of intercourse is carried out, such as women or those equivalent to them, experience delight and pleasure."[27] Or, in Jean de Jandun's redaction, "men and women and those equivalent to them." Whether, as in Pietro's own phrasing, readers are asked to focus on the object of intercourse, or, as in Jean's rephrasing, to picture an alternative to the male-male couple, women were a visible element in the characterization of men with homosexual desires. Yet, like the phrases "womanly actions and conduct" and "womanly nature and conduct," these phrases are frustratingly devoid of specific content.

On the one hand, Pietro's introduction of women into the picture somewhat deflected attention from the men; on the other hand, it firmly established a structural similarity between the two groups. He made clear that those with whom men had coitus could be either women or "those equivalent to them." Placed at the beginning of his explication of the *problema*, women became an unavoidable element in the reading of the rest of the *problema*, coloring the more elusive references, such as those just discussed. Whatever else might be meant or implied by "womanly," "feminine," or "effeminate," women's status as the objects of intercourse stands in the background, if not the foreground. If there is something coy about the phrasing of the equation, neither its context nor its logic leaves any doubt that the people being placed on a par with women are men of some sort. Like many interpretations of *problema* IV.26, Pietro's reading of "those with whom intercourse is had" will turn out to be less than conclusive, but it provides a baseline from which to evaluate the function of "women" in the medieval question about men. Furthermore, there is nothing about the concepts of women as opposites or objects that is peculiar to the textual tradition of the *Problemata*. They were, in fact, commonplace elements of the scientific and cultural environment. So too were women's passivity, imperfection, and insatiability, with which the remainder of this chapter is concerned.[28]

Passivity: Acted Upon

Associations of the feminine with the passive are not difficult to find in medieval society and culture. Legal restrictions on women's political, economic, and religious activities are well documented. Recent research has revealed, however, that such limitations were neither universal—statutory and customary law varied widely—nor, where they existed, entirely effective. Single women are found representing themselves in court, married women contracting indebtedness, religious women preaching. Even within the universe of ideals, the Virgin Mary, mild receptacle of the incarnation, becomes the Queen of Heaven and a powerful intercessor. And, as James Schultz has pointed out in the context of courtly literature, medieval bodies that were the objects of desire, insofar as they incited attraction, were hardly passive. Rather they were actively aphrodisiac.[29] Thus, as was the case in the *Problemata* tradition, the passivity of women was a prominent motif but hardly a monolithic system.

One environment in which the motif of passivity found ample play was the world of courtly and urban elites among whom texts with loose ties to academic authority circulated. The *Problemata* itself extended into these circles, as the history of its manuscripts and the French translation suggest. Far more widely disseminated, however, were a number of works that integrated learned traditions in natural philosophy, theoretical medicine, and practical medicine, and rendered them accessible for the enlightenment, entertainment, and (occasionally) use of a wider public. Transformed by rendition into vernacular languages, presented in manageable or attractive formats, adorned with inviting prefaces, and attributed to ancient or modern authorities, such texts lent shape, voice, and grist to norms, wishes, and anxieties concerning women. Monica Green has dealt at length with the content, history, and significance of several of these, especially the group of works, known as the "Trotula," on aspects of women's medicine.[30] Even in their most sober form, these texts rehearsed many of women's disabilities. According to one, women are cooler thus weaker, which condition (in combination with frequent childbirth) makes them susceptible to disease. Their poverty of heat affects their ability to process the kinds of superfluities to which *problema* IV.26 refers, and the substances that need to be expelled collect in the menses. With regular and moderate menstruation, the fragility of female health is mitigated, but many dangers to their health result from the all-too-frequent absence of a proper balance.[31] As Green has shown, the cultural valance of this group of texts evolved, as it circulated in numerous versions in various vernaculars as well as Latin.[32] Its tone was darkened considerably by association with the work known as the *Secrets of Women*, a compendium of material on reproduction and embryology falsely attributed to the famous Dominican scholar Albertus Magnus.[33] Drawing upon philosophical, astrological, medical, and other material, that work is infused with a fierce misogyny not present in most of its scholarly sources. As part of a body of knowledge expressed in a diverse literature of "women's secrets" it was entangled in a web of gender constructs and the erotic.[34]

The *Secrets of Women* did not travel in exactly the same circles as the *Problemata*, but its diffusion (like that of the "Trotula" texts) similarly illustrates the porosity of the membranes separating different elite sectors of late medieval society—learned and popular; Latin and vernacular; university, court, and professional environments.[35] Hence it underscores the broad dissemination of the ideas on which late medieval readings of *problema* IV.26 drew and thus modern readings depend. Likewise forming part of what Ann Blair has called the "low" tradition of the *Problemata* was another group of texts that circulated under the title "Aristotle's *Problems,*" referred to as "*Omnes homines*" to distinguish it from the work under consideration here.[36] Taken together, the representations of female and male bodies in these texts expressed a number of the notions and ambiguities found in readers' treatments of *problema* IV.26.

One recurrent theme is the uterus as a defining but inert organ. The pseudo-Aristotelian *Problems*, for example, treats it as a receptacle, and repeats the commonplace that fat women and prostitutes are sterile because their wombs are slippery. A too-cool womb may dilute the male semen, a too-warm one dry it up, but the female has no role in the creative aspects of procreation.[37] The author quotes the Pseudo–Albertus Magnus's *Secrets of Women* in his account of how the planets direct the development of the fetus.[38] A woman's blood is the material basis of the offspring, but it is thick and coarse, lacking in vital heat.[39] Yet this very passivity and incapacity is the basis of hidden activity—sometimes creative, sometimes destructive, always the object of the male gaze.[40] A properly constituted womb is an effective container, but it is also greedy (*avida*) in its desire to attract male seed; female blood is so crude that it rots easily into a poison, and for that reason can harm women, men, and embryos; and women themselves are in possession of powerful knowledge about their fragile receptacles and what causes the abortion of a fetus.[41] Selected and gathered together in works like the Pseudo-Aristotle's *Problems* and the Pseudo–Albertus Magnus's *Secrets of Women* such beliefs converge in a medieval sketch of the passive yet active female, weak yet empowered. The picture was all the more vivid because it drew upon the works of ancient authorities, respected by experts and lay people alike.

Those who approached Book IV of the *Problemata* would undoubtedly have had Aristotle's genuine work, *Generation of Animals,* in mind. Even if they had not actually read it, they would have encountered it cited often in works like the *Secrets of Women* and the pseudo-Aristotelian *Problems.* And Pietro made certain that his audience saw the connection between the non-reproductive sex discussed in *problema* IV.26 and some of the basic principles of sexual differentiation and sex roles developed in *Generation of Animals.* Chief among these was the idea that all procreation involved the interaction between an active, formative principle and a passive, material principle. Form was, of course, better than matter, and although among worms and the like the two principles coexisted in each individual, in higher animals they were separate. The form (and thus the male individual) was active; the matter (and thus the female individual) was passive. For Aristotle himself, this framework served to incorporate reproduction into a comprehensive scheme of natural causation which extended well beyond animate creatures and in which activity and passivity were abstract relations. But *Generation of Animals* also conveyed more concrete meanings. First, the reproductive substances produced by female and male reflected the distinction: the menses or female seed was cool, heavy, and sluggish; the male semen, as the medium through which the life force was conveyed, was warm, light, spiritous, and mobile. Second, in intercourse the male usually acted upon and penetrated the female (receptacle and receptive), though Aristotle noted exceptions.[42]

The natural questions that constituted the *Problemata* presented no general schema of female, male, and reproduction. But in the original text of IV.26 the distinction between active and passive was essential: "For what reason," it asks, "do certain people with whom intercourse is had enjoy it, some also acting but others not?" Later, it reiterated the standard philosophical terminology of acting (*agere*) and being acted upon (*pati*), saying that those in whom the moisture exits at the anus desire to be subjected; those in whom it exits both at the anus and at the penis desire to act and to be subjected. And medieval readers did not hesitate to link the active-passive axis to the man-woman axis from the outset, as is evident from a fourteenth-century manuscript with ties to the medical milieu of northern Italy. Following a standard medieval practice, one user inserted his succinct explanation of terms between the lines of Bartholomeus's Latin text: "For what reason do certain people with whom intercourse is had [Gloss: namely passively, in the manner of females] enjoy it, some also acting [Gloss: as males] but others not?"[43] Elsewhere, such abstractions were embodied. Boys destined to be shaped by their pleasurable experiences are said to be "subjected" or, literally, "placed underneath [*supponuntur*]," the proper location for a woman.[44]

The anonymous glossator sets up the equations that were the basis of many variations on a familiar theme: objects of intercourse are passive, and passivity is the mode of females; subjects of intercourse are active, and activity is the mode of males. In his own restatement of *problema* 26 Pietro too emphasized the language of activity and passivity. Bartholomeus used the verb "to have intercourse [*coire*]" in the grammatically passive form ("those with whom intercourse is had [*coitur*]"); Pietro echoed it in the passive as "those with whom intercourse is enacted [*peragitur*]." And he continued in the same mode, referring to the two distinct sources of pleasure as "by reason of acting or being acted upon." Indeed, he rendered temporarily invisible the natural philosophical and social difficulty that some enjoy both roles, asserting instead that the question is about "why certain people act and certain people do not, but are acted upon." This move underscored his interpretation of the objects of intercourse with "women or those equivalent to them." Thus, at the opening of the commentary on IV.26, the passivity of women is undisturbed; men who act (no matter upon whom) are exempt from being confused with women.

The crispness of these distinctions did not last, either in Pietro's own commentary or in the minds of later authors and readers. The simple refrain—male is active; female is passive—was sometimes accompanied by dissonant counterpoint that itself took various forms. Of these, the most prominent and destabilizing advanced a reciprocity between sexual partners that threatened the usual gender hierarchy, even as it confused the distinctions among men who engage in sexual acts with other men. If the usual medieval portrayal of a man and a woman engaged intercourse shows the man on top, exceptions are not hard to find (Figure 3).[45] Pietro defined sexual intercourse, the subject that unifies the various questions in Book IV of the *Problemata*, in terms of male and female, but not in terms of activity and passivity. Drawing on the Galenic theory that all individuals cook some of their excess blood into seminal fluid by a series of refinements, he said, "Coitus is a mutual action of male and female by which the superfluity of the third digestion exits."[46] The implicit parity between the partners is confirmed near the beginning of his treatment of *problema* IV.26, where he uses the word *congaudent* ("enjoy together") in his general discussion of the pleasures of intercourse.[47]

Figure 3. Above, an illuminated initial from a work on reproduction showing a conventional representation of a clothed couple, man on top and woman on bottom. The consequence of copulation is depicted in the lower half of the initial. Aristotle, *De animalibus*, IX, in a thirteenth-century French manuscript, Oxford, Merton 271, 65vb. Reproduced with permission of the Warden and Fellows of Merton College Oxford. Below, an illuminated initial from a work on the nature of semen, with an unconventional representation of a naked man and woman standing face to face. Pseudo-Galen, *De spermate,* in a thirteenth-century Italian manuscript, Biblioteca Malatestiana, ms. D.XXV.1, 163va. Laboratorio fotografico dell'Istituzione Biblioteca Malatestiana, photograph by Ivano Giovannini. Reproduced with permission of the Biblioteca Malatestiana.

Subtler evidence suggests that readers and copyists, in general committed to the analogy with male-female intercourse and to the active-passive dichotomy, were nevertheless a little uncertain how to phrase (and perhaps how to picture) the relations at issue. A cluster of variant readings suggests the blurring of "passivity" and, as a consequence, of the sense of the *problema* as a whole. The source of confusion or disagreement was a prepositional phrase, and it illustrates the minute but significant opportunities that manuscript culture presented to medieval readers. In most manuscripts of Bartholomeus's translation, the text speaks of those "with whom" intercourse is had.[48] Although the phrase does not seem ambiguous, Pietro rendered it so by asserting that "with whom [*cum quibus*]" meant "by whom [*a quibus*]." He would thus be taking the question to ask why "those by whom intercourse is had experience pleasure": the passive objects have disappeared in favor of neutral or even active participants. Is the question, then, about "those with whom intercourse is had," who are clearly like women? Or is it about "those by whom intercourse is had," who may be the active partners? Or is it about everybody? This unresolved dissonance may, in turn, help to explain the appearance in manuscripts of Pietro's second version of *aliquibus* rather than *a quibus*.[49] The word *aliquibus*, which is the plural of "someone" or "anyone," is grammatically equivalent to the *quibus* that it would presumably replace. The substitution would, in turn, rephrase the question as a whole to be asking why anyone (or perhaps some people) with whom intercourse is had experience pleasure. These variants do not express clear interpretive postures. On the contrary, they make the passivity of women and the men who are like them all the more difficult to locate.

The senses of hierarchy and mutuality thus coexisted in medieval readings of the opening lines of *problema* IV.26, in what commentators labeled as Aristotle's first question. When they turned to what they understood to be Aristotle's second question, focused on the differences between the active and passive roles among men, the comparison with women reemerged and took on more specific meanings. These hints of possible activity and mutuality do not mitigate the dark view of men engaged in sex acts involving anal stimulation. Nor does emphasis on the feminine and passive do much to lighten the burden on a hypothetical group of men who took the active role. Pietro did not hesitate to mark as wicked the men who undertake "sodomitical" behavior, such as (actively) placing the penis between the thighs of another. The comparison with heterosexual relations in the first question of the *problema* might convey a degree of normality, as Pietro's development of the natural processes of expelling superfluities has illustrated; but in Evrart's work, as we shall see, it simply provides a baseline for insisting on the unnaturalness of alternative forms of intercourse. In other words, men who perform that role in sex with other men benefit little from the generally positive and specifically masculine connotations of "active," in spite of being on top of their passive partners.

The entanglement of womanliness, passivity, and a certain type of man appears in these scholastic texts mainly in abstract forms like those just discussed—philosophical terminology, grammatical structures, and vague references to normative gender relations. Occasionally, however, something resembling a real body inhabits such clothing. In those cases, the language of the natural philosophers sometimes approximates that found in reports of judicial proceedings. For example, in a passage discussed earlier, Pietro represented the receptive partner's similarity to women in anatomical terms: "making friction around the anus and placing the penis into it as into a vagina."[50] The circumspect language of theologians concerning proper and improper vessels is translated into everyday (if still Latin) terms. Actions too take on greater specificity. Pietro speaks of one type of man that desires " 'to be acted' upon and rubbed," and another type that desires "to rub others with [his] penis and, with the penises of others in [his] anus, to be rubbed by them."[51] The Prague commentator's language includes active and passive grammatical forms. For him, the central question is simply, "Why are there certain men who want only to be rubbed, certain ones only to rub, and certain ones want to be rubbed and also to rub in intercourse?"[52] But he does not use terms like "active" and "passive" or "acting" and "being acted upon." And he alludes to women only as alternative partners.[53] On the one hand, the language and image of rubbing created a certain contrast with the heterosexual act, usually presumed to involve penetration as well as ejaculation.[54] On the other hand, neither in natural philosophy nor in court proceedings did the association of passive men with women depend on penetration, for Pietro was not alone in remarking on men who put their penis between the thighs of others.[55]

These examples, various and malleable, illustrate the adaptability of the passivity that connected the men of *problema* IV.26 with women, but they also point to its dispensability. Aristotelian metaphysical principles and reproductive theory, definitions of coitus, the notion of the female receptacle, ideas about male and female seed, canon law, and social expectations concerning coital positions were all building materials available but not required for the explication of the fourth book of the *Problemata*. Individual authors and sometimes individual readers chose to gather them up or to leave them by the wayside. Furthermore, although the cumulative evidence demonstrates the power of these connections, it also suggests that their meaning was not fixed—hierarchy could be mitigated by mutuality, for example, and objects could become subjects. In a similar way, the attribution of monstrosity to the men described in the *problema* sometimes evoked the "deformity" of women postulated in Aristotle's *Generation of Animals*, and sometimes a less specific (often still Aristotelian) notion of monstrosity. The physiological inadequacy of men with unnatural desires suggested systemic failure; their misshapen vessels suggested a birth defect.

Imperfection: Monstrosity and Menstruosity

Female passivity and female defect were closely linked within Aristotelian natural philosophy. To perfect the reproductive process, the active, form-giving male seed must dominate the matter from which the offspring will be shaped—the passive, form-needing female seed or menses. Although the masculine seed might in some sense be held accountable for any failure, "It is altogether better to suppose that the cause is in the [female-supplied] matter."[56] The most common, regular, and, indeed natural failure of this kind produces a female child, a "deprived" or "mutilated"—*orbatus*—male. Pietro d'Abano applies this same term not only to women in the context of *Problemata* IV.10 but also to the anatomy of men born with irregular sexual desires in the context of IV.26.[57] One fourteenth-century Italian scholar used a related word, *occasionatus*, often applied to women as failed progeny, to gloss the phrase "eunuchs and effeminates."[58] In this respect, women resembled other reproductive accidents called "monsters" or "defective animals,"[59] including multiple births, polydactyly, and hermaphrodites, even though women alone among these served a useful purpose.

But the conceptual universe of *Problemata* readers was not limited to academic theories. Like the idea of feminine passivity, the idea of female imperfection was deeply embedded in medieval culture. Although in some respects medieval gender roles overlapped and in other respects they can be construed as having been reciprocal, women's incapacity manifested itself in social practices from the ecclesiastic ban on female ordination to the legal impediments to female office holding. Whether grounded in their inadequate rationality or in the divinely and naturally authorized structure of the household, such limitations were often conceived in terms of a lack. Incompleteness shaded into failure, which shaded in turn into worse. Women could be dangerous to the social order, as was the case with prostitutes, for example; and they could be dangerous to men's and children's health, as well as their own. Indeed, in some contexts, their defects constituted a sort of monstrosity.

Like the varieties and implications of female passivity, the forms and functions of female monstrosity surface not only within technical Aristotelian natural philosophy but also in works like the Pseudo-Albertus's *Secrets of Women*, the Pseudo-Aristotelian *Problems*, and those influenced by them. Works such as these elaborated on various ways in which females are defective males, and in doing so they drew upon but reached considerably beyond philosophical principles of teleology and medical tenets about the role of intercourse in releasing superfluities. Women are phlegmatic, that is to say, cold and moist—the standard explanation for their inability to convert their superfluities beyond the stage of blood and into semen as males do.[60] Menses (the result of female inabilities and the cause of female complaints) and menstruation (the purgation of that failed and harmful substance) are defining features of women. Although these texts treat conception, gestation, and related subjects at considerable length, their discussions of those mysteries are often concerned more with the influences of the stars than the properties of mothers, who remain mere receptacles. In contrast, although other causes may be implicated, the main culprit in the formation of "monsters" is the matter supplied by the woman—its qualities and its quantity. Even the birth of twins requires that the material contribution of the female be specifically susceptible to this outcome, regarded as an abnormality in humans.[61] Accordingly, the procreative dimensions of female defectiveness would not seem at first to be relevant to the *Problemata*'s comparison with defective men. But because it was implicated in monstrosity, associated with sexual pressures, and understood as polluting, menstrual blood was one of the strongest lines of connection between women and the men whose bad pores produced irresistible impulses and perverted practices.

The explanations of defective offspring in this literature derive mainly from the same source invoked by Pietro in his discussion of monsters, namely, Aristotle's *Generation of Animals* (perhaps via the authentic works of Albertus Magnus). But other sites of authority came into play as well. *Secrets of Women* cites Avicenna for the view that "all specific types of monsters can be reduced to two general [causes]: . . . disobedience or insufficiency of matter."[62] The destructive intractability of the matter evoked in this passage merged with the pathogenic properties of imperfectly refined superfluities and the possibility of blockages, creating a fusion of menstrual blood and danger. This association colored, in turn, the cultural atmosphere in which the *Problemata* texts were produced and circulated. Furthermore, the notion of "disobedience" to the (male) formative principle suggests a certain threat to the social order implied by these notions about female physiology.

The *Secrets of Women* offered an especially venomous perspective on women's fluids and behavior, but its basic elements were readily available elsewhere. Although a necessary and productive component of the reproductive process, menstrual blood was also a dangerous impurity that needed to be purged, prone to pathogenic accumulation or putrefaction in women. Men should beware, since intercourse with a menstruating woman is harmful. Old women who menstruate can harm a child just by looking at it in the cradle, but so can those who do not menstruate, since the contaminated material remains trapped inside them.[63] Poor women, in whom the production of usable reproductive matter starts with coarse food, pose a particular threat.

These ways of construing and expressing the physiognomy of the feminine suggests how the *Problemata* literature was rooted in the larger social and cultural environment. *Secrets of Women* helps us to understand which notions about women were applied to the men of *problema* IV.26 both in terms of their currency and in terms of their prurience. The appeal to Greek and Arabic sources added authority, but ideas like the poisonous gaze of a menstruating woman grew like weeds in a variety of medieval European niches beyond the university curriculum.[64] Ideas about female imperfection thus had a role in naturalizing forms of exclusion and subordination, but they were not so widely accepted as to have become invisible. Christine de Pizan called the *Secrets of Women* "a treatise entirely of lies." Not only did she object (in another context) to the notion that she (and indeed every woman) was regarded as a "monstrosity of nature," she also made the imperfections of women, as portrayed in natural philosophy, one of the major targets of her *Book of the City of Ladies*.[65] Just as the sense in which men with naturally caused blockages possessed "monstrous natures," the sense in which females were defective males was not entirely settled.

The integration of female imperfection with the condition and characteristics of men who desire anal sexual stimulation is most vividly represented by the paradoxical danger that, if subjected to surgery, men who were already like women ("effeminate" and possessed of "womanly nature and behavior") might in some sense or other be turned into women. But this was not the connection that commanded the most attention of medieval readers. Evrart de Conty's French translation and gloss do not even mention it. Imperfection more generally did, however, constitute a node at which exceptional malformed males and ordinary flawed females were joined. Physiological theory and social disdain lent weight to the association. Evrart, for example, said men whose complexions were unbalanced and whose desires were not directed toward women were "of a depraved, effeminate, and imperfect nature."[66] Like many of the comparisons offered within the sphere of *problema* IV.26, the one with women was open-ended. The content of the shared imperfection comes into focus with the notion of monstrosity and with the further connection between the monstrous and menstrual blood.

The Latin translation of *problema* IV.26 does not use the word "monstrous," but medieval readers could easily accept their contemporaries' introduction of it into the description of the anatomical defects that might give rise to disordered desires. The innate irregularities resonated clearly with ideas about the natural process of reproduction and the factors that could derail it, as Aristotle had explained. Imperfections present in an individual from birth bespoke the incomplete domination of form (the active principle proceeding from the father) over matter (the passive principle proceeding from the mother). The result was something that was not fully possessed of its proper, defining essences—a male that was not completely male. Thus Pietro d'Abano invoked the association with monsters six times in his explication of the *problema*. He spoke twice of sperm traveling through the vessels to the place where it is destined to collect in accordance with this sort of man's "monstrous nature"; then he elaborated, saying that the place is "monstrously constituted"; next he referred the reader to the fourth book of the *Generation of Animals* for further information about "this type of monstrosity" and others; and later he returned to defend the notion that "such a distortion and monstrosity [can arise] from birth."[67] Neither Pietro nor any other author who treated the text mentioned women in this connection. Indeed, the only hint we have about what he had in mind when he spoke of "this type of monstrosity" is in his final use of the term. He pointed out the absurdity of Avicenna's refusal to acknowledge the natural causes "halubnathi" while being quite willing to do so for hermaphrodites, who are "more monstrous according to nature."[68] Aristotle too had raised the subject of hermaphrodites as individuals born with both male and female reproductive parts, outside the proper course of nature (*preter naturam*).[69] The gender disruption invoked by double-sex individuals is created not by a direct comparison with women but rather by the problematic female who lurks below the surface of the word "monstrous."

The subject of hermaphrodites was esoteric in the Middle Ages, and Pietro's reference to them highlights the bizarre aspects of monstrosity.[70] But the decadent and sometimes dreadful nature of women was ordinary. Pietro was simply doing his job as an expositor when he paused to explain why Aristotle's list of the superfluities that need to be expelled from the body included not only urine, feces, tears, and mucus but also blood. Why would blood, "the most natural humor," need to be expelled? As a physician, Pietro had at his fingertips a number of possible answers deriving from the principles behind the widespread prophylactic and therapeutic use of blood letting. He chose instead to offer an explanation that reflected the environment in which physiological misogyny thrived and notions of the evil eye circulated—an environment that encouraged attributing to women's bodies not just failure but also contamination. He took the position that what Aristotle meant was not good, nourishing blood but rather the sort that needs to be purged, "an unnatural superfluity mixed with many superfluities of other humors. . . . Menstrual blood and [blood] emitted by hemorrhoids is of this kind."[71] Furthermore, he went on to discuss the vessels involved in this purgation, namely, those connected to the "circle of the anus" and specific locations in the uterus. The Aristotelian text had mentioned blood only in passing; Pietro's digression paved the way for further associations. The appearance of the perianal region in the foreground as well as the references to the womb and to the "melancholic" blood that is involved in menses and hemorrhoids suggest a set of implications for the understanding of anal intercourse and its affiliation with women. Pietro had at first listed nosebleeds along with menses and hemorrhoids, but he dropped them in his revised version, and Jean de Jandun omitted them as well.[72] Both apparently found them out of place, and their absence further focuses the meanings of negative blood on the lower regions of suspect bodies. This prologue to the men susceptible to anal sexual pleasure thus announces the triad of the menstrual, the anal, and the melancholy. Later Pietro reinforces it with a reference to a passage in the *Generation of Animals* where anal bleeding in imperfect men is said to be similar to menses in women—except that the latter is natural.[73]

The immediate textual conditions created by Pietro's excursus on the word "blood" encouraged a reading of subsequent passages that brought the nature of the men with distorted desires into conjunction with the nature of their female counterparts. And so, at later references to "monstrosity" (or "monstruosity"), some scribes replaced the monstrous with the menstruous. Like the English words, the Latin made the substitution easy.[74] And in one case a reference to the *Generation of Animals* may have put a copyist in mind of the female contribution to reproduction (Figure 4). The system of scribal abbreviations widely employed in the production of books allowed for the suppression of the "o" and "e" that distinguished the two words. Thus *m̄s* could be read either as *mons* or as *mens*. In most of the manuscripts of Pietro's commentary on *problema* IV.26, however, the monster-related words are written out, rather than abbreviated, so a mental connection surely played a role in the readiness of at least six scribes to effect or accept the transformation.[75] And evidence from other sources shows that they were not alone. A manuscript of a different set of natural questions refers to Aristotle's view that "a woman is a defective man and a menses [*menstruum*] in nature."[76] There are other chapters within the *Problemata* that refer to monsters and menses where copyists seldom confused the words at those points. But in connection with an earlier

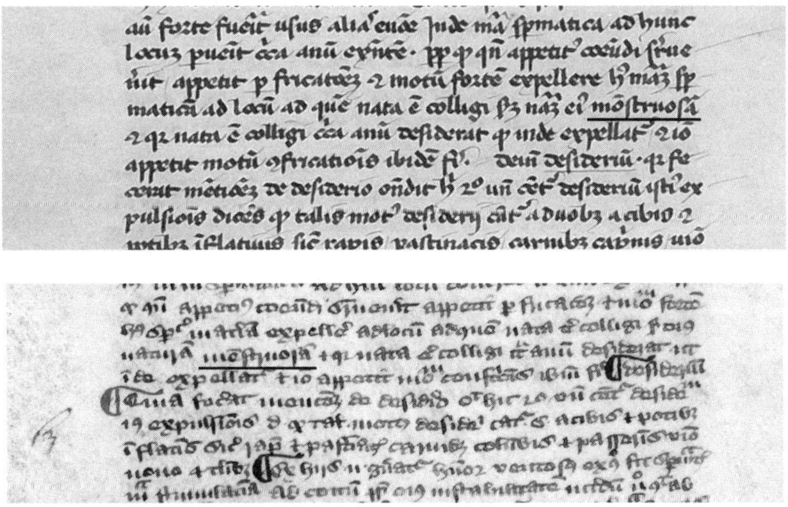

Figure 4. Above, a manuscript (Italian, 1381) of Pietro's commentary on *Problemata* IV.26 refers to "their monstrous nature"; below, another (Jean's version, French [?], fourteenth century) reads "menstruous" (underlining added). In the usual style of medieval abbreviation, the scribes have written *n* as a line above the vowels. The little hand ("maniculum") in the second example is one of the ways in which readers registered their interest. PA, Biblioteca Malatestiana, ms. D.XXIV.2, 36vb. Laboratorio fotografico dell'Istituzione Biblioteca Malatestiana, photograph by Ivano Giovannini, reproduced with permission of the Biblioteca Malatestiana; JJ, Peterhouse, ms. 79, 31va, reproduced with permission of the Master and Fellows of Peterhouse, Cambridge.

question on the birth of monsters, the same scribes read the "menstruous" into the monstrous.[77] The substitutions in IV.26 thus constitute a strand of the entanglement of women and womanliness (and, in turn, their polluted bleeding) with those men whose malformed pores deflect their pleasures.

The affiliation of woman's imperfection with blood drew together several aspects of her defectiveness: her status as a "monster" resulting from the failure of her father's seed to master her mother's menstruum; her inability to perfect her superfluities into fully refined semen; her identification with passive matter rather than active form; her contaminated and contaminating monthly purgation; and its comparability with hemorrhoids and thus with anal bleeding. For some readers another set of derogatory implications may have come to mind. A cluster of associations situated at the intersection of learned and popular traditions linked Jewish men to menstruation, hemorrhoids, and anal bleeding. Those who left their marks in the manuscripts made no reference to these notions, but they were widely available.[78]

As the collapse of the distinction between the monstrous and the menstruous illustrates, the properties of ordinary women and deviant men became inextricably intertwined. Insofar as the readers of *problema* IV.26 provide us with clues to the meaning of ascribing a "womanly nature" to the men in question, those are the traits with which they would seem to be invested. Yet that characterization poses problems on at least two levels. First, as mentioned in connection with the idea of a surgical correction, the men are said at one point to have inherently womanly natures and at another to be in danger of acquiring them. Second, while some of these traits, such as passivity and monstrosity, are unambiguously assigned to the men, and others, such as menstruosity and impurity, are clearly associated with them, still others have little or no presence in this family of texts. For example, although the spermatic moisture that is excreted around the anus is scanty, none of the manuscripts hints that it is bloody. Present in the *problema* itself or introduced into its explication, these traits extend and withdraw a tantalizing offering of specificity to "these men."

Insatiability: Little Women and Old Wives

Women's unsatisfiable sexual appetite was so well established as to require no substantiation. It found its way into Roman comedy and Christian patristic writings, courtly romances and bawdy student songs. It often conveyed a sense of women's fundamental weakness, their devious strength, or both. Thus the

Problemata tradition is not isolated in incorporating its multivalence, and Pietro d'Abano picks up a quotation on women's sexual appetite from the classical Greek poet Hesiod, "who said that women are especially lustful," to fortify his use of medical sources.[79] Like the ascriptions of passivity and imperfection to men with unnaturally constituted pores and desires, calling them "insatiable like women" both gives and takes away information about them. Medieval readers of that ancient phrase had no trouble making sense of it; yet it strains against the sense they made of the comparisons based on passivity. Compared to passivity and defectiveness, its social implications are the most visible; yet the scholars who considered insatiability did so in mainly physiological terms. In spite of these ambiguities, their explanations and their language do convey certain expectations concerning men whose behavior was not reliably heterosexual and "active." Voraciousness was connected to menstruality—Aristotle mentions that barren women whose menses collect but do not get expelled are prone to sexual incontinence.[80] And, like menstruality, sexual insatiability projected instability. Thus the disruptive powers of improper desires reinforced other aspects of the comparison between women and the always passive or sometimes passive men. But the dangers of unsatisfied sexual appetite are not themselves fully congruent with those linked to monstrosity and menstrual pollution. The latter features conveyed a physiological inertness that resists action, while the consequences of insatiability were more behavioral and, indeed, more active. That is, the characters, female and male, in this version of the story are on the move.

In the interpretation of the Aristotelian phrase "insatiable like women," Pietro d'Abano once again took the lead, and once again other readers responded in a range of ways. For some, analogy seemed to function as a medium through which the main subject could better be understood; for others, the phrase was apparently of special interest for what it had to say about women and girls themselves. The difference between two marginal notes is suggestive, if no more than that. In one manuscript of Pietro's commentary someone, probably the fourteenth-century Italian scribe himself, has made numerous annotations throughout the work, mainly labeling the contents. At this passage he notes: "Young women of fourteen years, and the others about whom this discussion is, seek intercourse."[81] No less elusive in his language than Pietro himself, this reader nonetheless sees clearly that the subject matter here includes "others," not just women. In contrast, seeming to lose sight of the men, one fifteenth-century reader wrote only a single comment in a full copy of Bartholomeus's translation: "Note that women are insatiable in intercourse."[82] Similarly, the author of the Bavarian version, which is close to being an abridgment of Pietro's work, incorporated the explanation of young women's appetites and omitted the subsequent passages that contain the other side of the equation—the men who are likened to them.[83] For yet others, the whole comparison itself was apparently not especially illuminating. The Prague commentator skipped this and all other references to women presented by the Latin text of IV.26, though he discussed the underlying physiology of men's persistent desire and frustration that the comparison was intended to explain.[84] Sexual appetite and indulgence were certainly an aspect of the cultural profile of at least some types of men who engaged in homosexual acts, but the examples just given suggest once again the difference between an association that is available and one that is invariably asserted. Some readers gave it some thought; others apparently did not.

The *Problemata* text itself is direct about this element in the comparison with women. The reference to female lust follows the brief discussion of the implications of surgery, with its remark that a person who was not a woman might be turned into one. Returning to the effects of improperly constituted bodies, Aristotle explained, "[Some] of the seminal excretion must therefore be diverted and moved elsewhere, on account of which [they are] insatiable, like women. For a modest amount of moisture does not manage to get out and is cooled off quickly."[85] Before moving to the next major topic, habits, this passage summarizes the results of the physical irregularities discussed so far: that "some desire to be acted upon in the anus and some . . . both to act and to be acted upon in both places."[86] The woman-like sexual appetite thus occupies the meeting point between the men's anatomy (blocked emissions) and the manifestations of their desires (dual sensitivity). Late medieval commentators explored two distinct aspects of this convergence. The first involved the physiological and behavioral implications of the anatomical structures, and depended on interpreting "women" as pubescent girls at the inception of their sexual activity. The second emphasized the nature and dimensions of female pleasures, and depended on interpreting "women" as sexually experienced adults. The first centered on frustration and appetite; the second centered on the double pleasures of emitting and receiving seed.

Young women are particularly humid, and may have narrow pores. These conditions make them prone to menstrual pain and to a buildup of their superfluities that causes a great desire for sex. Intercourse is therefore good for them. The author of the *Secrets of Women* notes that young women (*iuvenes mulieres*) are especially susceptible to "suffocation of the womb," the failure to expel harmful superfluities, and remarks: "These young women, when they have a superabundance of this material, seek intercourse a lot, on account of the abundance of the matter. Therefore, it is a sin with respect to nature to restrain them and to prohibit intercourse to him whom they love, though it be a sin with respect to morals—which has nothing to do with the question at hand."[87] The author here poses (and declines to resolve) a problem also raised by the belief that men's desire for anal stimulation was often innate: that the natural and the moral are not necessarily congruent. From the perspective of nature, a man (indeed, the male reader) would seem to have a sexual responsibility to a young woman in this condition. Permeated with such disreputable suggestions, the notion of young women's particularly voracious sexual appetite played an important role in the profile of the men who are like women.

Pietro regarded the similarity between the two groups as a logical corollary of the preceding description of anatomical irregularities and their physiological consequences. Glossing the phrase, he added a psychological dimensions, saying these men are " 'insatiable' and insane 'like women.' " And he asserted that Aristotle had in mind mainly young women whose seminal vessels, not yet fully opened, formed the link with the men whose vessels are permanently reduced: "And [they are] 'insatiable' just like 'women' and most of all young women [*iuvencule*] whose passages for this [expulsion of seminal fluid] are narrow. For this reason there is 'a small amount of moisture,' moved and liquefied by rubbing, that 'manages' to be expelled. This, because it is 'a small amount,' 'cools quickly,' not being able to be [fully?] expelled. On account of this they seek to be rubbed again, so that it may be expelled."[88] In reapplying the notion of insatiability to the specific phrases that describe the men with imperfect passages in the Aristotelian text, Pietro seals the connection between the two groups. The impulse to be rubbed and the minimal secretions refer back to his own earlier description of moisture that collects around the anus and the consequent desire for friction in order to release it. Some of the particulars are unclear, for example, how much moisture, if any, is emitted.[89] In spite of these difficulties, the passage points clearly to the shared features: an anatomical structure (narrow vessels) that creates a physiological predicament (the difficulty of adequately emitting spermatic material), which gives rise, in turn, to a behavioral pattern (the perpetual quest for friction).

Bartholomeus's translation speaks exclusively of desire (*desiderium*, *desidere*), with a more passive connotation of longing; medieval discussions of women's insatiability introduce the more active vocabulary of seeking (*appetitus*, *appetere*), conveying a sense of striving. Perhaps this sense of a frenzied quest helps to explain why one person who made a copy of Jean de Jandun's version read not just that "women are insatiable" but that "women are especially insatiable."[90]

Pietro's digression continues with what is essentially a gloss on his gloss, directed at elaborating his own word, "most of all [*maxime*]" young women:

> It should be noted that women are insatiable, especially young women around the age of fourteen, about whom basically the same discussion [is found] in the ninth book of [Aristotle's] *History of Animals*. Young women most of all hurry to the practice of sexual acts [*venereorum*]. When they begin, having experienced sexual activities, they become incontinent. The cause is that at this time, because of the descent of the spermatic humor and menses, the genitals begin to thicken and the opening of the vulva [begins] to be closed, and its softened labia get plumper and become hairy. And then they begin to be mad for intercourse. They do not, however, emit [anything] in [their] desire. And however much more they have intercourse or rub themselves with their hand, so much more do they seek, because the humor is attracted but not emitted by such rubbing together. And with the humor, heat is attracted and windiness arises too. But, since the body is cold and its pores are also closed, it does not quickly emit seed.[91]

Here Pietro confirmed his conviction that the insatiable women mentioned by Aristotle were adolescents.[92] Here too he explained the relevance of madness: although insanity always implies the loss of reason, this specific form manifests itself as a frenetic quest for sexual gratification—enhanced by the very impossibility of attaining it.

Like the real Albertus Magnus, who had written about the sexual physiology of adolescents a few decades earlier, Pietro saw their desires and behaviors as distinct from the procreative process. He thus assumed that, for those who have not yet attained full reproductive capacity, intercourse and masturbation were functionally equivalent.[93] Just as he observed that the "sodomitical act" could be performed in any one of several ways, he likewise presented friction by intercourse and by the hand as equally serviceable (if ultimately insufficient) in girls' quest for satisfaction. Apparently, with respect to their unbridled desire, the two groups would resort to any available mechanism of release. To a certain extent, this perspective on the result of blocked vessels suggests that, even when acts of male-male intercourse are explicitly at issue in medieval texts, the crux of the matter may be not object choice per se but rather the pursuit of particular pleasures. In the context of this comparison to female adolescents, the defining feature of men who engage in such activities—especially the subgroup of those who both act and are acted upon—may be their transgressive promiscuity, rather than their involvement with male partners. In the context of *problema* IV.26 as a whole, however, the specific focus on contact between male anus and penis prevents the subject from dissolving entirely into a matter of general sexual excess.

As a formally constituted excursus, the passage on female insatiability illuminated two sources of tension at once: the main subject of the *problema* and the new subject interjected by the comparison. The latter—the fourteen-year-old girls—have momentarily moved into the spotlight. At this point Pietro's first version and thus Jean de Jandun's redaction include the quotation from Aristotle's *Generation of Animals* that highlights the dangerous social consequences of these symptoms of female adolescence: "around this time they need a chaperone."[94] With its sober analysis of the anatomy and physiology of young women's sexual frustration and its agitated references to rubbing and insanity, the emerging portrait of men who suffered similar symptoms fed upon and fed a particular strain of misogyny. At the same time, it enhanced a disdainful perspective on men who sought anal stimulation. These people who furiously rub and constantly seek to be rubbed were problematic not simply for their own cycle of frustrated desire but also for the threat they posed to others—even, according to one reader, the king and prophet Solomon, whose virtue was otherwise inviolate.[95] Beyond the sense of moral miasma by which illicit and disruptive desires might spread, here is a threat of social disintegration perpetrated by the active pursuits of groups elsewhere represented as weak and passive.

The particulars of medical and scientific learning had led Pietro to propose that men's structural and behavioral similarities with women depended on interpreting the female side of the equation as pubescent. (The consensus among the manuscripts was fourteen years old, though one manuscript of the Bavarian *Problemata* says nine.)[96] But not everyone was sure the comparison was so limited. A scribe who compiled an index of Pietro's exposition included an entry for young women around twenty-four in addition to one for young women around thirteen.[97] Another reader, referring to the need for a chaperone, noted in the margin, "When a woman needs a guardian lest she commit an abuse with another man."[98] The use of the word "woman [*mulier*]," which carried a sense of maturity, and the mention of "another" man make it clear that, for this reader, the danger to be guarded against was adultery by married women. This spontaneous reaction ignores the anatomical and physiological logic of the text to which it is appended; it responds instead to a more general cultural logic, according to which sexually experienced women were also prone to insatiability. Older women, especially celibate women such as widows and nuns, were susceptible to the retention of superfluities that required release by means that included intercourse and genital massage.[99] The menstruous females who figured in the ominous passages of the *Secrets of Women* were not adolescents. And readers had a wide range of cultural resources to draw on as they imagined these voracious and disruptive women, whether the adulteresses conjured and condemned from the pulpit or figures like Chaucer's wife of Bath. When, in his prologue to the *Problemata*, Walter Burley alluded to the insatiability of the womb, he was making a biblical reference, albeit one in harmony with the physiological notion that the womb desired and attracted the male seed. Pietro cited Galen on the same point: "In the womb [*vulva*] the appetitive power is strong."[100]

Perhaps precisely because his explanation of the comparative mechanics of desire depended on the anatomical and physiological peculiarities of puberty, Pietro himself did not explore these other possible meanings of "insatiable like women." The Prague commentator, however, seems to have been doing just that when he concluded his far briefer treatment of *problema* IV.26 with: "Note that women have greater pleasure in intercourse extensively. This means that women experience pleasure for a longer time than men, since they get pleasure in receiving seed from men and in emitting their own seed. But men more experience pleasure in intercourse intensively, that is by a more intense pleasure than women, although it lasts for less time than [in a] woman."[101] This author has combined two ways of comparing women's and men's sexual pleasure that were common in scholastic natural philosophy and medicine.[102] The first, according to which men's pleasure is greater with respect to its intensity, whereas women's is greater with respect to its extension, would not usually be taken to support the special emphasis on the female appetite. As the author expresses it here, however, the sense tilts in that direction. He qualifies the man's advantage ("although") but not the woman's. At the same time, he explains the woman's advantage ("a longer time") and not the man's.

More important, the Prague author adds a second layer to the claim. Women get two kinds of pleasure—from receiving and emitting seed—whereas men get only one. Positioned at the end of his discussion of *Problemata* IV.26, this otherwise unexceptional reference to the pleasure of receiving seed takes on new significance. It suggests a source of pleasure for the receptive partner that is not referred to in—or even consistent with—the Aristotelian text, with its emphasis on the expulsion of superfluities: receiving the semen of another man. In this respect, the insatiability of mature women and of comparable men manifests a curious combination of activity and passivity. It is not unsatisfied desire but rather experienced pleasure that leads mature women to actively pursue sexual contact. In this they resemble not the men with blocked pores but those who have come to their desires through habit. Unlike the former, the latter are capable of satisfaction. Furthermore, active and passive sources of pleasure can be combined, as they are in sexually active women. The Prague commentator, who had mentioned that the men with no physical impediment sometimes have intercourse with women too, thus multiplied the forms of pleasure involved beyond those in the original text and beyond Pietro's comprehensive commentary, at the same time as he brought out further ambiguities of activity and passivity.

Conclusion: Who Is Like a Woman and How?

The force of comparing the "men with whom intercourse is had" to women depends above all on the assumed contrast between women and proper men. Men and the male principle are active, whereas women, in their passivity, constitute the paradigm of "those with whom intercourse is had." Males are perfect both as products of the reproductive process and as fully formed and physiologically capable adults, whereas women fall short on both counts, imperfect from birth and menstruous as adults. Finally, men's sexual impulses are naturally temperate, whereas women's are inordinate. For a man to be reduced to the "nature and behavior of women" was thus doubly degrading. First, it was a radical breach of the order of nature—monstrous. Bad enough for a natural object to be so defective as to resemble an entirely different one; worse still for it to resemble its own opposite. Second, since the partners in the male-female dichotomy were of unequal value, for a man to resemble a woman was for him to have the characteristics of "something worse." The fact that the men of *problema* IV.26 were not actually women trapped them in this betrayal of their proper kind; that they were indeed men disgraced manhood itself.

As entertained by readers of the *Problemata*, all of these features of the comparison resonated with other forms of cultural expression. Women were, in principle at least, a disappointment as heirs and a cause of strife in vernacular literature. And sexually transgressive men were tarred with the attribution "womanly," whether they were the subject of religious polemic against sodomy or on trial for seducing or coercing young men. The tropes of female passivity and defectiveness were deployed by even the most powerful female writers, as when Heloise pleaded that nuns need a more lenient monastic rule than men—and that Abelard should write it. Like the municipal regulation of prostitution, the warning in the margin of a manuscript about adultery and the anxieties about King Solomon convey the impression that the sexual demands of women posed problems not only for personal rectitude but also for the stability of a well-ordered polity. Comparable threats were posed to late medieval municipalities and states, when rulers and inhabitants feared the literal and figurative consequences of becoming a Sodom.

The coherence of this picture suggests how powerfully men who had sex with men were stigmatized by their assimilation to women. If the "womanly actions and behavior" were not specified, the attributions of passivity, deformity, and sexual insatiability in the *Problemata* literature nonetheless lend concreteness to the elusive category of "these men." Their bodies are deformed, their sexual postures are passive, and their desires are endless. Yet even in the context of *problema* IV.26, which poses the question in the most circumscribed terms—those of susceptibility to anal stimulation—this formula is only part of the story. First, the objects of the inquiry do not constitute a homogeneous group. There are those whose site of pleasure is the anus and those whose sites of pleasure are the anus and the penis. And the causes of their dispositions are also divided into two types, the innate physical source and the acquired psychological source. The interpretations that commentators and readers offered of the resemblance to women did not apply equally well to the differences encompassed by these variables. Second, as has already been suggested, the fundamental elements of the description are not altogether consistent either internally or with each other.

Who among these men are deformed? Who passive? Who insatiable? This is a point on which the struggles of readers to understand Aristotle's text and arrange its elements of nature and habit affect but do not clarify their interpretations. The notion of deformity applies most clearly to those with completely or partially blocked passageways. Yet in the polemics surrounding sodomy, the term "unnatural" is not limited to birth defects; the psychology of habit undermines the distinction between innate and habitual sources; and the corrupting effects of habit suggest a behavioral deformity that parallels the physical one.[103] The ascription of passivity refers most securely to the receptive partner—to the possessor of the vaginal, anal, or intercrural receptacle. This category would therefore include members of two groups: congenital and habitual participants in homosexual acts. That correspondence too is somewhat confused, however, since either group may include individuals who like both to act and to be acted upon. With respect to passivity, are those men always like women? Sometimes like women? Somewhat like women? Though the use of words like "effeminate" and "eunuch" might provide insight into this problem, the places and ways in which "woman" and "womanly" occur in the context of *problema* IV.26 do not permit any further specificity. Finally, the characteristic of insatiability, like that of deformity, at first appears to apply only to those with a physical deformity (not a habit), since it depends on a structural impediment to proper ejaculation and thus to satisfaction.[104] Indeed, for the same reason, it might be the disposition only of those with fully blocked pores and the consequent exclusively passive desires. After all, other men, namely, those who have passageways open through the penis, are in fact capable of expelling spermatic superfluities. Are those whose pleasures come from both roles also sometimes insatiable? In some respects insatiable? Somewhat insatiable? The language of the commentaries would suggest that, whatever the physiological rationale, all such men are marked by inordinate sexual appetite, for they seek to be acted upon and to act, to be rubbed and to rub. Those whose desires arise from habit, and who are thus excluded from anatomically caused insatiability, arrive at the feminine state by another route. The very mechanisms of habit formation—experience, pleasure, memory, imagination, and desire—themselves constitute an endless cycle of appetite, as the Aachen commentator asserted: "Habit is already nature in them, because they almost always rub those parts . . . or want to be rubbed by others."[105] Thus, although the comparisons with women occur in the context of specific natural philosophical arguments that explain only certain subsets of the men included in the *problema*, the traits they identify are diffused into the larger set. Such is the force of the comparison with women, operating both within and beyond the *Problemata* tradition, that it consumes everyone in its path.

Nor are the associations necessarily weakened by the disjunctions and paradoxes they contain. Not only are there too many distinctions among the men who rub and are rubbed to allow unilinear accounts of their causes and natures, but, as this chapter has argued, the various associations developed in the texts and the margins drew upon distinct lines of reasoning and cultural assumptions that were not always congruent. Those who are most clearly passive have the clearest anatomical imperative to seek sexual stimulation actively; those whose deformed vessels render them incapable of pleasure and those whose experiences have given them a taste of pleasure pursue it with equally insane fervor. For all the distinctions among the causes and manifestations of their erotic impulses, the men of *problema* IV.26 are all, in some way, "equivalent to women." And yet, even if they have pores like adolescent girls and colons like vaginas, even if they have (or are in danger of acquiring) the nature and behavior of women, they embody a transgression: they are nevertheless men, not women.

Chapter 4

"Beyond the Boundaries of Vice"
Moral Science and Natural Philosophy

Like the fifteenth-century sodomite priest Johannes Stocker, Arnaud de Verniolle spoke in his own defense at trial. Accused of multiple acts of active and passive intercrural sex with multiple youths, as well as of posing as a priest, he appeared before an ecclesiastical court in southern France in 1323. "Asked if he told anyone or believed that because his nature required him to satisfy his lust either with a man or a woman it is not sinful to have relations with men or women, or that these may be minor or venial sins, he replied that his nature inclined him to commit sodomy, although he always believed sodomy is a mortal sin."[1] The Inquisitors may be asking about his "nature" in the common sense of his general sex drive; he himself seems to be referring to his individual nature that directs him toward particular acts. In addition, a witness testified that Arnaud had justified the emission of semen when men lay together as promoting health. Although he knew some Latin, there is no reason to assume that he was familiar with the technical medical theories concerning superfluities or with the philosophical distinction between universal and individual natures. And, as other portions of the trial record show, he held (or at least expressed) a somewhat garbled view of the relevant religious doctrines. Nevertheless, Arnaud's language, like Stocker's, incorporated and adapted multiple discourses, which in turn reflected the layers of understanding and of power inherent in his particular situation. Both men were searching for cultural resources with which to explain themselves in the face of legal and moral opprobrium.

As the gendered language of the *Problemata* texts makes clear, more than academic natural philosophy was involved in interpreting and explaining men like Stocker and Arnaud. The word "sodomite" itself, which appeared not only within Pietro d'Abano's commentary but also in the margins of manuscripts, formed a bridge to a different cultural and social area, where it was colored by associations with heresy and political disorder.[2] When readers of the *Problemata* contemplated questions of accountability, they generally did not venture into the domains of religious doctrine. Their apparent restraint was due in part to the institutional structures and political dynamics of universities, in which powerful faculties of theology (especially in the North) and of canon law (especially in Italy) defended their prerogatives. But some of those who read and made remarks on the Aristotelian work were themselves clerics or theologians, and Walter Burley did not hesitate to tell the theologians to mind their own business, saying that their books contained dirtier material than the *Problemata*'s collection of questions about sex.[3] Nevertheless, most followed two more familiar and comfortable paths when confronted with questions about values that emerged from the subject matter of *problema* IV.26. One, treated in this chapter, was to have recourse to an alternative, complementary philosophical domain, specifically Aristotelian ethics; the other, treated in the next chapter, was to bypass philosophy and to express their opinions and feelings in any number of textual gestures—omissions, asides, slips of the pen.

Appeals to Aristotle's *Nicomachean Ethics* in commentaries on the *Problemata* acknowledged both a distinction and a relationship between natural philosophy and moral philosophy. On questions about sex, the two disciplines might differ, but the citations conveyed the idea that the answers would still be found somewhere within the framework of a secular, academic science. Do the dictates of nature have any bearing on the praise and blame attached to particular acts or individuals? Do innate conditions, diseases, or madness insulate an actor from responsibility for prohibited behavior and its consequences? What about habits? The claim that natural science should investigate even what is ugly and vile rested on bracketing moral questions: natural philosophers could simply assert that the causes of whatever is found in nature are the proper subjects of philosophical investigation. Pietro exercised his prerogatives under those terms, offering a set of emphatically natural explanations for the disturbed desires. Yet, as though to concede or even assert the inadequacy of that approach, *Problemata* commentators turned to moral science. This crossover highlighted for late medieval readers the volatility of the subject matter and it pointed to a specific and well-developed body of thought—Aristotelian ethics—within which it might be considered. At the same time, the bridge they created between the two philosophical disciplines directs modern readers to the difficulties or conflicts they perceived.

At one level the points at which Pietro d'Abano referred to Aristotle's *Ethics* in the course of his comments on Book IV, *problema* 26, provide an introduction to the specific issues medieval readers encountered when natural explanations of homosexual pleasures (passive, active, and ambiguous) were measured against a system of values. At another level, the management of the relationship between two texts of secular philosophy suggests a larger project concerned with preventing a direct confrontation between the natural and the moral sciences.[4] His first quotation comes in the context of his digression on sex as the greatest source of pleasure, where he invokes the general principle that the experience of pleasure reinforces and improves a behavior. The context is a positive one, in which pleasures serve the individual and the species. The second occasion for a quotation from the *Ethics* produces a very different tone. At the opening of his discussion of those whose desires are grounded in a habit, Pietro declares that "those who do those abominable things [are those] whom Aristotle calls 'beyond the boundaries of vice' in the seventh book of the *Ethics*." On this point, apparently, moral philosophy alone was not sufficient for his purposes, since he continued by quoting both Roman law and Avicenna's assertion that all nations condemn pederasty. Later, turning in a cooler voice to explain how habit becomes a kind of nature, Pietro quoted Aristotle to pose a possible objection. In the *Ethics* the Philosopher had pointed out that one cannot teach a stone to go upward by throwing it up; in response Pietro made a distinction which allowed him to claim that Aristotle actually supported his position. And finally, as he reached the end of the *problema*, just before his closing references to astrological authorities, he again brought the *Ethics* to bear upon the *Problemata* by pointing to a specific mention of male-male copulation in the former: "Truly Aristotle says in the seventh book of the *Ethics* that the cause of this abominable act is complete bestiality . . . and corruption of the soul."[5] The specific functions of these references within Pietro's scholastic arguments have been treated in Chapters 1 and 2. In this chapter they serve as a point of entry into the domain of moral science as it was understood by the readers of the *Problemata*. And, as Pietro's invocations of the *Ethics* suggests, the meaning and status of nature is once again in play.

Understanding moral virtue (and thus true human happiness and thus in turn the basis of a well-ordered polity) is the central project of Aristotle's *Ethics*. Among other things, Aristotle asks whether virtue and its opposite, evil or vice (*malitia*), are in us by nature; examines the roles of desire, pleasure, pain, and reason; considers the characteristics of self-restraint and its failure; and sorts through various related categories, including intemperance, incontinence, and bestiality. As Pietro's references suggest, the text could be deemed relevant in a number of ways to male homosexual practices as framed by the *Problemata*. These included the relationship between nature and habituation and questions about the moral culpability of people overtaken by forces beyond their control. In addition, Aristotle's brief mention of men who have sexual intercourse with other males gave a clue about where to place them on the map of bad behavior. The *Ethics* thus formed an important dimension of the textual environment within which medieval readers approached *Problemata* IV.26. It had a bearing both on the way they evaluated the sexual irregularity treated there and on the larger question of how far natural philosophy could be extended into the realm of human behavior.

The *Nicomachean Ethics* in the Late Middle Ages

The science of ethics was the leading member of the so-called practical sciences, which also included politics and economics. Although other ethical systems, notably that of the ancient Stoics, influenced medieval philosophical thought on the subject of the good and felicitous life, the ethical writings of Aristotle defined and constituted the field for late medieval intellectuals. Even in antiquity, several versions of Aristotle's writings on the subject circulated, and the textual situation was further complicated by the time they reached the Latin West, translated at various times by various scholars. Two versions arrived mediated by Arabic authors. But the Latin text that came to dominate the university scene was the translation from Greek by Robert Grosseteste, bishop of Lincoln, made in the 1240s. He also translated a composite Byzantine commentary on the work that was known to a number of Western European commentators.[6]

Grosseteste was one of many philosophically oriented readers of Aristotelian ethics, which took a place in the curricula of European arts faculties. Because it dealt with questions of critical importance to theologians, they too often discussed it, especially in commentaries on Peter Lombard's *Sentences*, a cornerstone of theological studies.[7] Thus, unlike some of Aristotle's other works, notably the *Metaphysics*, that were often met with theologians' suspicion or even condemnation at the time of their introduction into the nascent universities in the early thirteenth century, the *Ethics* met with general approval.[8] It dealt with the kinds of virtue accessible to pagans and manifested in social relations, so its arguments and conclusions needed to be carefully meshed with and subordinated to the Christian virtues. That project was greatly advanced by Augustine and continued by medieval scholars. This domestication of Aristotelian ethics was, for the most part, successful, but commentators trod on some dangerous territory.[9]

Physicians, trained in Aristotelian philosophy in the arts faculties of universities and participants in late medieval urban and court life, were among those who took an interest in the *Ethics*. The thirteenth-century Bolognese practitioner and professor Taddeo Alderotti, for example, played a role in the preparation of an Italian translation of one version of the work.[10] Acquaintance with its ideas reached beyond university circles thanks in part to their prominent place in a philosophical compendium by Brunetto Latini (1220–95) that was well known during the fourteenth and fifteenth centuries in Italian, Catalan, and especially its original French versions.[11] In the late fourteenth century Charles V of France commissioned a French translation of and commentary on the best Latin version of the *Ethics*, while an early fifteenth-century Latin translation made its way into royal and aristocratic libraries in Spain, and a Castilian compendium circulated widely.[12] No doubt because of its subject—right and rewarding conduct in the secular world—and because of the values it extolled, such as strength, magnanimity, justice, and friendship, the work seems to have appealed to ruling elites.

Indeed, the *Nicomachean Ethics* was one of the most widely read of Aristotelian texts, and thus readers of the *Problemata*, having themselves passed through the environment where it was taught and discussed, were likely to have been familiar with it. A rough measure of its importance can be taken by the number of medieval manuscripts that have survived. The Grosseteste translation has come down to us in more than two hundred copies, more than the main translation of Aristotle's *Metaphysics*, more than twice as many as its own companion text, the *Politics*, and more than four times as many as the *Problemata*. Its numbers rival those of Aristotle's logical works, which were required foundational texts for all university studies.[13] Hence, just as Pietro's invocation of Avicenna's *Canon* would have resonated even beyond his medical audience, his citation of the *Ethics* called up a familiar body of theory and controversy for those readers oriented toward philosophy and theology and beyond. Book VII of that work was not an especially contested site, compared to sections of the work that fueled controversy, for example, about whether unclouded reason led inexorably to right judgment and right action, and what the implications might be for free will and divine grace.[14] In posing the problem of individuals burdened with constraining conditions, the seventh book did, however, raise general questions about the limits of individual responsibility; and in placing the inclination to male-male sexual conduct among those conditions, it provided a specific opportunity for addressing sodomy in the context of moral philosophy. Commentators on the *Ethics* itself often paused briefly to reflect on the larger issues, but they seem to have had little to say about the specific case of homosexual desire.

About seventy commentaries or sets of scholastic questions on the *Ethics* by known authors have come down to us from the thirteenth through the fifteenth centuries, in addition to numerous others by anonymous scholars.[15] A full survey of their treatment of the relevant passages in Book VII that deal with incontinence and bestiality in relation to vice would yield a richer understanding of medieval readings of same-sex pleasure. Such an inquiry is beyond the scope of this book, but, in what follows, a small sampling provides a sense of the ties and distinctions between the *Ethics* and the *Problemata* in terms of natural dispositions and acquired habits. Because they were chosen to illuminate certain narrow questions about the extent and limits of blame under specific circumstances, the works considered here are not necessarily typical or orthodox. Most particularly, on the points in question here, they hardly reflect the introduction of divine grace into the discourse surrounding ethics and the possibility that virtue might be divinely infused.[16] They do serve, however, to provide an additional context for the medieval discussions about the men with whom intercourse is had. Pietro d'Abano was more inclined than most of his contemporaries to frame subjects that had moral and social implications in terms of natural processes.[17] But his appeal to the central textbook of scholastic moral philosophy for support made sense within the range of late medieval interpretations.

Natures (Essential and Accidental) and the Limits of Ethics

Much of Aristotle's argument in the *Ethics* concerns the acquisition of a state (*habitus*) of moral goodness by practice—through the habit of acting. Unlike, say, reason, virtue is not something preexisting in humans simply waiting to be actualized. Thus medieval Aristotelians took as axiomatic the Philosopher's dictum that "none of the moral virtues is in us by nature."[18] That this premise could be understood as consistent with the Christian doctrine of Original Sin may have made it plausible to theologians, but medieval philosophers' understanding of it followed Aristotle's reasoning.[19] "Nature," in this sense, was an aspect of a being's essence, part of its very definition, without which it would not be what it is. If virtue could, as they believed, be acquired and developed (or lost), it could not be regarded as something inherent, fixed, and permanent. Thus, he concludes, we must not regard virtue as something that is in humans by nature or as something in conflict with nature. For in neither case could it be instilled in a person—nor, as he says elsewhere, could it be the subject of praise

Medieval readers developed implications of these principles that bore on the question of the moral status of the subjects of *problema* IV.26. Thomas Aquinas (1225–74) concluded his exposition of this passage of the *Ethics*, "Therefore, habit [*consuetudo*] has nothing to do either with those things that are according to nature or with those things that are contrary to nature."[20] The acquisition and thus the discourse of moral virtue are in play only in those domains in which right conduct can become ingrained by being practiced. Thus, if nothing inherent in us can qualify as virtue, because it is not (as virtue is) a state produced by repeated actions, it is also the case (as with the stone repeatedly thrown upward) that a person cannot be trained to behave in a way that is contrary to some fundamental aspect of human nature. Neither Aristotle nor his medieval commentators acknowledged (much less attempted to resolve) the tension between this position, articulated in Book II, and a later allusion, in Book VII, to unnatural desires acquired by habit upon which Pietro played. Rather, Aristotle's argument, which emphasizes the higher human faculties and the exercise of choice, presents a dichotomy between natural (that is, necessary or automatic) action and voluntary action. This perspective thus appears to postulate that some things can be changed or overcome and others cannot. In that context, there is at least one sense in which an aberrant property, such as a physical disability that inheres in an object or a person, does so with the same fixity as a normal property, so that it is no more subject to alteration than is the (natural) heaviness of a stone. Such constraining conditions, as the influential Muslim scholar Averroes had explained in his *Ethics* commentary, are in us "involuntarily" and are not included within the purview of ethics.[21]

Thus Christian Aristotelians also understood the domain of ethics to exclude behaviors that were the result of a person's general or particular "nature." In the first case, what was beyond choice and control was something essential for membership in the human. In the second case, a fixed individual trait—what scholastics called an "accident"—constituted a contingent feature of an individual's nature. The ethical consequences of recognizing the intractability of individuals' conditions are explicit in Aristotle's work: "No one reproves people who are shameful [*turpes*] from nature but those who [are] from inaction and negligence. Similarly, . . . surely no one will reproach a person who is blind by nature or from an illness or a wound, but rather [that person] will be pitied, whereas everyone certainly will blame a person who [is blind] from drinking wine or some other incontinence."[22] This principle is undoubtedly what Walter Burley had in mind when he cited the *Ethics* in the prologue to his *Problemata*, saying: "For no one ought to be blamed or praised for natural things."[23]

The mutually exclusive relationship between the results of a trait inherent in a person due to necessary causes (nature or illness or wound), on the one hand, and those actions that are properly subject to blame, on the other, set the stage for discussions about what actions may and may not properly be qualified as morally good or bad. It remains to be seen whether a person's sexual nature can be exempted from the axis of praise and blame on this principle. The blockage of seminal pores was, after all, compared to blindness; yet some of the men in question might be more similar to those who drink themselves blind. Though there is no reason to believe that Burley had *problema* IV.26 in mind, a comment he made in his own survey of Aristotle's *Ethics* leaves the door open. Paraphrasing the passage just quoted about "shameful things that result from nature," he allows for the possibility of an innate flaw of spirit as well as body: "For example, no one should reprove [someone] blind or shameful from birth, but rather he should be pitied."[24]

Aristotle's statement that virtue is not in us by nature bears not only on the *Ethics* and the *Problemata* as individual texts but also on the relationship between the disciplines to which they were attached. There are two distinct sources of action, the moral and the natural (whether universal or individual). They are not to be approached with the same philosophical questions, language, or methods. Natural philosophy, after all, is about nature; ethics is about conduct that does not arise from nature. In the eyes of some readers, the disciplinary divide was crucial. For example, Heinrich von Friemar (ca. 1245–1340), a theology professor from Thuringia whose commentary on the *Ethics* was widely disseminated, added reasons to support Aristotle's exclusion of the natural as a source of virtue.[25] If goodness were in us by nature, he argued, then it would follow (absurdly) that virtue is the subject not of moral philosophy but of natural philosophy.[26] That is, there would be no separate ethical causes of behavior, only natural causes. For Heinrich, whose interests were pastoral as well as theological, the line drawn between the two domains served decisively to exclude virtue and vice from consideration by the natural sciences; but for those whose interests ran to the sort of natural philosophy entertained in the *Problemata*, the line between the disciplines opened up rather than closed down serious questions about the reach of nature and of the branches of learning dedicated to it. Are not good and bad conduct indeed sometimes explicable in natural philosophical terms? Like Pietro's treatment of the passage in Avicenna, in which that prince of physicians dismissed homosexual desire as a nonmedical issue, the distinction of the moral and the natural in the tradition of the *Ethics* was the beginning, not the end, of a story. This difference in perspective may help to explain why certain passages in the seventh book of the *Ethics* took on greater significance in the context of *Problemata* commentaries, where the possibilities of natural causation were central, than in the context of *Ethics* commentaries, where they were tangential.

"Insofar as Nature Is the Cause"

Although the proper subject of the *Ethics* was moral goodness (*virtus*) and (by implication) its opposite, vice (*malitia*), Aristotle saw the state of virtue and justness as accessible only to the best male citizens. To highlight the particular qualities of true virtue and provide a framework for evaluating weaker or more common folks, including women, children, and many men as well, he treated other levels at which behavior may be regulated. These formed a hierarchy of distinct axes along which conduct (*mores*) could be measured. For those who are able to achieve prudence, temperance is possible, as is its opposite, intemperance. People who are not prudent can nevertheless be continent if they restrain their bad impulses, and are incontinent if they cannot. But in some people, the higher, human faculties are lacking or destroyed, so that their conduct is ungoverned, or rather it is governed by animal inclinations, and they are simply "bestial." Aristotle gives some colorful examples, collective and individual: the cannibals of Pontus and a woman who tore open the bellies of pregnant women.[27] Bestiality in this sense (as distinguished from the practice of copulation with animals) can only successfully be opposed by "a virtue that is beyond us, best qualified as heroic or divine."[28] As a consequence, Aristotle continues, it makes no sense to describe gods in terms of moral goodness or beasts in terms of moral evil. "Rather the former is more honorable than virtue and the latter is of a different category from moral evil."[29] Although Aristotle later complicates this taxonomy, the exemption from the moral axis accorded here to bestiality is the principle to which Pietro appealed in one of his quotations from the *Ethics*. This exemption is by no means an expression of acceptance—bestiality is no less to be reviled—but the text suggests the possibility that it is removed from the ordinary burden of culpability. As the views of Jean Buridan will illustrate, the superhuman moral strength required to conquer certain terrible impulses raised questions about the limits of human reason and will.

Thus, in addition to virtue proper, the ancient text examines various responses to desires and pleasures—continence and incontinence, temperance and intemperance, perseverance and weakness (*mollitia*). And in the course of doing so, Aristotle makes distinctions among the objects of desire as well as the ways of dealing with them, mentioning, for example, that some pleasures have to do with necessities, such as food and sex, while others, such as those associated with victory, are optional. Medieval Aristotelians saw appetite in a fundamentally positive light, because in general its object was good or produced a good result. Pietro elaborated on this idea in his explication of the word "desire" in *problema* IV.26. And, invoking the same notion of the teleological functions of desires, he quoted from the *Ethics* to explain the proposition that sex produced the greatest of all pleasures. In that same context, however, Pietro acknowledged in passing that the appetite might pursue an apparent good rather than a real one. Medieval scholars encountered that pitfall in Aristotle's *On the Soul* and again in the *Ethics*, where it appears by way of introduction to a list of appalling behaviors: "Some things are pleasurable by nature, and of these some [are so] absolutely and some according to the types of animals and people. Others are not [pleasurable by nature] but some become [pleasurable] on account of passions, some on account of habits, some on account of destructive natures."[30] Here Aristotle's language moves beyond the general category of "bestiality" to suggest the types of human conditions that played a role in explanations of homosexual pleasure in *problema* IV.26.—anatomical defect and ingrained habit. Furthermore, it illustrates what appears to be a paradox concerning the role of nature in defining what is good and explaining what is not. With the phrase "destructive natures [*perniciosas naturas*]," it invokes the distinction between essential (universal) nature, by definition good, and accidental (individual) nature. This is the same tension echoed in Pietro's formulation, "monstrous nature."

Medieval commentators made this point explicit. For some, like Thomas Aquinas, pleasures whose fulfillment served no providential purpose posed a fundamental challenge to the logic of Christian teleology.[31] But for some other readers of the *Ethics* the stakes were apparently not so high. John Buridan (ca. 1292–ca. 1358), writing on the *Ethics* in Paris not long after 1310, when Pietro finished his *Problemata* commentary, offered an explanation that might almost serve as a gloss on Pietro's phrase "their unnatural nature." He distinguished depraved natures from "universal" ones: "Pleasures are called 'outside of [*preter*] nature' that give pleasure to living creatures because of their particular depravity. Therefore I concede that such things are suitable and thus natural for such depravities, but it does not follow therefore that they are natural in the unqualified sense [*simpliciter*] or fitting in the unqualified sense—just as it does not follow that something having a bitter taste when it is spoiled is therefore fundamentally bitter."[32] The contrast between these two levels of nature was equally clear to scholars reading the *Physics*, of whom Buridan was one, and they applied it without difficulty to the majority of natural questions posed in the *Problemata*. But the binary distinction was insufficient for scholars reading the *Ethics*, leaving less than fully determined the moral status of the individuals and groups featured in *problema* IV.26.

One complication that emerged was the multiplicity of natural conditions. At the level of possible objects of desire, natural and therefore appropriate pleasures included not only those common to all living creatures but also those proper to particular kinds and even to particular types within a kind. Albertus Magnus, like Buridan one of the commentators on the *Ethics* most committed to natural philosophy, explained Aristotle's passage on natural and unnatural appetites this way, quoting bits of the text, but adding layers to the concept: "And with 'some things are pleasurable by nature' [Aristotle] says that [they are pleasurable] either 'in the unqualified sense' to all—such as food in general and such things to all animals—or 'according to the types of animals,' since one food is naturally pleasurable to some [kind] and another to an ass and another to a human. 'And' similarly according to types of 'people,' since some [foods] are naturally pleasurable to melancholics and others to cholerics."[33] Some of the appropriate objects of desire are thus universal, others are common to all human beings, but others naturally appeal to only some specifically constituted people—according to Albertus's example, those having a particular physiological constitution.

"Sodomites" are clearly not included among these benign human varieties. They belong to the other half of the distinction, which follows neatly: some pleasures are *not* natural; some objects are *not* naturally desirable. Yet of the three circumstances that Aristotle cited to explain unnatural impulses, the first, passions, might originate from natural causes; the second, habit, is, as Pietro explained, very similar to nature; and the third, "pernicious nature," is by definition a "nature." Indeed, Aristotle was explicit both about the ways in which nature or natures can be the source of abominable behavior and about the ethical implications of the constraints arising from such sources. Having described cases involving matricide and cannibalism caused by sickness and madness, as well as apparently lesser offenses, such as eating raw meat, he continued: "Insofar, therefore, as nature [is] the cause, no one would say that these people are incontinent, any more than [one would call] women [incontinent] because they do not lead but are led. Similarly for those in a sickly condition because of a habit. So, therefore, each of these things is outside the boundaries of vice, and like bestiality."[34] Buridan explained in his discussion of this passage that, in order for moral categories like virtue and vice to apply, the subject must have unclouded reason, confront the force (*impetus*) that opposes right reason, and possess the strength to enforce the reason over the temptation. People who, like children, have weak or fragile natures cannot be held properly accountable for their actions.[35] The reference to children invoked a commonplace of social, legal, and religious culture that was manifest, for example, in their incompetence in civil law and the relative indulgence with which authors of penitential literature prescribed corrections for them. The reference to women, as suggested in Chapter 3, provided an equally accessible, though significantly different, reminder that social standards and values were contingent on the state and status of persons. Like a woman, a man "who, in his body, has experienced womanly things," was held incompetent to represent anyone in a legal proceeding, unless he was raped—thus suggesting that the appetite, not the act, was decisive.[36]

Early commentators on this passage in the *Ethics* had no difficulty with the notion that some individuals might be constrained in a way that restricted their responsibility for what, under other circumstances, would be sinful or criminal acts. These might be "bad," but they were not properly understood as moral evils.[37] Even those commentators on the *Ethics* who had extensive interest in natural philosophy, however, did not go so far as to suggest that such natural causes might in some sense neutralize the unnatural. In this light, Pietro's appropriation of Aristotle's phrase "beyond the boundaries of vice" to develop a point of natural philosophy may have been bold; it was not, however, incomprehensible to medieval readers either of his *Problemata* commentary or of Aristotle's *Ethics*. Jean Buridan's importation of terms from physics, such as "force" (*impetus*), into his scholastic questions on the *Ethics* signaled his natural philosophical approach to the psychology of desire and restraint.

In this view, the desires that did not need to be resisted (though they ought to be restrained and measured) were those pertaining to the essence, the very definition of a human being—as an animal that grows and reproduces, for example, or as a rational and social animal. Those that ought to be resisted, insofar as they were attributable to "nature," do not arise from nature as the defining essence of a permanent kind, nor from nature in the sense of the ordinary course of things. Rather they arise from contingent nature—what another commentator called "errors of nature,"[38] that is, traits inconsistent with those very essential, defining features. Such errors are attributable to accidents, irregularities, and disruptions in the ordinary course of things. "It happens," according to Albertus Magnus, "that what is contrary to a well constituted nature somehow happens to a person naturally. . . . So [pleasures] of this kind are not natural in the sense that they are consistent with the natural principles of the species. They are, however, natural for the individual . . . just as it is natural for a particular person that he has six fingers."[39] (Albertus is alluding to the same passage in Aristotle's *Generation of Animals* that Pietro cited on the subject of birth defects.) Later, in a similar context, Albertus reminds his audience that monsters do occur, and that their characteristics are natural to them.[40] The irregular seminal pores and passageways of the men described in *problema* IV.26 were, in Pietro's view, exactly this sort of naturally occurring (and familiar) transgression of the natural.

This distinction between "natural" in the full, universal sense and "natural" in a more individual and contingent sense is thus found not just in natural philosophy but also in ethics. And it is what made it possible for medieval readers to accept the notion that there existed individuals whose natures produced behavior that, no matter how appalling, was not vice properly speaking and was therefore not subject to moral blame properly speaking. It thus constituted an important element in the exculpatory framework to which Pietro appealed when speaking of men who took the receptive role in intercourse with other men. But the moral philosophy of Aristotle did more than leave an opening in the wall between the natural and the unnatural by offering medieval authors the occasion to reflect on depravities that were themselves naturally produced. The very text that established the distinction between the two senses of nature also suggested that the line between them might not be entirely clear. When medieval commentators reflected on Aristotle's formulation of the relativity of natural desires, they entered a territory in which the healthy shaded into the pathological, the fully natural into what was natural to some individuals only because of a defect. Of course one species of animal had different natural desires from another—they had appetites for different foods, for example. But what did it mean, in the context of fully natural desires, that different types of humans had different appetites?

The unnatural natures had been set not against a single, unified norm but against a range of human natures, none of which was unnatural. According to medieval understandings, groups and individuals regularly differed according to land of habitation, age, pregnancy, and other circumstances.[41] The most common gloss on Aristotle's otherwise unexplained reference to "types of people" was that suggested by Albertus Magnus and quoted above. His elaboration of physically distinct human types—melancholic, choleric, and, by implication, phlegmatic and sanguine—followed familiar and uncontroversial principles of medicine and natural philosophy, based on the ascendency of one of the four humors: black bile, yellow bile, phlegm, and blood. The natural status of the complexions (as they were called), each leaning slightly toward one humor or another, was grounded firmly in the notion that healthy individuals were not all constituted in exactly the same way. For astrological, environmental, or other reasons people were born with slightly different humoral mixes. Furthermore, in the course of a lifetime, it was expected that the humoral balance would shift: people in their prime tended rather more to the sanguine, those in old age to the melancholy.[42] This system of underlying complexions was different from the related system of pathology, according to which many fevers were characterized by an excess of yellow bile, many respiratory ailments by an excess of phlegm, and so forth. Aristotle and his commentators were separating natural desires from unnatural, and people with various humoral constitutions—these different types of human being—are unambiguously on the right side of the line by both ancient and medieval lights. They are not sick. Thus the objects of their desires are not, in and of themselves, suspect. Setting aside that one can desire an honorable object too much, there is no harm in a choleric's appetite for cucumbers. Ambiguity creeps in nonetheless. At their most fundamental, the complexions represent naturally caused innate traits that deviate from the ideal human essence—a perfect balance of all the humors. Though not strictly speaking pathological, the condition of each group is short of perfect physiological moderation, and for that reason the desires themselves reflect and confirm a deviation from perfect equilibrium. Thomas Aquinas uses as examples the choleric (naturally hot) and the phlegmatic (naturally cold), explaining that, "among men, naturally cold [foods], which temper their complexion, are pleasing to cholerics, and hot ones to phlegmatics."[43] Something immoderate calls for rectification. Albertus Magnus puts the problem more forcefully at another point in his close reading of the *Ethics*: "The melancholy humor 'stings,' so the body of 'melancholy people' is always sort of corrupt and thus 'always needs medicine' against pain, namely, pleasure."[44]

That nature's productions were not perfect or uniform was not a source of concern to medieval intellectuals, who had no reason to expect they would be. For them, nature was not structured according to static essences nor governed by inexorable laws.[45] On the contrary, the prodigious and sometimes quirky profusion of individuals within one type was comprehensible from numerous perspectives, ranging from the religious belief that the Fall of Adam and Eve had introduced imperfection into the divinely created world to the philosophical belief that, in the domains of coming to be and passing away below the changeless spheres of the perfect heavens, the intractability of matter or the interaction of separate processes or the influence of the celestial motions often caused inconsistencies or failures. Some deflections from the ideal happened frequently enough to be recognized as subtypes. Of these groups, some, such as females, though they occurred as a result of accidents, served some good purpose within the larger natural order. Others, such as melancholics, did not contribute to any higher goal but represented no serious qualitative or quantitative infraction of normalcy. And still others, such as hermaphrodites, deviated in a way or degree that placed them in the category of monstrosities.[46]

The existence of these gradations further dulls the distinction between the natural and the unnatural, in spite of the fact that medieval authors often exhibited confidence about what belonged on which side of the divide. As interpreted by them, the moral dispensations that resulted from Aristotle's characterization of virtue rested on the delineation of moral science as concerned with a subject altogether separate from the workings of nature. (Even good behavior, if it was the direct result of natural compulsions, did not qualify as *morally* good.) But the separation between the natural and the ethical was based on two not entirely congruent premises. The first was the fixity of a thing's fundamental nature. Just as a stone, essentially heavy, cannot be trained to move upward, some essential human characteristics cannot be altered. The second depended precisely on the mutability of fundamental nature. Not only do deviations from what is essentially human in fact occur at birth or as the result of sickness or habit, the acquisition of virtue itself is not a natural process. Becoming a mathematician involves the realization of qualities that are in us potentially from birth, needing only to be actualized. Becoming virtuous, however, involves the inception of something new within us (whether earned or divinely infused). The two perspectives came together in the passage quoted, where Aristotle explicitly compared the behaviors that result from a fixed natural type, women, with behaviors that result from adventitious deviations producing an individual nature, a sickness, or a habit. When nature is the cause, as in such cases, there is no question of culpability—"no one would call them incontinent."[47] Some medieval authors worked to resolve the incongruities of these two lines of analysis, notably by introducing principles of free will. Nevertheless, the cloud of meanings and operations of "nature" in medieval Aristotelian moral philosophy introduced some uncertainty about the moral responsibility of those deemed defective, infirm, or even constitutionally weak. Thus it obscured the logic of the epithet "unnatural," so often applied to sodomites.

Aimed at delineating the specific characteristics of virtue by distinguishing it both from its opposite and from states or behaviors that resembled it, this section of Aristotle's argument did not condone the consequences of unnatural desires. Nor did medieval commentators like Thomas Aquinas read it in a way that suggested any such implication. Pietro and some subsequent readers of *Problemata* IV.26, in contrast, took this qualified acknowledgment of natural causation ("insofar as nature is the cause") in the *Ethics* as an invitation to mitigate the opprobrium aimed at deviant sexual behavior or to extend the territory of natural philosophy or both. What had seemed absurd to the theologian Heinrich von Friemar—that morality might be the purview of natural philosophy—was rather inviting to those who favored explanations from the other side of the disciplinary divide.

"What We Call Sodomy"

The questions addressed in Book VII of the *Ethics* were all the more relevant to readers of *Problemata* IV.26 because it explicitly mentioned male-male sexual contact, and it did so in the context of a passage elaborating on the concepts of bestiality and incontinence. Aristotle illustrates the force and inexorability of unnatural desires, giving some particularly appalling examples of their consequences. But these abominations keep company with some behaviors that, however deeply ingrained they may be, are merely annoying at worst. As mediated by the thirteenth-century translation, the passage reads:

> I call pleasures bestial such as that attributed to the woman who tears open pregnant women to devour their children; or certain

> wildmen around Pontus, some said to enjoy raw meats, some the flesh of men, some to serve each other [their] children at a banquet; or what is said about Philaris. These are bestial. Some become [thus] on account of sicknesses and madness for certain things, for instance sacrificing and eating one's mother and [eating] the liver of a fellow slave. Some are due to sickness or from habit, for example, pulling out one's hair and chewing one's nails. In addition, [eating] coal or earth. And along with these [the practice] of sexual acts with males.[48]

Scholastic expositions of the *Ethics* dutifully explained the classical references in this passage and elaborated on the anthropological lore concerning the practices of peoples in distant lands. Albertus Magnus, for example, named the woman who craved fetuses and added that people in Pontus also eat raw fish.[49]

The Byzantine commentary that often circulated in Western Europe along with the Aristotelian text had highlighted this particular reference to sex. Having attributed the eating of coals or dirt to humoral imbalances, the anonymous author assumed that much of Aristotle's subsequent discussion of causation was directed specifically at the "bad inclination" to "abuse males" and repeats the phrase "corruption of males" three times in the course of his exposition.[50] His expansiveness represented an encouragement to Latin scholars to engage the subject. But medieval commentators largely declined the invitation: few offered much explication of the final example in this passage. Their opinions and attitudes must therefore be read out of a few clues in the phrasing and organization of their glosses. Grosseteste's Latin translation of the *Ethics*, quoted above, speaks literally of "things venereal with males [*venereorum masculis*]."[51] Albertus Magnus offered the paraphrase "males sleeping together [*concubitus masculorum*]," which he had borrowed from the translation that accompanied Averroes's Arabic version, and Thomas says "intercourse of males [*coitu masculorum*]."[52] Each, in other words, had taken a phrase that distinguished the subject from the object and replaced it with a phrase conveying mutual action. Also deriving from an Arabic text was a simpler, shorter Latin compendium of the *Ethics* that had achieved a certain currency by the end of the thirteenth century. According to its paraphrase of the passage in question, "Pleasures resulting from a bad nature are, for instance, sleeping with males and other disgraceful things of this sort." Not only did the author of this version omit the point about being outside moral categories, he employed the very term that Aristotle had disallowed—"vice [*malitia*]"—to characterize cruel acts that arise from sickness.[53]

The medieval textual responses deriving from these Greek and Arabic traditions, slight and occasional though they be, do betray an impulse to stress the reprehensible nature of this particular behavior. Since the purpose of the passage in which the phrase occurs was to illustrate bestiality, the word "abuse" in the Byzantine commentary would seem redundant. Yet, echoing it, the late medieval moralist and preacher Heinrich von Friemar underscored the tone of condemnation, speaking of "the abuse of sex acts with males [*abusus venereorum cum masculis*]."[54] Nicole Oresme (ca. 1320–82), the prominent scholar who wrote a French commentary at the behest of King Charles V, ventured a slightly more substantive foray into interpretation. He translated Aristotle's phrase as "abuse of carnal pleasure with male children" and glossed it with "This is what we call sodomy."[55]

Given the orientations and audiences of these authors, this fragmentary evidence suggests that where the philosophical text of moral science, the *Ethics*, approached the more social project of teaching conduct, the language and tone became moralistic. That rhetoric has no specifically scriptural or religious associations but rather retains a general or even secular character, with the exception of the identification of "sodomy" by Oresme, whose book belonged to the same French environment as Evrart de Conty's. Brunetto Latini, whose work was written in one vernacular and quickly translated into others, referred to "sleeping with males and other dishonorable things."[56] More academically oriented treatments (all in Latin), such as those of Albertus Magnus, Thomas Aquinas, and Jean Buridan, did not expand upon the reference to men who desire intercourse with males in connection with the *Ethics*. The division between authors with more general audiences and those more narrowly directed is not absolute, however. Albert of Saxony (ca. 1316–90), whose career at the universities of Paris and Vienna centered on the development of natural philosophy, singled out this particular unnatural desire from the others at the same time as he retained its natural causes: "Sometimes it happens to some people on account of soul, body, and a corrupted [*perversa*] and extremely base complexion."[57] A systematic survey of *Ethics* commentaries and their authors might clarify the extent to which broader and more diverse audiences were more likely than philosophy students to encounter this kind of invective.

Both the general reticence and the occasional murmurs in the face of this Aristotelian passage are significant. There are no signs of any attempt to suppress or evade this example of the bad things some people are inclined to do.

No doubt it seemed reasonable and appropriate to include it, and, in any case, these were Aristotle's words, to be treated with respect. For the most part, the glosses represent routine explications of a phrase that was not at all difficult to understand. Whatever a reader's or a listener's view of homosexual acts, the question more likely to present itself to a medieval audience was "Who was Philaris?" As with the *Problemata*, so with the *Ethics*, sodomy was not a prominent issue within the larger scheme of the work. Nevertheless, the internal dissonance of the passage itself, with its apparent lack of proportionality among the elements (from eating coal to eating people), along with the evidence we have from elsewhere about the volatility of the subject, would suggest that the paucity of discussion was more than coincidental. Albertus Magnus is unusual in alluding to the problem that would seem to present itself: sodomy is a sin and thus, by definition, morally culpable. Among the propositions he undertakes to discuss relating to the *Ethics* is: "It is false that no one would call the people to whom these things happen 'incontinent.'" By way of explaining a possible objection to Aristotle's position (an objection he ultimately rejects), he observed, "Sleeping with males is called 'the sin against nature,' therefore it would seem that nature could in no way move [a person] to it."[58] But his response sidestepped this particular dilemma. Instead, it focused primarily on a different example from the same segment of Aristotle's text, namely, the extent to which women, who are naturally weak, are morally answerable, adding an analogy to the meaning of horoscopes. At the end of the various arguments and counterarguments, he dismissed the objection about astrology (and, in an added phrase, the one about men sleeping together), as irrelevant to the central question. Mark Jordan has noted Albertus's curious reluctance to resolve the challenge concerning sodomy that Aristotelian ethics and natural philosophy posed, in spite of his thorough command of both. And Thomas Aquinas's evasion is even more suspect, considering the implications of "outside the boundaries of vice," and given the way in which, elsewhere in his work, the unsettled position of sex between men in his taxonomy of vice depended on a peculiar convergence of the bestial, the cannibal, and the sodomitical.[59]

In addition to placing him within the matrix of disapproval, Oresme's French version of the *Ethics* touches on another aspect of male-male sexual interaction that comes into play in the *Problemata* commentaries: the question of age. His position that the males in question are boys, not men, occurs in his translation, rather than in his glosses on it. Thus, in principle, this experienced and self-conscious translator based his choice of words on the narrow linguistic determination that the text intended *enfans masles* rather than simply *masles*.[60] The original Greek spoke of sex acts with τοῖς ἄρρεσιν, which does not suggest boys. The Latin that Oresme and his contemporaries were reading gave *masculus*.[61] That word does indeed sometimes refer to male children, especially to sons, but that sense is normally set up by the specific context in which it occurs.[62] Thus in Exodus, when the order is issued to midwives dealing with newborns, killing all the *masculi* refers to boys. But in Genesis, when the Vulgate makes reference to killing all the *masculi* of a neighboring people, the victims are adult males—the women, children, and cattle being taken captive.[63] Oresme may have had in mind a subsequent reference in the *Ethics* to "habits acquired in childhood," though the phrase does not in itself imply intergenerational influence. The anonymous Byzantine commentary, which he cites elsewhere, refers to those who are, from the time they are young, accustomed to corrupting males, in which case both partners would be young. That is, the practices he has in mind are more likely to involve interactions among boys than the initiation of younger men by older ones. On the other hand, according to the same text, the process involves associating with bad men, presumably adults.[64] Oresme may also have been basing his own choice of words on an anthropology of sorts, reflecting what he knew of ancient Greek erotics from literature, or sharing Pietro's characterization of contemporary practices, or conceivably even alluding to Avicenna's mention of the taboos and legal prohibitions against intercourse with children among many peoples.[65] In any case, his reading accords with the structure of the original text insofar as both convey the notion of an implied male actor and speak of those whom he acts with or upon, in contrast to Albertus Magnus and Thomas Aquinas, whose phrasing had weakened the presumption of asymmetry between active and passive partners.

In the end, these treatments of a single phrase linking the *Ethics* to the *Problemata* do less to define the acts and actors than to gauge its significance for its readers. Their level of interest tended to be slight; their tones diverse and often elusive. For example, the motive and meaning of a scribe who intervened in Heinrich von Friemar's commentary is difficult to discern. He enhanced Heinrich's phrase "abuse of sex acts with males [*abusus venereorum cum masculis*]" to read *abusus venereorum cum ipsis masculis*. The added *ipsis* (a demonstrative pronoun) highlights the word "male" and has the effect of adding an exclamation point to the phrase.[66] At the very least, the verbal gesture serves to call attention to these men, in the same way that a marginal "nota bene" might do. The stress may be pointing toward Aristotle's causal explanations: sickness or habit can account for hair-twisters, nail-biters, and even sodomites. Or it could (either alternatively or simultaneously) be underscoring the incongruity of the juxtapositions: infanticide, cannibalism, matricide, twisting hair, biting nails, eating dirt, and "even" males having sex with other males. But where, on this broad scale, does the last case belong? For a variety of reasons, including the tone conveyed by Heinrich's term "abuse" or "misuse," the person who copied the work probably had in mind something graver than a compulsive tick. Jordan has observed that the sense of sodomy in Aquinas's writing "swings from a middling sin of *luxuria* to the eponymous sin against natural teleology."[67] This passage of the *Ethics* poses a special challenge in that regard, since, short of confronting the lack of proportionality between hair-pulling and cannibalism, which commentators declined to do, it fails to specify the position of sex between males along the axis of disorder at the same time that it naturalizes the various disorders to which it alludes. Thus, in the end, the inclusion of men apparently akin to those who figure in *problema* IV.26 was not what made Aristotle's *Nicomachean Ethics* relevant to Pietro and the other expositors of the *Problemata*. Rather it was the work's more fundamental engagement with questions about the relations among nature, behavior, and responsibility.

First and Second Natures

For medieval readers of the *Problemata*, the ambiguous status of male-male intercourse within the array of unnatural desires was illustrated by but not limited to the passage in which it was mentioned in proximity to infanticide and eating dirt. It was compounded by the more philosophical problem that, no matter how abominable they were, their perpetrators could not be called *morally* bad, any more than a bad doctor or a bad musician is bad in an ethical sense.[68] The attribution of causes added yet another layer of complexity to the way in which the *Ethics* illuminated (or obscured) the implications of *problema* IV.26. In the single, brief passage that Pietro cited and that has been under consideration here, the Aristotelian text mentions: (1) unnatural desires "some on account of passions [*passiones*], some on account of habits [*consuetudines*], some on account of destructive natures [*perniciosas naturas*]"; (2) bestial people who eat human flesh "on account of sicknesses [*egritudines*] and madness [*maniam*]"; (3) those who twist their hair or bite their nails, "[because they are] sickly [*egritudinales*] or from habit [*consuetudine*]"; and (4) after mentioning eating dirt and sleeping with males—whether referring to them in particular or summarizing the whole passage—things that "happen to some by nature [*natura*] and to some from habit [*consuetudine*] to which they have become accustomed from childhood." Finally, after stating that those for whom the cause is nature (*natura*) are outside the limits of vice, Aristotle adds, (5) that the same goes for those in a sickly state (*egrotative*) because of a habit (*consuetudinem*).[69]

From the point of view of moral philosophy, these distinctions raise the possibility of different degrees of responsibility, depending on the source of the behaviors. For example, should those with bad habits be blamed while those with innate defects be exempt from moral judgment? Asking which is worse, evil resulting from nature or from habit, Jean Buridan concluded that the former is "naturally" worse, since it is incurable, whereas the latter is morally worse, since changing it is possible and lies within our powers. Elsewhere, however, he makes an exception for those habitual inclinations that it is simply not within our power to change.[70] The subject of unnatural desires, the characterization of bestiality, and the refinements implied by the wide variety of possible causes all add urgency to the question "What, in fact, are the boundaries of vice?" The limits of vice (and virtue), as construed in the *Ethics*, delineate the limits of moral science. Since the candidates for exemption involve "nature" and "sickness," the territory outside the borders of vice belongs, by implication, to the realms of natural philosophy and medicine. (The citizenship of "habit" is more difficult to determine.) Seen in this light, the case of men who have sex with men is one example among many of the relationship between natural and moral philosophy, sister disciplines separated by a thin, flexible, and permeable membrane.

Modern commentators on Aristotle's work have tried to sort out which types of act belong to the category "bestiality" and which ones are attributable to which causes. For example, Gauthier holds that bad nature explains the fetus-devouring woman, the people who eat raw or human meat, and the ones who serve each other their children; that defects occurring after birth (including illnesses and madness) explain the man who sacrificed and ate his mother and the one who ate his friend's liver; that habit explains pulling out hair, chewing nails, eating charcoal or dirt, and sleeping with males. His own translation (of the Greek), however, unravels the neat order. For example, immediately after the last set of behaviors it adds, "And these practices are sometimes the fruit of nature and sometimes the fruit of habit (as for those who have been abused from the most tender childhood)."[71] Jordan reads in Aristotle a distinction between the bestial (Gauthier's group following their bad natures) and those prompted by sickness or habit.[72] The difficulties of the Greek were compounded by the Latin version on which medieval authors depended, not only in the variable iteration of the causes but also at the point where all appear equivalent in relation to the limits of "vice" or "incontinence" properly speaking.[73] Heinrich von Friemar apparently found the standard text confusing, for he resorted to the streamlined Latin compendium translated from Arabic: "Note that the members of this enumeration are more clearly expressed in the Arabic translation, which says, 'Some pleasures are natural, some bestial and some savage, and of these some arise by reason of the season, some by reason of illness, some through habit, some on account of bad natures.' "[74] Following the same authority, Brunetto Latini limited bestiality to the infanticides and cannibals, whereas Albertus Magnus and Nicole Oresme seem to have seen the whole range of causes and examples as referring to bestiality.[75] Elsewhere, Albertus divides the field into two types that seem to differ more temporally than causally: "Either what is converted into a nature of this kind, and such are bestial pleasures; or what is still moving toward a disposition of this kind, and such are certain pleasures related to sickness."[76] Sickness, presumably, is reversible, whereas a monstrous nature is beyond repair.

Thus, whatever the ancient or modern consensus about this passage, the medieval responses constituted and revealed diverse readings of the scope of "bestiality" and the permutations of its causes. Not surprisingly, they did not agree on the status of same-sex intercourse within this tangled matrix. The phrasing of the passage was sufficiently loose that commentators differed not only about the inclusion of homosexual acts in the category of bestiality but also about the cause to which Aristotle attributed it. Indeed, by the uncertainties in its treatment of unnatural desires, the seventh book of the *Ethics* seemed to pose from a different perspective the very question contained in *Problemata* IV.26: are men born this way or do they get this way by developing a habit? In keeping with their general reticence about the phrase "sleeping with males," individual commentators generally declined to acknowledge the issue; taken as a group they highlighted the puzzle, rather than resolving it. There does seem to be agreement that biting nails, twisting hair, and eating inappropriate substances are the result of bad habits, and both Nicole Oresme and Heinrich von Friemar put sleeping with males in that same category.[77] Brunetto Latini, on the other hand, takes it to be the only example of an unnatural desire caused "by a bad nature," following, as he usually does, the Latin version of the Arabic summary. The Latin version incorporated in Averroes's commentary referred to two distinct types of men who sleep together: "This happens to some of the men as the result of a bad nature; to others, in fact, as the result of habit."[78] The Byzantine commentary likewise divided those who "corrupt males" into two groups with inclinations arising either from a depraved nature or from a habit acquired early in life. Thus both these authors, who exercised considerable influence in late medieval Europe, appear to have been familiar with the *Problemata* and to have made the same connection as Pietro between the two Aristotelian works. Albertus Magnus, who knew the Latin versions of both the Byzantine text and Averroes's work, was nevertheless vague about the cause or causes. Thomas Aquinas, in his *Ethics* commentary, held that hair-twisting, nail-biting, coal-eating, dirt-eating, and sleeping with males can all arise from either cause—nature or habit. In his theological work, however, he seems to attribute cannibalism and intercourse with beasts or males to habits.[79] Thomas's construction of this passage figures, in turn, in his association of bestiality and cannibalism with sodomy. For Albert of Saxony, the whole list of behaviors was attributable to bad habits, yet, at the same time, he indicates that homosexual inclinations in particular sometimes result from a bad "soul, body, and complexion"—which would seem like a gloss on "bad nature."[80]

The significance of individual commentators' choices is hard to gauge, especially since they are working within textual traditions, including some that may be lost to us. Nicole Oresme and Heinrich von Friemar were the authors who attached harsher rhetoric to the behavior itself, Oresme being the only commentator to utter the word "sodomy." They did associate it with trivial failings and thus might seem to have been downplaying its gravity. On the other hand, identifying it as a bad habit might have suggested to these authors a greater degree of responsibility. Brunetto Latini, who qualified sex between males as "dishonorable," attributes the desire to nature, thus perhaps according it a sense of greater inevitability.[81] Albertus Magnus, who—if only for the sake of argument—reminded his audience that sleeping with males was a sin against nature, took some of the stigmatized behaviors to be graver than others. After his exposition of the various murderous and cannibal figures, he continues (as usual, quoting snippets of Aristotle): "And some people are this way on account of some other 'illnesses' 'or' on account of 'a habit,' just as some eat their 'nails' and pull out 'hairs' and eat them, and certain other things that are less unnatural than the aforesaid, such as the eating of 'coals and dirt' and the 'sleeping together of males.' And all these things 'happen' on account of the aforesaid causes."[82] Albertus's categorization of what was so often called the "sin against nature" as one of the "less unnatural" mentioned by Aristotle does not help to clarify its status, and neither does his lack of correlation between the gravity of the offenses and the types of cause to which they are ascribed.

The most obvious reason for the lack of distinctions based on causal factors is that Aristotle clearly states that *all* such forces are sufficient to place a person "beyond the boundaries of vice." But even if innate defects and grave physical or mental illness can clearly impede the sort of autonomy required for the categories of virtue and vice to obtain, how can habit do so? Indeed, it is precisely by practice, by becoming accustomed to certain ways of acting, that one achieves the state of virtue according to Aristotle. Thus habit (in the sense of being a state achieved by certain practices) would seem to be at the very center of the ethical. Late medieval attempts to resolve this paradox, such as they are, rely on both moral and natural science, especially in the interpretation of Aristotle's remark that the habits outside the limits of vice are acquired from the earliest childhood. The acquisition of virtue is a practical discipline, dependent on choices and experience. Children develop their will before they develop the capacity to make ethical choices. A youth may quickly grasp the principles of geometry, but acquiring a prudent disposition takes a long time, a virtuous one even longer. The same holds for the development of a truly vicious (as distinguished from immature and intemperate) character. In this sense, children, like animals and women, are excluded from the discourse of ethics.[83]

The training of the young, which is the subject of the latter part of Aristotle's *Politics*, aimed at preparing each individual appropriately for a future role in society. But if children could be said to have anything that might be called "virtue," it was of an imperfect kind and less attributable to them than to their teachers.[84] Upbringing, like innate qualities, could help to determine how easily and to what extent a person could attain the condition of virtuousness. (Aristotle did not expect laborers, disabled by birth, education, and way of life, to acquire full virtue or happiness.)[85] Nevertheless, the appetites, actions, and habits of youth are not themselves proper objects of moral science. Against this background, therefore, habits acquired by the young relate to the categories of virtue and vice differently from those acquired by adults, who have come into full rational and ethical capacity.

Since, in the passage about those whose acts could not be called morally bad, Aristotle's text had explicitly mentioned habits formed in childhood, and because children are, in any case, not ethical (or unethical), medieval commentators on this passage were not concerned with how the habits were formed—though they discussed the general subject elsewhere, and Pietro d'Abano did incorporate the sequence of experience, pleasure, memory, and desire into his commentary on the *Problemata*.[86] The notion that habits could have such a grip that a person had no control helped to make sense of their association with both nature and illness in the context of unnatural desires. Pietro had cited Cicero and Hippocrates to the effect that people's deeply ingrained habits become very much like their natures; he could also have referred (again) to the seventh book of the *Nicomachean Ethics*. For although Aristotle held that "it is easier to change a habit than one's nature," he nevertheless observed, "A habit is difficult [to change], since it becomes similar to nature."[87] There were limits to this similarity. Albertus Magnus, for example, took the position that habit did not attain the stability of nature.[88] Nevertheless, it was this phenomenon—the way in which habit can become second nature—that seemed relevant to the medieval commentators as they considered the three causes of unnatural pleasures: nature, sickness, and habit.

Thomas Aquinas entertained the idea of a virtual equivalence between nature and habit when he discussed Aristotle's example of the stone that moves downward, no matter how often it is thrown up. He introduced the teleological language of appetite (commonly invoked in the medieval physics of natural motions) and explained that inclinations were "either from nature that inclines [one] toward that which is appropriate for oneself, or from a habit that is turned into nature."[89] In that context he was emphasizing the close association between the ethical and the habitual. In the later context of unnatural desires, however, he preferred to assimilate (bad) habits not to bad natures but rather to the third basic cause, illness: "To some people this happens from the nature of their bodily complexion, which they received from the beginning. To some people, truly, [unnatural desires] occur from habit, namely since they have gotten used to this sort of thing from childhood. And it is similar with those who fall into this from a bodily illness. For a shameful habit is almost a type of mental illness [*aegritudo animalis*]."[90] Oresme was probably following Aquinas when he made the same link between habit and illness, saying, "Some bestial people are this way by nature, others from sickness, others from habit, which is a sickness of the soul."[91] This notion of a mental illness surfaces in Pietro d'Abano's commentary on *problema* IV.26, where he quotes the *Ethics* for the last time. The formulation rang true with scribes who repeated it in margins of manuscripts, and with the Bavarian commentator who followed Pietro almost word for word on this point: "Aristotle says in the seventh [book] of the *Ethics* that the cause of this abominable thing is bestiality of all sorts and a corruption of the soul. On account of which, not without reason, he wishes it to exist outside the boundaries of all vice."[92]

In these commentaries, habit seems to resonate between its status as something very like nature (and thus fixed) and something very like illness (and thus not necessarily permanent), both potentially exculpatory. Aristotle's own disinclination to make distinctions in connection with unnatural desires seems to have discouraged hard and fast divisions. After discussing the example of women—clearly a case in which innate nature places limits on moral culpability—Aquinas concluded, "And the same reason seems [to apply] to those who are in a sickly state, that is, who have a corrupt disposition because of a bad habit, which also overpowers the judgment of reason in the manner of a bad [*perversa*] nature."[93] The lines of causation—sickness, habit, nature—turn out to form a web.

"Action Proceeds from the Will"

Although Jean Buridan had suggested that faults arising from habit were more properly subject to moral condemnation than those arising from nature, his observation that habits, unlike nature, were under our control did not define the medieval perspective on the ethics of the unnatural—or even his own. It was insufficient not only because Aristotle had grouped nature, sickness, and habit together; and not only because some habits (especially those instilled in the young) were very close to being nature; but also because medieval Christian thinkers were not entirely prepared to concede that even those compulsions arising from nature were beyond human control. The doctrine of free will was an essential element in the economy of salvation, and although it was empowered by and dependent on divine grace, its claims in the domain of human conduct were strong in the late Middle Ages. Thus Albertus Magnus would assert bluntly at one point in his *Ethics* commentary that, whatever a person's natural condition, "the will nevertheless remains, since however much nature inclines [a person] toward doing something, action still proceeds from the will."[94] Indeed, in many respects, the classical notion of human reason or the rational soul was regularly read to be in harmony with the Christian notion of the individual will and the immortal soul. Whether seen from an Aristotelian or a Catholic perspective, for example, this substance was the defining principle of the human being, not instilled in the course of embryological development but infused into a new person from the outside. The correspondences had their limits, however, and one significant concern of Christian commentators on the *Nicomachean Ethics* was to reserve a distinctive role for the will. In particular, with natural appetites generally tending toward appropriate ends and with human reason capable of discerning what is right, was there any choice, any freedom, remaining for the will proactively to embrace (or fail to embrace) providentially ordained goodness? Or was the process leading to proper behavior essentially automatic? Attempts to resolve such questions were complex and subtle, and they affected the way scholars evaluated the sources of human choice. At one point in his theological masterwork, for example, Thomas Aquinas explained that the will can usually, but not always, dominate the appetite; that the will is *in* the reason; and that the reason rules the passions—but not as a despot.[95]

The challenge of natural reason to Christian views of moral responsibility, and thus of answerability in the afterlife, helped to distinguish and to contrast both the subjects of moral philosophy and theology and the disciplines devoted to their study. *Ethics* commentaries were one site for exploring these boundaries.[96] One question was whether the secular (indeed, pagan) philosophy of ethics, with its rational psychology and dispensations based on nature, would align itself with natural philosophy or whether, with its sights set on a higher good and its implications of moral judgments, it would gravitate toward theology. This was the predicament that Heinrich von Friemar had in mind when he suggested that to accept the inevitability of natural forces was to reduce the science of human behavior to a branch of natural philosophy, thus eliminating a separate science of ethics.[97]

Even so, in the context of *Ethics* commentaries, some medieval scholars did entertain the possibility that what was natural might also be necessary, that innate desires might preclude moral choices. They universally rejected absolute determinism, but they were not certain that human freedom could always overcome the forces of nature. The paradigm for natural necessity in this period was the radical version of judicial astrology, according to which the configurations of the heavens exercised inexorable control over the course of human events. Thus formulated, this position had no actual defenders, but it served to dramatize the dangers of according too much power to forces of the created world. Albertus Magnus judged faith, because of its divine origin, to belong in theological, not philosophical, treatises.[98] He nevertheless mentioned in his *Ethics* commentary "the error of the horoscope casters who say that people are naturally thus or thus, according to the signs of their nativity, which is not only against the faith but also against all civic order and good behavior [*civilitatem et bonos mores*], as Cicero proves."[99] This comment, like the reminder immediately following it that sleeping with males is a sin against nature, represents not Albertus's final position but an element in his dialectical reasoning toward a more nuanced judgment. For clergy and laity, educated and uneducated shared the conviction that celestial bodies did indeed exercise significant influences over terrestrial ones. Horoscopes and astrological consultations at critical moments (a marriage, an illness, a battle) were commonplace. Similarly, no book on the rearing of princes was complete without a section on physiognomy, the science by which a ruler could judge a person's character from physical (especially facial) features. Thus the question was not *whether* human actions were marked by the stars or whether a person's innate nature translated into appearance and behavior, but rather by what means and to what extent. Pietro d'Abano's concluding appeal to the role of the planets in the explanation of men with deflected desires was thus not simply an indication of his personal interest in the subject, but rather an obvious allusion. It called up both the commonplace confidence in celestial influences and the thorny (and equally commonplace) questions of their strength and moral significance.

The responses to these questions that make their appearance in connection with the unnatural natures of the seventh book of the *Ethics* seldom allude to theological or doctrinal truths, but they bear nevertheless upon a generalized notion of free will. One standard argument is that of Albertus Magnus just quoted—that nature inclines a person to behave in a certain way but action proceeds from the will; another is that nature affects the body, not the mind or soul.[100] But, as is already evident from the conviction that the blind are not to be blamed for what they do not see and that children cannot be held accountable for failing to control their appetites, medieval authors regarded those positions as incomplete. As Aquinas observed (citing the *Ethics*), "Natural acts are not in the power of the natural actor, since nature is limited to a single [end]. And thus, although there may be failing [*peccatum*] in natural acts, there is not, however, blame [*culpa*] in them."[101] What is and is not in the power of human beings? When are we (merely) natural agents? In the context of medieval Aristotelian ethics, the key to answering these questions resided in the capacity of reason to prevail over appetite, and this emphasis was especially strong where unnatural desires were in play, because of their association with bestiality. Philosophically speaking, the foremost distinction between humans and beasts was, after all, reason.

Medieval opinion was not, however, univocal on the matter of defective reason. In his theological work, Aquinas took the position that, although a passion, such as anger, or a physical disorder can hamper reason and thereby impede the will, only the full loss of reason constitutes real compulsion.[102] Only the truly beast-like, then, would qualify for exclusion from the strict moral sense of "incontinence" or "evil."[103] In his *Ethics* commentary, however, Aquinas advanced a less absolute standard and took a more psychological approach, using some of the same concepts later employed by Pietro. Due to a bad nature, he argues, people weighed down by unnatural desires have mostly immediate mental images and memories and very little rational judgment. "This small amount seems to be almost nothing. Nor does it happen easily that [this] small power of reason would repel strong desires."[104] John Buridan, whose scholarly production centered on natural philosophy, went a step further, not only echoing the psychological concepts like "appetite" and "judgment" familiar to medieval scholars from Aristotle's treatise *On the Soul* but also introducing the terminology of contemporary physics, such as "force" and "resistance." His arguments thus illuminate the medieval struggle to mediate between natural coercion and free will and, at the same time, the expansion of natural philosophy into the neighboring discipline of moral philosophy.

Buridan's version of Aristotelian physics posited that for an object such as a stone to move in any way other than its natural, downward fall, it must not only have some sort of impetus imparted to it but also encounter some resistance from the medium through which it passes. (The resistance is required because, without it, there would be no reason for the stone not to reach its end point instantaneously, that is, to travel at infinite speed.)[105] In just these terms of physical dynamics, Buridan interpreted what was required for moral dynamics: "First, right reason and second, the impetus of a passion opposing that reason is required. . . . One more thing is required for continence and incontinence properly understood: the efficacy or strength of both the reason and the passion."[106] The sense of this model is that if the reason or will has nothing to resist and thus to triumph over, then a good act cannot properly be called virtuous. In this particular passage, Buridan is speaking of continence, so he represents the passion or appetite as the active element and the reason as the more passive medium, resisting rather than initiating action. That is, self-restraint must hold the fort against the onslaught of desire. This comparison to the physics of motion sets up a way of looking at the clash between the will and unnatural appetites that highlights not the absolute claims of one to prevail over the other but the question of their relative strength in particular cases. After all, in some instances a strong push can move a stone through a slightly resisting medium, in others a strong resisting medium will prevent the motion of a weakly impelled stone.[107]

The result of Buridan's reasoning is not moral relativism in the modern sense but an ethics in which the evaluation of an individual's acts hinges in part on the properties of the individual; specifically, it hinges on the relative strength of the passions or appetites on the one hand and the reason or will on the other. Any uncertainty is not about what is good but rather about the capacity of some people to attain it. Building on Aristotle's own comment about the limits of what can be expected of women and alluding to the accepted indulgence with respect to children, Buridan likened those with unnatural pleasures, whether occasioned by nature or by habit, to those recognized to be morally disabled—women, slaves, and children. Quoting from Aristotle's *Politics*, he concluded that "'Everyone should be judged but not in the same way, but rather for each person according to his own role.'"[108]

In a separate argument, Buridan investigated the limits of human moral capacity more generally, including the powers of reason and will. As part of his commentary on Book VII of the *Ethics* and its references to beasts and their unreason, this discussion of incontinence (by nature or habit) bears upon his conclusions concerning the potential power of unnatural pleasures, to which he makes direct reference. Pointing to Aristotle's introductory remarks on bestiality at the beginning of the seventh book, he poses the question "whether heroic virtue or bestiality is possible for a human being."[109] Among other things, he is asking whether a person in the throes of unnatural desires can (heroically) resist them. He resolves the question by saying that it is indeed possible in principle for mortals to exhibit the divine sort of virtue necessary to overcome an inherent depravity, but it cannot be expected. As in the case of women, so in the case of any actor, much depends on one's place in society. Those who are well disposed by birth and upbringing can perfect virtues and thus attain a moral status beyond what we normally suppose to be human limits. "Others, however, are worse born, lacking basic prudence, and they have bad [*pravos*] behavior, naturally inclining to vices. It is not in any way possible for the badly born to achieve such a degree of excellence as the well born. Or, if this is possible because of the freedom of the will, . . . it is, however, difficult and never seen. . . . From these things, therefore, it is sufficiently clear that bestiality and heroic [virtue] are possible for human beings—not for all, but heroic [virtue] for those who are well born and bestiality for those who are badly born."[110] Buridan's nod to free will here is perfunctory: far from being required, overcoming a bad natural disposition is exceptional and rare. Indeed, even among those who are well born, only those with the best upbringing can achieve a heroic state.[111]

As Buridan's earlier reference to the *Politics* makes clear, in other contexts this reasoning served to confirm a social hierarchy and exclude large categories of people from political participation. Here it is brought to bear on the unfortunate carriers of unnatural desires. As he explained it, Aristotle's exclusion of the heroic and the bestial from the proper discourse of virtue and vice, and thus from the concern of ethics as a science, was based on the almost unattainable conditions that would make the heroic possible and the likewise almost insurmountable conditions that disabled the reason and will of the depraved. Combined with his ethical physics, according to which the forces of appetite were measured against the resistance of reason, simple arithmetic would confirm that, for some, it was a losing battle. Without denying free will, he let it be swept aside.

Buridan's representation of the logic and psychological necessity that placed the consequences of unnatural desires outside the boundaries of vice was by no means generally accepted. It was, nevertheless, one interpretive perspective open to readers of the *Problemata* and Pietro's commentary on it. Walter Burley, the only late medieval scholar known to have written about both Aristotelian works, placed "those who get pleasure in coitus of males" under the causal heading of "habit" rather than nature or illness.[112] But these men, like the others, lack sufficient "universal reason," the higher faculty that would make it possible to grasp the general concepts necessary for ethical judgment. Thus, Burley summarizes:

> Concerning objects of pleasure that are against nature, there is no continence strictly speaking. Animals, which have imagination and memory of individual things but not universal [reason], are not said to be continent or incontinent. Then [Aristotle] argues thus: Those people who have just a little universal reason that does not suffice for resisting strong desires are not called continent or incontinent, since not only those who are continent but also those who are incontinent in the strict sense have strong universal reason. But people who are like animals in vice [*malitia*] related to objects of pleasure against nature do not have strong universal reason that can resist strong desires, whether that vice arises from a corrupt nature or from a sickness or from a habit. Therefore people of this kind are not continent or incontinent. And consequently, concerning objects of pleasure that are against nature [*contra naturam*], there is no continence strictly speaking.[113]

Then, having used the term "vice [*malitia*]," Burley goes on to say that it does not (strictly) apply in such cases, citing the same phrase with which Pietro brought his commentary on *problema* IV.26 to a close: "To have any one of these, namely pleasures relating to objects of pleasure against nature from sickness or habit, is completely 'beyond the boundaries of' (read: human) 'vice' and just like bestiality."[114] For what distinguishes beasts from humans is precisely their lack of reason. At the same time as it offered a qualified dispensation from ethical judgment, the association with animals resonated both with religious polemics about sodomites and with Avicenna's prescription of beatings for the sufferers of *halubnathi*.

Once removed from the ethical axes not only of virtue and vice but also of continence and incontinence, and thus of praise and blame, the dynamics of unnatural desires and pleasures were potential objects of natural philosophical inquiry: the anatomy, physiology, and astrology of unnatural natures; the psychology of unnatural natures and unnatural habits. As will become amply clear in Chapter 5, not all scholars with an interest in natural processes were eager or even willing to accept the transfer of sodomites from the jurisdiction of ethics to that of natural philosophy. For some, resisting that shift involved insisting on a different interpretation of Book VII of the *Ethics*. The physician Evrart de Conty, while acknowledging that some things were beyond the powers of mortals, nevertheless asserted of men who desire sex with men: "There is no doubt that they can very well withdraw from this very mad and very abominable persuasion . . . if they very strongly will and very virtuously resist their inclinations from the very beginning. And for this reason Ptolemy (in the prologue to the *Almagest*) says that the prudent master the stars. And thus Aristotle says elsewhere (in the seventh book of the *Ethics*) that we are lords of all our deeds from the beginning of our life until the end."[115] Evrart, who was fully familiar with Pietro's commentary, rejected his reading of Aristotle by citing (somewhat misleadingly) the very book from which Pietro had drawn the phrase "beyond the boundaries of evil."[116] In keeping with this interpretation, Evrart made it clear that he found the topic of *Problemata* IV.26 an inappropriate one for natural philosophy.

Conclusion: "Less Than Vice" but "More Horrible"

Within the context of Aristotelian ethics, neither the possibility of associating homosexual acts with nail-biting nor the latitude to interpret "beyond the boundaries of vice" as an exemption from responsibility offered a significant cultural haven for men who found themselves described in *problema* IV.26. Walter Burley struggled with the limitation on the word "vice" or, more literally, "badness," entailed by Aristotle's argument. First he applied it to unnatural desires, then he excluded people with the very same desires from the spectrum of moral good and bad, and finally, just a few lines later, he reintroduced evil by implicitly postulating a broader, higher, or deeper sort of *malitia*: "In all these things there can be badness [*malitia*] beyond the boundaries of human vice [*malitia*]."[117] For Aristotle, this would have pointed to "bad" musicians, but Walter seems to be getting at something else. As Pietro d'Abano's rhetoric has already suggested, his quest for explanations within the parameters of natural philosophy did not cancel his stern disapproval.[118] Even those scholars inclined, like Buridan, to represent resistance to unnatural desires as virtually impossible for some people did not erase their associations with cannibalism and matricide. Rather, late medieval intellectuals were likely to agree with Aristotle's evaluation of those who, having lost the powers of reason and choice, have lost the very principle of human virtue: "If bestiality is less than vice, it is, nevertheless, more horrible."[119]

Most commentators on the *Nicomachean Ethics* did, however, note that nature was an elusive standard in moral science, and some confronted directly the difficulties involved in squaring moral responsibility with necessities effected by birth, illness, and even early formed and deeply ingrained habit. Acknowledging the force that these might exercise on an individual opened the gate to explanations of human behavior based on the known principles of the natural world. To varying degrees, commentators regarded this possibility as a danger and took measures to maintain the perimeters. They at least nodded to free will and at most asserted its ability to prevail. In addition, they cordoned off the area in which nature could oppose itself to virtue. When "nature" was taken in its general and primary sense, natural forces, natural order, and natural reason were in complete harmony with each other, and all in turn conformed to moral goodness, divine Providence, and Christian faith. Only local and contingent low-level causes could produce the singular and futureless deformities that might escape the otherwise powerful teleology and stern accountability of moral virtue. Speaking in his *Summa of Theology* about questions raised by Book VII of the *Ethics*, Thomas Aquinas summarized the views of many later commentators already encountered above:

> For it happens that in a given individual one of the fundamental aspects of the species' nature is corrupted. And thus what is against the nature of the species becomes natural to that individual by accident—just as water [cold by nature], having been heated, may cause warming. Therefore it happens that what is against the nature of man . . . may become inherently natural for this particular man, because of some corruption of nature existing in him. And this corruption can be either with respect to the body . . . or indeed with respect to the soul—for instance, on account of habit, some take pleasure in eating people, or in intercourse with animals or males, or other such things that are not consistent with human nature.[120]

The isolation of what Pietro later called a "monstrous nature" is intended to preserve the integrity of the essential nature of the human species and, indeed, of the providential orders of Creation and society.

Yet the nagging problem of the exceptions, the accidents, the escapees did not disappear. The philosophers' own struggles to adjudicate between the natural and the moral attests to this residual trouble. So too does the reaction of the bishop of Paris. In 1277, Étienne Tempier, whose see included jurisdiction over the University of Paris, issued a list of philosophical propositions which it was an excommunicable offense to assert, because they were judged to conflict with various tenets of faith. Many of the most important of these had to do with the nature of the individual human soul, a subject to which Greek and Arabic learning had contributed a great deal of excitement and controversy. Several others illustrated the problem of drawing boundaries between the natural philosophical interest in discerning necessary physical causes and the theological imperative to sustain the autonomy of God's will. Thus, in the bishop's opinion, it was improper to maintain that God, should He wish to do so, could not move the universe in a straight line, because it would leave an empty space.[121] Most of the errors Tempier identified in connection with ethics had to do with the relation of the Aristotelian temporal virtues and their benefits to the Christian virtues and their benefits. Several condemned propositions concerned sex and appear to have targeted simultaneously the mores of Paris students and the perceived threat of philosophical naturalism or even determinism.[122] One, the opinion "that simple fornication is not a sin," may have been a subject of scholastic fun among the young men of the arts faculty, whose formal public disputations sometimes addressed such questions as whether it is worse to fornicate with a prostitute or a respectable woman.[123] Another speaks to the intractability of the problem discussed here by rejecting the attempted solutions of various authors. It categorically bars nature from intruding on the regulation of morals and it likewise bars natural philosophy from interfering in matters pertinent to salvation. According to Tempier, it is a manifest and detestable error to hold "that the sin against nature, in particular the abuse in the course of coitus, although it is against the nature of the species, is not, however, against the nature of the individual."[124]

The bishop's insistence is notable in two respects. First, although "abuse in the course of coitus" seems to have an ominous indeterminacy, like Pietro's use of the word "sodomy" in connection with *problema* IV.26, the context makes the target clear. Aquinas's formulation is evidence that Tempier is almost certainly alluding to the discussions surrounding "intercourse with males" in the seventh book of the *Ethics*. Searching for trouble spots at the arts faculty of the university, he did not fail to notice the inflammatory possibilities contained in the delimitation of the territory governed by the discourse of virtue and vice—Aristotle's concession that some abominations were beyond the reach of moral condemnation. Christian judgments also made provisions for exemption, but those were to be determined by canon law, not by science: no one can have a nature that goes against nature; no one can be a natural sodomite. (Sodomy by second nature would, of course, be even less acceptable.) The second significant dimension of this specific condemnation is that it apparently had no effect.[125] Some of the challenges posed by the condemnations of 1277 and others like them may have led scholars to rethink specific philosophical problems, so, for example, the absolute power of God became a framework within which some speculated about what it would be like if God decided He did indeed want the earth to move.[126] But scholars commenting on the *Ethics*, even those at Paris like Buridan, did not turn away from the notion that individual natures could deviate from the norm, where unnatural desires were concerned. John Buridan and Heinrich von Friemar, writing after the condemnations, no less than Albertus Magnus and Thomas Aquinas, writing before them, reserved a role for natural causation and explanation in the dynamics of the pleasures called "unnatural." And Pietro d'Abano, who arrived in Paris not much more than a decade after the bishop's proclamation, could conceive of Book VII of the *Ethics* as a justification for the natural philosophy of *problema* IV.26, even if some, like Evrart de Conty, rejected it.

Men who desired and derived pleasure from anal stimulation by other men were wrong. From the point of view of anatomy, physiology, astrology, and psychology, they were either born wrong or they got to be wrong through experience. They did not fully conform to human nature. From the point of view of moral philosophy, they were wrong in another sense, and it was in that context that the phrase "against nature" was most persistently invoked. Within each domain, there were arguments about the exact causes and implications of what was wrong, and, in addition, the two disciplines intersected as scholars imported vocabulary and textual authority from one to the other—Pietro d'Abano citing the *Ethics* to explain the *Problemata*; Jean Buridan referring to the *Physics* to explain the *Ethics*. Such collaboration was accompanied by competition, as the conflicting views about Avicenna's "mental, not natural" have illustrated. But in addition to domestic quarrels among branches of Aristotelianism, the men of *problema* IV.26 were the subject of less systematically formulated opinions and even feelings. One could be a serious and well-read intellectual and still abandon the formal rhetoric required of a good scholastic reader to express one's views. It was an accomplished fourteenth-century scholar who systematically glossed a copy of Bartholomeus de Messina's Latin version of the *Problemata*. Toward the end of IV.26, at the point where the age of habit formation is discussed, he offered his own way of interpreting "beyond the boundaries of vice," writing: "Sodomites are in the extremity of vice. Aristotle in the *Ethics*."[127] To dismiss the comment as a misreading of Aristotle would be to pass up a chance to look through the philosophical discussions of men with unnatural natures to the more individual and even personal responses of readers beyond. These are the subject of the next chapter.

Chapter 5

What's Wrong?

Silence, Speech, and the *Problema* of Sodomy

Evrart de Conty's challenge on the subject of moral accountability—that "we are the lords of all our deeds"—illustrates the scholastic debate surrounding the subject of male homosexual pleasure. At the same time, his rhetoric illustrates medieval reactions detached from the structures of systematic philosophy. While conceding some degree of natural causation, Evrart had no doubt that men could defeat their abominable inclinations and therefore that those who did not were morally culpable, whereas Pietro d'Abano, Walter Burley, Jean Buridan, and others placed such heroic resistance beyond the reach of ordinary men and therefore deemed irrelevant the moral scale of praise and blame. But Evrart raised another question about *problema* IV.26 for himself and his contemporaries: Should Aristotle—and should we—even be discussing this subject? He then declined to treat the contents of the *problema* fully, referring the interested reader to the Latin text and commentaries for a fuller treatment. The gesture is ambiguous, like that of the translator, Gerard of Cremona, who introduced the term *halubnathi* into the West when he either failed or refused to find a Latin expression for the word Avicenna had applied to the condition of sexually receptive men. The text of *problema* IV.26 was not the only text in the *Problemata* that Evrart found awkward (*pas bien seans*) to discuss in the vernacular, but he reveals grounds for his reticence that go beyond the difficulty of translation:[1] "This subject . . . is not nice [*gracieuse*] to deal with in French or in any language, because it is abominable even to nature."[2] He abbreviates the *problema*, declines to name specific parts of the body, and laces his prose with phrases like "this very mad and detestable will."[3] Although it is consistent with his reasoned rejection of a permissive interpretation of the *Ethics*, Evrart's demur and his language signal something different and more culturally diffuse. Just as the texts that developed around the Aristotelian *Problemata* convey the explanations its medieval readers entertained for the phenomena related to sex between men, they also reveal a less obvious but no less urgent medieval dialogue about whether such a subject should be discussed at all, especially within the frameworks of natural philosophy and medicine. In addition they offer clues, albeit often hard to interpret, about individuals' attitudes toward men susceptible to these particular pleasures. Why do some scribes see and copy "the fault [*vitium*—written ııcıııııı] of their disease" and others "the origin [*initium*—written ıııcıııııı] of their disease"?

The frank and open explanations of male homosexual pleasure proffered by Pietro d'Abano and his successors also present a challenge to modern scholars who have documented the extent to which the forms of "sodomy" confronted in the *Problemata* were unspeakable, even lethal, yet sometimes spoken.[4] The very existence of these medieval discussions suggests the possibility that certain cultural and social spaces were at least partly shielded from the powerful religious and judicial forces of stigma and suppression. Natural philosophy and medicine appear to have enjoyed a dispensation to address sexual experiences—even those judged inimical to the natural and moral order of things. In some ways, the situation is not surprising. The narrow world of medieval Christian elites was neither ideologically uniform nor institutionally united. Medieval universities, for example, hosted competition between Franciscans and Dominicans and between arts faculties and theology faculties. At the same time, their patrons—towns, bishops, kings—pursued various and not always consistent interests regarding their curricula, organization, personnel, and learned opinions. Scholars enjoyed varying degrees of latitude, but even at the institutions with the most stringent regulation and even when doctrines considered heretical were at issue, the prosecutions of controversial academic positions were not necessarily fatal to a scholar's career, much less to the scholar himself.[5] Given this environment, as well as the cover of Aristotelian authority, the anatomical, physiological, psychological, and astrological reasons for men's susceptibility to anal stimulation may not have been especially dangerous territory when compared to theologically more urgent questions, such as those about the nature of the soul or the eternity of the world.

As Evrart's ambivalence illustrates, however, the absence of monolithic and systematically enforced censorship did not insulate scholars from the troubled meanings associated with "those with whom intercourse is had."

Permeated by the multiple resonances of gender and spilling over into controversies about moral responsibility, even the briefest and calmest treatments of the *problema* could not escape the burdens of its subject. For that reason the text itself and Pietro's widely read commentary disturbed medieval scholars. They were subject to external constraints and internal conflicts concerning what might be spoken of as well as what should be said when speech was permitted. Like the *Problemata* as a whole, the thirty-two chapters about sex that constituted Book IV, especially the text of *problema* 26, all presented difficulties to authors and readers. Many of the *Problemata*-related texts and manuscripts that have proved useful to illustrate discussions within natural philosophy also contain evidence of the various ways medieval scholars reacted both to the men described and to the very project of explaining sexual desires and pleasures. Based on their responses, both considered and spontaneous, this chapter examines how medieval intellectuals articulated and managed these challenges, and argues that both individual circumstances and personal sensibilities help to explain the variety of ideas and attitudes expressed. Some claimed unequivocally the propriety of writing about homosexual desires and acts; others imposed silence on themselves and on the text they held in their hands. Furthermore, the responses we can detect convey attitudes ranging from discomfort to distaste to horror, in language that often makes it impossible to distinguish among the religious, philosophical, medical, and social frameworks within which the men and their condition were stigmatized.

The comments of Pietro d'Abano, Jean de Jandun, Walter Burley, and Evrart de Conty will provide some of the evidence, and the fact that we know something about their lives and writings will help to contextualize the ideas and attitudes elicited by the Aristotelian text and Pietro's commentary on it. But most of those who expressed their views are anonymous—copyists, owners, readers—some leaving us as little as a squiggle in a margin, others leaving us a piece of their mind. They signaled disapproval or defiance, embarrassment or amusement as they dealt with the group of thirty-two questions on coitus or with the specific one on anal sexual stimulation. The *Problemata* and its parts did not necessarily require any comment, qualification, or justification from these members of its audience. And whatever they may have thought, the majority of those who read the texts, expositions, and abbreviations of the *Problemata* did so without recording any explicit remarks or visible indications of hesitation. Thus, the spontaneous marks, words, omissions, and other signs that do appear in the manuscripts reveal (however obliquely) something about a few individuals confronting the work in specific circumstances. If they also reflect something about medieval thought and society in general, it is by registering a range of perspectives on the same material rather than by converging on a single norm.

Speaking of Sex: Gravity

The awkwardness began at the level of the Aristotelian work as a whole. As the various prologues and briefer remarks indicated, the fragmented subject matter of the compilation and the nature of a *problema* as a curiosity or puzzle risked casting any engagement with the material in a disreputable light.[6] Because of its irregular nature, commentators took pains to point out that the work could also be useful, sometimes highlighting the medical sections as evidence. Still, some were forced to admit, one encountered in the text undignified and vile material, so that (even before arriving at Book IV) a reader might feel some uneasiness. No doubt for that reason, when Walter Burley came to explain the purpose of the *Problemata* in his general preface, he took a defensive posture: "The [book's] final cause is to know the causes of the assembled problems and to contemplate the various marvels in them. Nor should [anyone] avoid them on account of a certain baseness, since contemplating base things ought not to be rejected. For, according to Aristotle in *On Animals*, XIII, in anything, no matter how base, something in some way divine and wonderful is discovered."[7] Burley reiterated this argument—and the reference to Aristotle—in the justification he inserted at the beginning of the group of questions on sexual intercourse. Immediately after his assertion that "nothing natural is shameful," he declared: "We are not made base by contemplating base things but by doing or willing them. *Ethics*."[8] Yet, when he faced *problema* 26, he abandoned his stalwart defense of philosophical discourse and became evasive. In this respect, Burley exemplifies a pattern that emerges in the margins of Book IV: readers found more, as well as more direct, ways to interact with the other questions about sex than they did with the subject of men having sex with men.

The significance of readers' decisions to accept or protest the presentation of the sodomite as a legitimate object of philosophical inquiry thus emerges when set against the ways the same audience responded to other sexual topics. Desire and pleasure, intercourse and the emission of seed were respectable subjects when framed within the context of reproductive theory, as was the case in Aristotle's *Generation of Animals* and the pseudo-Galenic treatise *On the Seed* (see Figure 3); or when addressed in the medical context of promoting of procreation, as was the case in medieval treatises on sterility.[9] Generally accepted as well was the medical principle that male and female semen, like all other bodily superfluities, had to be expelled to maintain the health of the individual, even though, in specific instances, this imperative could be subject to moral or social objections. Unmoored from such contexts, however, learned works that enumerated and explained sexual parts, feelings, and experiences, though some circulated widely, were frequently the object of disapprobation.[10] Christine de Pizan, the sophisticated and prolific writer (whose father was, like Evrart de Conty, a physician in the service of Charles V of France), roundly condemned one such text; and scribes sometimes made individual decisions about what was too indecent to be copied into the medical manuscripts they produced.[11]

As Pietro d'Abano explained at the opening of his commentary on Book IV, however, the respectable subjects of reproduction and regimen were not always (or even often) in play among these thirty-two questions about sexual desires and acts: "[Coitus] is engaged in for the sake of pleasure, as is usually done; and sometimes for the sake of producing children; and also occasionally for the sake of health and weight loss."[12] The assertion struck one medieval reader as remarkable, and he called attention to it by making a mark in the margin of a manuscript.[13] Book I of the *Problemata* had attracted a medical audience, and some undoubtedly expected practical advice on health and procreation in Book IV. In the space available in a margin of the Aristotelian text itself, toward the beginning of Book IV, someone recorded a test to determine whether a woman is pregnant.[14] But neither the *Problemata* nor Pietro's exposition obliged their expectations. "Only among humans," he had said earlier, "are more individuals moved to intercourse for pleasure than for producing offspring."[15] This observation was not new, but usually it was expressed as a prelude to a discussion of procreation: pleasure was providentially designed to promote coitus, since rational beings might otherwise shrink from it;[16] it was, as Evrart said, "ordained for the conservation of the species."[17] Pietro transformed the motives for pleasure by placing his remark at the head of Aristotle's series of questions, few of which had much to do with reproduction. That is, the men discussed in *problema* IV.26 were not alone in violating the teleology of sexual intercourse.

Perhaps because of this unusual content, the textual history of Book IV as a whole shows signs of trouble, starting with the problem of its title. Bartholomeus de Messina had rendered a word derived from "Aphrodite," the

Greek goddess of carnal love, ἀφροδίσια, with a word derived from "Venus," the Roman goddess of carnal love. Thus the heading in almost all manuscripts of his Latin translation reads, "About Venereal Matters [*Circa venerea*]," and points toward the theme of pleasure. Pietro, however, changed the title to "On Sexual Intercourse [*De coitu*]," bringing it into conformity with the conventional name for chapters found in medical works, where coitus, along with such factors as food and sleep, was treated as a fundamental variable in the maintenance of health. Ignoring this label's poor fit with topics like the stimulative effects of horseback riding, not to speak of IV.26 about men who have sex with men, his introduction defined narrowly and in physiological terms the domain he was about to treat: "Coitus is the mutual action of male and female in which the excess material from the third stage of digestion [i.e., the seed] is expelled."[18] Almost all the *Problemata*-related texts that follow Pietro's retain his title for Book IV.[19] Pietro's shift from the pleasures of *venerea* to the mechanics of *coitus*, along with his definition of intercourse as a male-female act, paved the way for Evrart de Conty to characterize the group of questions as those dealing with the "procreative act" (*oeuvre de generation*) in spite of their actual subject matter and in spite of Pietro's acknowledgment that reproduction was seldom on the minds of the actors.[20] Philosophical and medical French was just in the process of being developed, so Evrart had considerable latitude in his choice (or invention) of terms.[21] His use of discreet language at this point foreshadows objections he was to raise when he arrived at *problema* 26. In different ways, both he and Pietro gave Book IV as a whole a deceptively respectable veneer.

The protean title of Book IV suggests that sex in general, as presented in the *Problemata*, was a site of unease, a surmise borne out by both the direct comments and the indirect gestures of those who dealt with the work. If Pietro and Evrart had, by their word choice, brought Book IV into the dignified folds of health and reproduction, respectively, even Pietro had to acknowledge (quoting a famous pseudo-Aristotelian work) that coitus was responsible for "the destruction of virtues, transgression of law, and feminine behavior [in men]."[22] Although he was quick to add that the warning applied only to excessive sexual activity, those who read the following chapters were not all won over to that measured perspective. Reactions to the very beginning of Pietro's introductory remarks bear witness to an interest in the moral and social dimensions of the subject. "Note how coitus is an infraction of the law," observed one annotator;[23] and in a manuscript of Burley's epitome containing very few marginalia, near the beginning of Book IV a reader has written the phrase "because of the destruction of the conduct (*morum*) of the soul."[24] Such concerns underscored the question of whether scholars should be studying such a subject. If it was articulated most clearly by Evrart, who proceeded nevertheless to say something, it was expressed most emphatically by an anonymous English scholar, who refused to do so. In the copy he made for himself of Walter Burley's abbreviated *Problemata*, he skipped from the set of questions on drunkenness (Book III) to the set of questions on exercise (Book V), calling attention to his decision and stating his reasons: "Book [IV] has been completely left out on account of the shamefulness [*turpitudinem*] of the material."[25]

That book owner was rejecting Burley's own outspoken refutation of just such calls for the suppression of sex within academic discourse. In the special prologue composed for this group of *problemata* he had offered an animated justification for including sexual subjects within the purview of natural science.[26] What Pietro had implied by the fullness of his commentary on Book IV, Burley asserted directly: natural philosophers can and should search for the causes of any phenomenon that occurs in nature. His argument acknowledges the disreputable character of this object of study. The very heading of his prologue refers to the subject's "indecency" (*inhonestatem*).[27] But he defends it as a source of illumination. He takes up once again the theme he had developed in his prologue to the work as a whole, reiterating his earlier reference to Aristotle and adding another. And, as though for emphasis, he prefaces those appeals to authority with the sweeping statement about the inherent dignity of natural phenomena quoted at the opening of this book: "Nothing natural is shameful but all things that exist in the world [are] clean. Contemplating base things should not be spurned, since, no matter how base, something wonderful and divine is contained in anything, according to the Philosopher, *On Animals*, XIII. We are not made base by reflecting on base things but by doing or willing them [according to] *Ethics*."[28] This argument has two prongs. The first is that (apparently) vile objects are not without dignity. As most of Burley's readers would have known, the objects Aristotle was referring in his *Parts of Animals* were lower organisms, such as worms and flies. These are, the Philosopher held, disgusting to the senses but meaningful to the philosopher. In the context of Burley's prologue to the thirty-two questions about sex, the evocation of Aristotelian ethics has a target different from Pietro's use of the Aristotelian phrase "beyond the boundaries of vice." Burley is using it not to calibrate the moral burden on sodomites themselves but to defend the propriety of the subject matter as subject matter and the legitimacy of philosophical speech.

Burley paired his defense of the innocence of Book IV's subject matter with a second line of argument: a defense of those who undertake the study of such objects. The two approaches are not entirely consistent, since if the topics were truly respectable there would be no need to justify reflecting upon them. But, as the reactions of his readers will show, Burley had reason to address all possible objections. The second strategy, like the first, moves from the *Parts of Animals* to the *Nicomachean Ethics*. As read by medieval scholars, Aristotle had stressed in the work on animals that, to a philosopher, anything in nature can reveal its causes and, through them, its Creator. This principle gave weight to the poetic sentiment with which Burley had introduced the work as a whole, "Fortunate he who can know the causes of things."[29] But he supplemented the philosophical justification with a moral one: the determinant of good and bad resides in intentions and actions. Furthermore, "It is good to know evil but [it is] evil to do evil, since by knowing good and evil one can better be preserved from evil."[30] He thus concedes that this section of the work does contain something he regards as "evil," but, far more explicitly than Pietro, Burley raised the protective banner of philosophy (and propriety) as he ventured into the territory of Book IV.

Speaking of Sex: Levity

Indeed, the tone of Burley's defense is outright combative, as he adds a polemical challenge onto his rational stance. In this mode he makes even clearer that he is asserting the prerogatives of the natural philosopher—and against whose encroachment he is defending himself and his fellow Aristotelians. Burley himself was a doctor of theology, and, insofar as can be determined, there was not a simple divide between lay and clerical readers of the *Problemata*, but sexual behavior, especially the nonteleological types treated there, was one area prone to skirmishes. Tensions arose around the divergence between doctrinal and medical perspectives on sexual abstinence, for example,[31] and Burley sensed them here, at the opening of Book IV. Since he had read Pietro's commentary closely, he could have availed himself of the excuse that only the wrong kind or amount of sex caused moral harm, but he declined to employ it. Rather he confronted head-on a possible source of objection to the natural philosophical investigation of human sexual behavior and feelings. The truth about natural phenomena, he declared, should not be suppressed out of hypocrisy, "nor on account of those who abhor shameful talk [*turpiloquium*] in public more than [they abhor] behaving shamefully in private."[32]

Lest the reader have any doubt about who his target was, Burley pointed directly at those lecturing on the Bible—theologians—or, as he put it, those "in whose books are to be found things more base than in medical and natural philosophical books."[33] Specifically, he had in mind points at which the anatomically ambiguous term *vulva* (usually rendered as "womb" in these contexts) appears in the Vulgate.[34] He claimed he himself would be embarrassed to discuss such things openly in public. Crude banter about sex flourished in medieval academic environments, as elsewhere, and here Burley avails himself of the tone of levity that surfaced occasionally as a counterpoint to the gravity of voices calling attention to shame. It is likely that the passages, which he identifies by their unoffending lemmas, were well known not only to biblical scholars but also to their adolescent students, for whom references to subjects like the insatiability of the womb would have provided an echo of general misogyny and a source of prurient fun. Although he said he was offering these prefatory remarks lest anyone be offended by the material, Burley expects no quarter from these particular hypocritical detractors who fail to distinguish between a lustful and a philosophical approach. Pointedly quoting scripture, he accuses the theologians of habitually seeing the speck in someone else's eye and ignoring the log in their own (Matt. 7:1). If his book should fall into their hands, he expects them to ascribe their own failings to innocent people, including, presumably, the author himself.

Whereas Pietro's introduction to Book IV had acknowledged the potential negative effects of sexual activity itself, Burley's prologue addressed a different question: Was it bad to discuss sex? By appending to his philosophical defense an extravagant attack on a real or imagined opposition, he asserted the practical and institutional claims of natural philosophy against any encroachments, whether based on ethics (which he co-opts) or attempted by theologians (whom he mocks). If Burley had specific antagonists in mind, they are impossible to identify on the basis of his words or his career. He may have been aware of bishop Tempier's condemnations of certain propositions about sexual behavior discussed in Chapter 4. If so, he did not hesitate to contest them. And although comparable prohibitions were promulgated at Oxford, none can be connected to Burley or to Book IV of the *Problemata*. But the pious gesture of the person who ostentatiously omitted that section of the work suggests he was not entirely fabricating opponents in order to make fun of them.

The manuscripts of the epitome that Burley had based on Pietro's commentary circulated among English scholars, many of whom had documented ties to Oxford; many of whom had an interest in natural philosophy or medicine or both; and some of whom, such as a Franciscan from Canterbury, had clerical connections.[35] Although the volumes containing some copies also included medical works, the English audience for the *Problemata* seems to have consisted mainly of those with an extracurricular interest in natural philosophy. With the exception of the copyist who omitted Book IV, the fourteenth- and fifteenth-century Englishmen who produced or perused the surviving manuscripts of the work left no indication that they rejected Burley's claims about the possibility of reflecting upon sexual subjects in a dignified manner. At least one of them, however, although he appears to have accepted the presence of the questions about sex, preferred not to reproduce the satirical tone of the challenge to theologians. He respectfully named the "sacred scripture," which Burley had referred to as "their books." And omitting some of the accusations against hypocritical detractors, he humbly solicited potential critics: "Should [readers of this work] see anything worthy of blame, may they correct and improve [it]."[36] Later, at a point at which Burley had referred to intercourse between a man and a "woman," the copyist substituted "wife."[37] If English scholars exhibited reticence regarding sexual subjects, in northern Italy, where the *Problemata* was more associated with medical circles, readers appear to have been less cautious. Burley's work with its jibes at theologians did not reach them, but the surviving manuscripts of other *Problemata*-related texts offer a picture of a more jocular approach than that of Burley's more conservative audience. One anonymous adaptation of Pietro's commentary with clear ties to a medical environment offered this aside on the pleasure associated with sexual intercourse: "There's so much pleasure [in it], it's a wonder that theologians don't designate it a sin."[38] Here the presumption that the theologians have an interest in regulating sex is a source of fun, not of caution.

Indeed, one feature of the marginalia associated with Book IV (and not with other books) in copies of *Problemata*-related texts is their playfulness. Humor is difficult to interpret across chronological and social distances, and clerical literature not only sustained repressive sexual ideologies but also sought to regulate speech intended to provoke laughter.[39] Nevertheless, medieval academic culture has left many signs that educated Latin readers entertained jokes and expressions of prurience, satire, and curiosity in connection with sexual subjects—not least in the margins of manuscripts.[40] The subject of sexual intercourse, when formally illustrated in medical and literary books, was usually represented by a picture of a man and a woman in a bed, decorously covered or even clothed, and the exceptions often conveyed the immorality of a specific act, for example, when adultery was at issue.[41] Yet even the standard, sober picture could be deployed in a humorous manner. In a thirteenth-century copy of one of Aristotle's works on logic is a picture of a copulating couple. As Michael Camille has explained, the image provided a mnemonic device for the principle that the relationship between one thing and a second thing can give rise to a third. Someone objected to the joke, however, and erased it.[42] In Aristotle's *Generation of Animals*, at the chapter on human reproduction, a thirteenth-century illustrator has illuminated the top half of the initial letter with a couple, dressed (respectably), having sex (disreputably) in a garden (see Figure 3). In the bottom half of the letter is a woman rocking a baby in a cradle. At the bottom margin of the page, the same dramatization of cause and effect is played out by a peasant couple (Figure 5).[43] Comparable tonalities occasionally resonate in *Problemata* manuscripts. The copyist who wrote the following had just completed work on the Bartholomeus translation (though such colophons were attached to many other types of text): "Praise be to You, Christ, this book is ended. May the scribe be given a beautiful girl for his pen."[44] (In medieval Latin, "pen [*penna*]" would have formed a pun with "pain" or "trouble [*pena*]" and would have had a sexual connotation as well.) Although the majority of manuscripts associated with the *Problemata* contain no marks that can plausibly be interpreted as gestures of fun, enough readers registered their reactions to suggest the possibility of laughter about Book IV, "On Sexual Intercourse."

Much of the evidence is nonverbal. In an early fourteenth-century Italian manuscript of Bartholomeus's text someone (probably the mid-fourteenth-century owner) has drawn a naked man, prone, at the top of the page that includes Book IV.4–8.[45] The copy of Pietro's work in which a reader called attention to the dictum that coitus was a "transgression of law" belonged to a fifteenth-century Parisian physician who was responsible for at least some of the annotations in it. At the question about the cause of erections (IV.23), someone, probably the scribe himself, underlined the following statement: "Those having hanging testicles are not potent in intercourse." And, in a wordless gloss, he drew a pair of testicles in the margin.[46] These interventions are cryptic, but they cannot be construed as belonging to the ordinary business of scholarly reading or as expressing opprobrium. The same can be said of the way Johannes of Krakow, a student in Prague, made use of his red ink and the small empty spaces that remained after he had finished making

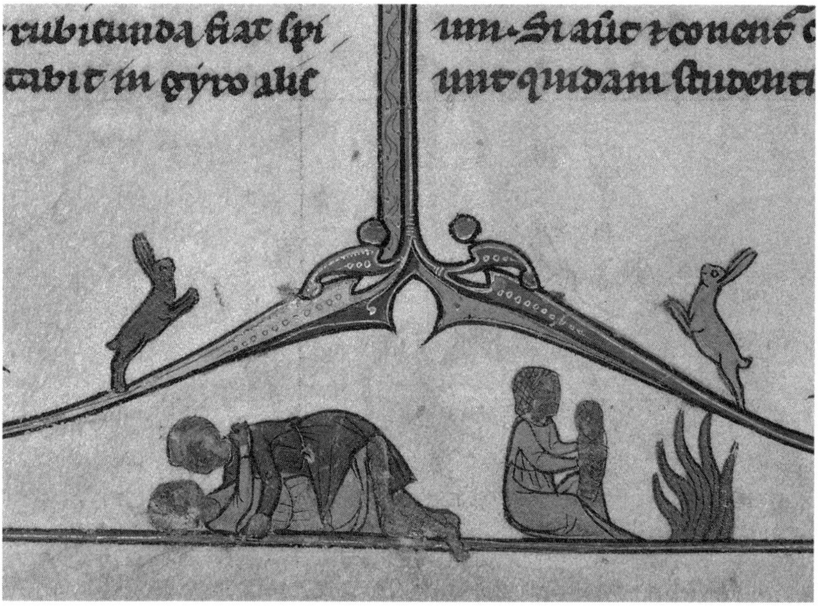

Figure 5. In a humble environment (and in the bottom margin), a peasant couple has intercourse as fertility-hares leap across the page. As a result, the woman rocks a cradle. See also Figure 3 (top). Aristotle, *De animalibus*, IX, in a thirteenth-century French manuscript, Oxford, Merton 271, 65v. Reproduced with permission of the Warden and Fellows of Merton College Oxford.

himself a copy of the Bavarian *Problemata*. In 1406, he wrote "Margaretha" on one page (between the chapters on sexual arousal from horseback riding and people beginning to smell bad when they reach sexual maturity). On the facing page (between the chapters on the difficulty of having intercourse under water and the reasons coitus is the source of the greatest pleasure), he wrote "Melchisidech."[47]

Playfulness of this sort is clearly distinguishable both from the righteous refusal to discuss the material and from the defensive insistence that it should be discussed. Yet all three postures suggest that sex in general was a problematic subject, whether because it was "indecent," "shameful," "vile," or "destructive" of laws, morals, and masculinity or because it was so much fun that it ought to be called a "sin."[48] The various perspectives were in dialogue with each other, as attested by the person who revised Burley's work to give it a more respectful tone and the person who erased the visual joke about Aristotelian logic. In *Problemata* manuscripts that contain marginalia, other sections of the work (especially those on medical theory and practice) are far more heavily annotated than Book IV, but only Book IV elicited so many humorous gestures. These did not, however, extend to *problema* IV.26. Although, in other contexts, the male ass is the butt of humor, sodomy does not seem to have been funny in the textual environment of the *Problemata*.[49] Or, perhaps more accurately, whatever their attitudes and responses, readers of philosophical and medical texts did not see fit to express their amusement on this especially stigmatized topic. Whether men having sex with men was a joking matter varied by time, place, genre, and social locus. On the one hand, Michael Rocke gives examples of wit and satire among elite men of Renaissance Florence.[50] On the other hand, Helmut Puff has pointed out that, in spite of the presence of "sodomite" clerics in the archival records, such figures are absent from the ribald vernacular genre of the fabliaux, which is laced with "narrations of a sexually rampant clergy."[51] Banter about men's desire for, access to, and acts with women surfaced more easily, aided perhaps by the participants' familiarity and comfort with the vocabulary of misogyny.

Problemata IV.26 and the Language of Stigma

As attitudes toward the subject of sex suggest, some scholars found it easier than others to talk or joke about what was often defined as the "mutual action of male and female." When they confronted a question that explicitly or implicitly (depending on the specific text) featured the anal stimulation of one man by another, the production of amusing asides dropped off precipitously. The mood in the margins becomes darker and more judgmental. One sign of the cultural and perhaps affective difference between the stigma attached to male homosexual pleasures and the frisson associated with other sexual impulses is visible in readers' marks in manuscripts at question IV.26. The passage in the various versions and treatments that occasioned the most significant activity in the margins, some of it lighthearted, was the one that compared certain men's unnatural desires to young women's excessive sexual appetites—one of the salient traits of the human female.

Some of the attention attracted by this section of the *problema* was inarticulate, for example, a pointing hand;[52] some was neutral, for example, the restatement of a phrase from the text;[53] some was more emphatic, for example, multiple index entries.[54] But other readers were more expansive. In one manuscript, two different readers left notes (the second perhaps inspired by the first): "When a woman needs a guard, lest she commit adultery"; and "The Emperor Solomon [says], 'Place a chaperon over a lustful girl [Ecclesiasticus (Sirach) 42:11].'"[55] The intended tone of these interventions is impossible to establish. The comment about infidelity resonates with medieval social concerns, while mentioning a specific sin; the one about the chaperon quotes scripture but would also have brought to mind contemporary lore about Solomon's susceptibility to the attractions of women. Like much medieval misogyny, such comments can function simultaneously as earnest warnings and as jokes against women told in the textual company of men. More unambiguously meant to be amusing is a brief note in the margin of another manuscript about a woman who was accustomed to having intercourse thirty times a day but was limited to ten.[56] The lines about insatiable women in the Aristotelian *Problemata* and its derivatives accorded readers the degree of latitude they experienced in other parts of Book IV, where comments ranged from moralistic to amused. This site in the text was thus the point of intersection between, on the one hand, the disdainable and derisible associated with (inevitably sexualized) women and, on the other, the despicable and unspeakable associated with men engaged in homosexual acts.

Bartholomeus's translation of the Greek text did not mandate the introduction of strongly derogatory language. In the course of presenting the questions to be answered, it describes the men of IV.26 as "effeminate [*effeminati*]" and their condition as "soft [*mollis*,]" having been constituted in a manner inconsistent with nature (*extra naturam*)—far from a neutral representation, but by no means vituperative. And several authors added nothing to this mildly negative language: the Erfurt commentator mentions effeminates and eunuchs; the Prague commentator omits even those words. Other authors and their readers, however, reacted more strongly, either by speaking out or by forbidding speech. Pietro d'Abano, in spite of his naturalistic posture and his restrained position on moral culpability, drew upon powerful value-laden discourses at a number of points in his exposition. When he referred to the "monstrous nature" or the "error of formation [*peccatum plasmationis*]" in men whose irregular anatomical configuration causes semen to build up in the wrong place, he was speaking in concrete terms of a physical malformation. Such phrases nevertheless participated indirectly and implicitly in the aura of stigmatization. And when the subject was those in whom homosexual appetites and practices develop as a result of experience and those who enjoy both the active and the passive roles, his language became unambiguously condemnatory. They acquire their condition as the result of a "twisted and abominable habit [*pravam consuetudinem et obscenam*]," it is a "wicked way of behaving [*profana operatio*]" and represents "an extremely shameful deviation [*prevaricatione turpissima*]." Indeed, this is the point in the exposition where Pietro, who had previously referred to them using phrases like "such men," introduced a Latin noun for his subjects: "These men are sodomites [*sodomite*]"; they are "corrupted by the polluted sodomitical vice [*scelerato vicio sodomitico depravati*]." They seek and enjoy a "wicked act [*actum profanum*]," whether they practice their "abominable sodomitical action [*nefandum opus sodomiticum*]" by anal penetration, intercrural friction, or (in his first version of the work) masturbation. And having declared them to be "beyond the boundaries of vice," Pietro reminded his readers of Avicenna's assertion that men having sex with boys is regarded as "filthy [*fedus*] among most peoples, and legally prohibited."[57]

With such expressions, Pietro highlighted the unique status of Book IV, *problema* 26. As the phrases accumulate, they achieve polemical force, and his many readers could not have been unaffected by his tone. Yet its influence was ambiguous, in the sense that his language did not provide a specific basis for condemning the men and inclinations in question; and it was limited, in the sense that many who came after him found their own terms in which to express their disapprobation or, indeed, to withhold it. The ambiguity resides partly in Pietro's word choice but largely in the cross-fertilization of learned discourses in the Middle Ages. Pietro and most members of his audience were at least somewhat conversant with the languages of classical literature, Aristotelian philosophy, Galenic medicine, Christian theology, civil and canon law, and aristocratic culture, even though they did not have specialized knowledge of all of them. When he uses the word "sodomite" it is undoubtedly tied into a religious (perhaps doctrinal) nexus of meaning. The same is certainly true of "vice" (*vitium*), which could be translated as "sin" when it accompanies "sodomitical." But, in other passages, *vitium* may well have more general meanings ("failure," "defect," "crime") found in Virgil or Cicero and common in medieval Latin. The term *peccatum*, which occurs in Latin scripture and medieval theology with a sense of "guilt" or "sin," may color but does not fully govern Pietro's use of it to describe an anatomical condition, where the sense of "error" may apply. Words like "shame" and "indecency" are even harder to pin down, unless they derive from quotations, as is the case with Pietro's adoption of "filth [*fedus*]" from Avicenna. Even then, the associations are layered: that word occurs in the Latin translation of an Arabic work on Galenic medicine and relates to the moral and legal imperatives of many unnamed peoples.

The indeterminacy of the logic and ideology that underpin Pietro's invective does not weaken its force any more than the juxtaposition of contradictory traits dampened the force of the ascription of womanliness. Indeed, the integration of so many discourses added scope and strength to the stigmatization of homosexual pleasures and the men in whom they inhered. This effect is visible elsewhere in the history of *problema* IV.26. For example, Evrart de Conty, closing his commentary on it with a flourish, declared that the men in question are "very perverse, very unnatural, very effeminate, very miserable, very ill-fortuned," for which reason "a certain poet" wrote that "Nature had Genius excommunicate them."[58] Here the concept of excommunication is ecclesiastical, and the image is ultimately derived from the sober Latin condemnation by the twelfth-century theologian Alain de Lille.[59] But as Evrart's audience knew, his immediate source was *The Romance of the Rose*, a late thirteenth-century French poem with the secular theme of love.[60] Furthermore, this passage in Evrart's commentary also incorporates a citation of a philosophical text (Aristotle's *Ethics*) and the term "sect," borrowed from a medical text (Avicenna's *Canon*), not to speak of common words like "miserable." The reactions of others who encountered the *problema* may have been built on less eclectic foundations, but for the most part their testimony is too brief and fragmentary to pinpoint the learned and vernacular paraphernalia contained in the cryptic expressions of their individual attitudes. Taken together, however, the shorter commentaries, marginalia, and variants create a picture of an intellectual and moral environment in which various cultural strands were tangled.

Only some of the authors of the other versions expressed their views directly. Those who prepared the shortest summaries incorporated no value-laden language at all.[61] Even the author who had wondered why theologians had failed to declare sexual attraction a sin declined to embellish his brief summary of IV.26. The lengthier Aachen summary of Pietro's exposition leaves out most of his negative language, probably for reasons of brevity, reproducing only the term "sodomites" as Pietro had applied it.[62] And the still longer Bavarian *Problemata*, which also follows Pietro's closely, reproduces very few of his disparaging phrases. Its author retains only "unspeakable men [*nephandi*]" and a single reference to bestiality and the "corruption of the soul [*perversio anime*]." He even goes so far as to omit references to "sodomites" yet, without specific prompting from his source, he declares the habit "extremely bad" (*pessisma*).[63] In contrast, even in this milder text, and even before he reached the passages on habit-based behavior where this negative language occurs, one scribe was apparently already leaning toward further derogation. Where the original text had spoken of small (*parvos*) testicles, he wrote that these men had bad ones (*pravos*). In the conventions of Latin paleography, the abbreviations for the two words are similar but distinguishable, so the copyist was presented with the opportunity but not the imperative to mix them up.[64] On the basis of this admittedly slender evidence, we might place this author and this scribe on the more philosophical or secular end of the spectrum, compared to those for whom the concept of sodomy loomed large.

Problemata IV.26 and the Limits of Speech

The various types and intensities represented in these ways of presenting *problema* IV.26 occur in many other *Problemata*-related texts. But at least as significant are various gestures with which authors and readers convey the impression that the taint that marked the men themselves also stained the act of speaking about them. Such gestures were more frequent and more forceful where sex between men was concerned than in the context of other perceived sexual irregularities. While readers' pens were busy highlighting insatiable women, authors and readers alike were busy obscuring sodomitical men, whether by omission, evasion, or protest. Different individuals appear to have reacted differently to the subject matter, and different contexts may have influenced their decisions about what to say and what not to say. The scribe who prepared one fifteenth-century manuscript of Pietro's work left no signs of hesitancy in the body of *problema* IV.26. But the one who prepared its detailed table of contents labeled it, "Why both people having intercourse get enjoyment from that act."[65] Perhaps he had not read it or perhaps he was motivated by a sense of decorum or by a desire to hide the more inflammatory content.

The person who composed the Flemish commentary simply skipped IV.26, as did one person who made a copy of Burley's work.[66] And one attentive reader of Bartholomeus's Latin glossed all of the questions in Book IV except three: on monstrous births, on why intercourse is so pleasurable, and on the causes of some men's enjoyment of anal stimulation by other men.[67] There were various reasons for omitting specific questions. Pietro himself occasionally passed over one, saying that it had been treated elsewhere, and the texts derived from his naturally followed suit. But no such excuse applies in the *problema* about men who enjoy sex with men.[68] One plausible explanation for its occlusion comes from a student who made a copy of the Prague version around 1400. Protesting in the same terms as the Englishman who cut out Book IV altogether, he passed over the main substance of IV.26, saying, "I do not wish to write many things on account of the shamefulness [*turpitudinem*]."[69] The connotation of the word (and thus the content of the sentiment) is difficult to locate, but it seems to indicate more than embarrassment, perhaps even pollution. It does not have a specifically religious ring, but neither does it have the protective polish of philosophical discourse. Since the other extant copy of the Prague *Problemata* discusses in especially great detail men rubbing and being rubbed,[70] and since the reluctant scholar appears to have been making a copy for his own use, the omission seems to be an honest expression of an individual's feeling. Either the acts under consideration or the act of writing about them, or most likely both, carried too heavy a social, moral, or religious stigma for him, even if they had not done so for the person who made the book from which he was copying.

Although he gives us a plausible way of understanding the omission of IV.26 in the Prague version, this student himself did not eliminate the *problema* altogether. As he said, there were many things he did not want to discuss, but he did copy what amounted to an aside at the end of the section: the commonplace that women had greater pleasure in intercourse extensively, men greater pleasure intensively.[71] Subsequent readers of his manuscript would have had the impression that the subject of IV.26 (although in some way embarrassing or even taboo) had to do with male-female copulation. Nor would they have been alone, for the obfuscation of the men with homosexual desires was as common as their erasure: many texts gave the impression that the *problema* involved men having sex with women. And this view was not implausible—if one read no further than the first line of the *problema*.[72] Pietro had opened the door to this approach in his elaboration of the general principles governing the buildup and release of seed, and Jean de Jandun had reoriented the question itself by referring to those who engaged in intercourse as "a man and a woman or those equivalent to them." Since Jean's redaction had tracked Pietro's full treatment of the *problema* in all its anatomical specificity, readers who got beyond its first lines could not remain uncertain for long about its subject. Nevertheless, in the Vatican *Problemata*, an abbreviated fourteenth-century Italian text, the central question was rendered almost invisible. The sole surviving copy of this work suggests that its author belonged to medical circles and was writing to make the long and difficult Aristotelian *Problemata* accessible to a learned audience by paraphrasing not just Bartholomeus's Latin but also elements of Pietro's commentary.[73] Although he was the same person who wondered why theologians had not objected to the substance of *problema* IV.15 on sexual pleasure in general, he declined to confront the apparently inflammatory issues of IV.26. Following Jean's formulation, he posed its two basic questions: (1) why "a man and a woman or those equivalent to them" enjoy sex; and (2) why some people enjoy both the active and passive roles and some engage in the act alone. But his summary of the answer addresses only the first of these: "in every emission of a superfluity, joy occurs." That is, it covers only the initial step of the explanations offered by Aristotle and other commentators. The author stops there, declaring, without further elaboration, that Aristotle has demonstrated this point "at length and very clearly."[74] Whereas Evrart de Conty invoked the awkwardness of Aristotle's Latin text as one reason he would have preferred not to deal with the *problema*, this author invoked its transparency as the reason he had no need to—with the result that he says nothing about blocked seminal passages or the acquisition of habits, much less about parts of male bodies rubbing together.

Since the full versions of Bartholomeus's translation and Pietro's commentary circulated in the same northern Italian intellectual circles as that text did, this truncation of IV.26 in the Vatican *Problemata* cannot be construed as the result of external censorship. Nor can it have been a mistake, given the author's access to both of those sources and the care he took in preparing his summary of them. Like the student in Prague, he decided not to say any more than his sense of propriety or conscience permitted. Similar conditions obtained in Paris at the monastery of St. Victor, an academic powerhouse with a superb library. By the end of the fifteenth century, it possessed at least two full copies of Bartholomeus's Latin translation[75] as well as an abbreviated version acquired for it by a bibliophile prior, in which the full text of IV.26 reads: "Why do a man and a woman similarly [*proportionaliter*] take delight in carrying out coitus? Three reasons: first, a superfluity is emptied out by the act, as is clear from what has been said; third, they experience pleasure in the emission of other superfluities, as is clear from defecation and urination. Wherefore the proposition follows."[76] In many respects, this version resembles the fourteenth-century Italian text that summarizes only the first part of the *problema*, but in the former work an uniformed reader would at least have had cause to speculate about the meaning of the phrase "a man and a woman or those equivalent to them [*eis proportionati*]." Here, however, the proportionality referred to is no longer between a male-female couple and some other couple but between a man's pleasure and a woman's pleasure in a mutual act.

The suppression may not have been a gesture of disapproval—the author was determined to produce a short text—but he could have proceeded in a different manner, as did someone in Central Europe. That person, the author of the Erfurt *Problemata*, included a brief discussion of "the act" (as he called the subject of *problema* IV.26). He chose to skip rather than to emphasize the first, general question about sexual pleasure and to omit any mention of women in his answer. Instead he addressed only "Why do some take pleasure actively as well as passively in coitus?"[77] His decision to see and confront the phenomena of homosexual desires, acts, and pleasures underscores the significance of those authors who decided not to.

The most striking example of this pattern of obfuscation occurs in the work by Walter Burley. He too expressed the goal of brevity. But the passages that purport to convey the meaning of *problema* IV.26 repeat substantial elements of his summary of an earlier question (IV.15), as one (perhaps puzzled) scribe noticed.[78] Burley himself said he was relying on Pietro's commentary, and he certainly understood the issues posed by the *problema*. Furthermore, he was the one who had asserted that it was the philosopher's job to examine anything that occurs in nature, "no matter how base," and to seek its causes. By his account, however, the question posed by Aristotle at IV.26 was simply, "Why do those engaged in intercourse experience the greatest pleasure?"[79] And to answer it he proceeded to give a list of a dozen answers, including the pleasure of the flesh, the beauty of women, the preservation of the species, the fulfillment of the nutritive soul, and health—especially that of women suffering from menstrual retention. As the Prague student was to do decades later, he concluded his answer with a commonplace which reinforced the impression that the subjects interacting were female and male: "A women gets pleasure both from conceiving and emitting, a man gets pleasure only from emitting."[80] Tucked away in an aside, introduced with "Nota," after what appears to be his answer proper, Burley's readers at last encountered the other topic lurking in IV.26: "The desire for coitus comes from four things, namely from nature, windy foods, the imagination, and the habit of having intercourse. Effeminate men are insatiable and insane in having intercourse, like women. They are cured by diuretic, laxative, and emetic [medicine], just as by hunger, thirst, sadness, staying awake, beating, and labor."[81] In this final passage, Burley has rejoined Pietro's commentary. The idea that such men are insatiable like women comes from the Aristotelian text itself; the idea that they are insane comes from Pietro. And, although he is writing an epitome, Burley becomes expansive on the subject of treating the barely visible condition in question. To Avicenna's punitive remedies he adds thirst and work; to Pietro's rejection of a surgical approach in favor of a pharmacological one he adds the particular categories of drugs to be used.

In the end, without being specific about the phenomena or their causes, Burley offered some concrete information, and, if they reached that point, most readers would probably have grasped its import. But to the extent that Burley's text served as a reference work (as his insistence on brevity and its existence in an alphabetical form would suggest it did), those scanning the text or looking for a particular topic would never have guessed the contents of *problema* IV.26. Nothing in his wording suggests a reason for what he omitted, what he included, or the order in which he arranged the material. His use of the term "effeminate men [*effeminati*]," which occurs in Bartholomeus's Latin, rather than the term "sodomites" preferred by some others is consistent with his defense of philosophy against the encroachments of theology. But it still left the subject unnamed and undefined, since (as Pietro had reminded his readers) excessive sexual activity of any kind could render a man "effeminate."[82] Thus Burley's own treatment of the question would seem to indicate that, contrary to his protestations, there were things in nature that at least bordered on the unspeakable. Like other texts that obscured the men who have sex with men, his version of *problema* IV.26 placed sex between a man and a woman in the foreground. That subject was clearly more acceptable. Even if some, like Burley himself, thought it might require defense and others, like Evrart, thought it might require delicacy, male-female intercourse was still not only speakable but also a potential target of fun.

In contrast to the insatiable women who represented the lighter side of sex, the men at the heart of *problema* IV.26 thus stopped some natural philosophers, medical authors, and interested readers in their tracks. This was the case not only with the active glossator in northern Europe who inserted comments only on what came before and what came after it[83] but also with the reader who, coming upon Pietro's statement that "rubbing between the thighs of boys [is] what many do today," scratched out the words "many" and "today."[84] For others, it was not enough to turn aside from or obscure what ought not to be spoken of. In various ways, they uttered their objections. An Italian scholar studying the Aristotelian text singled out IV.26 for special attention. (In the same manuscript it is also the only *problema* in Book IV that someone has divided into sections, as though it was to be the basis of a lecture.) Where Bartholomeus's Latin reads "condition acquired by habit [*ex consuetudine passio*]," he has written simply "twisted [*prava*]."[85] Similarly, in a manuscript of Pietro's work at the passage about what men did in Aristotle's time and his own, someone, probably a physician or medical student, has condemned the deeds: "Note: shameful things." And at the passage about those who like both to rub and to be rubbed, that same person has also condemned the words: "extremely shameful and filthy to speak of."[86]

Unlike Pietro and Evrart, most participants in the scholastic conversation about *problema* IV.26 do not address issues of religious doctrine or Aristotelian ethics, contenting themselves with the general vocabulary of disdain or disgust. But a few do align themselves with one discourse or the other. Such was the case with a reader of the Bartholomeus translation who otherwise had little to say yet took the trouble to write, next to the passage on forming sexual habits, "Sodomites are in the extremity of vice, [according to] Aristotle, *Ethics*."[87] Others who dealt with *Problemata*-related texts expressed themselves in language that seems more closely tied to Christian doctrine. In a different copy of the Aristotelian work, a scribe has made a substitution that echoes the religious language surrounding sodomy: for "outside nature [*extra naturam*]" he has written the more familiar phrase "against nature [*contra naturam*]."[88] And at the point at which most manuscripts of Pietro's commentary declare such men are "corrupted [*depravati*]" by the sodomitical vice, two copyists read "damned [*dampnati*]."[89] Even the scribes who prepared indices for Pietro's work, marking tag words in the margins and composing alphabetical tables of them, made choices indicative of their own perspectives. For example, Pietro used the word "sodomites" once and the word "sodomitical" twice. But the compositor of one such table found no fewer than six occasions to enter related terms under the letter *S*, including "sodomy," which he took to be the subject of the *problema* as a whole, and "those who have been sodomized [*sodomitati*]."[90] In addition to ignoring (or rejecting) Pietro's distinction between those who are born with blind pores and "sodomites," this person has opted to invest all those involved with the burdens of religious rhetoric. A different alphabetical index contains only one entry referring to IV.26, but its author reveals a similar perspective, registering "sodomitical sin" (*peccatum sodomiticum*) as the sole entry for that *problema*.[91]

Few of these commentators, copyists, and readers are explicit about the cultural or social origins of their sense of shame and their reluctance to speak. Undoubtedly for at least some the constraints came from more than one source. Commenting on a general work about regimen, the prominent Bolognese medical theorist and natural philosopher Taddeo Alderotti did not hesitate to include the emission of seed as a factor in an individual's health. He regarded it as "shameful," however, to discuss "other types of intercourse, involving the manner of having sex and the diversity of bodies having sex." Since there were so many of them and since sins were involved, he thought it best to leave such matters to others and he mentions both theologians and moral philosophers.[92]

Conclusion: Silence and Speech According to Evrart de Conty

Evrart de Conty was in no way typical of those who addressed the questions about sex in the *Problemata*. He wrote in French, was attached to a court, and dealt with the work at greater length than anyone but Pietro d'Abano. Yet, in other respects, he was representative of those authors, scribes, and readers. He was a physician with extensive training and interest in natural philosophy; he lived in Paris, one of the centers from which the *Problemata*-related texts were disseminated, maintaining ties with the university there; and he drew extensively on Pietro's exposition without entirely subscribing to his interpretations. Because of the length of his comments on Book IV, *problema* 26, Evrart's work brings into fuller view many of the themes and debates surrounding it that are only hinted at in the fragmentary evidence drawn from highly abbreviated works, textual variants, and marks in the margins of manuscripts.

As his position on moral responsibility and the *Ethics* has made clear, Evrart was the only author to develop an argument fully consonant with contemporary church doctrine on homosexual behavior and with the theology of free will. On the other hand, the terms in which he expressed his position on moral responsibility resonated more with literary, medical, and philosophical discourses than with religious ones. Even his allusion to "excommunication" was a metaphor borrowed from *The Romance of the Rose*. He spoke of men with a depraved "will" but appealed more often to "reason"; he cited not scripture but rather Aristotle, Avicenna, Ptolemy, Cicero, Ovid, and Jean de Meun.[93] In this respect, the attitudes he conveys in his treatment of *problema* IV.26 illustrate the integration of various genres and cultural domains in the world of late medieval intellectual elites. They also reflect a particular form of ambivalence regarding sexual pleasure. As he says near the opening of his treatment of IV.26, nature has made it especially great because it produces beneficial results. When undertaken properly (by a man, of course), "with a woman, in the manner consonant with nature and reason," it serves not only the individual, through the expulsion of superfluities, but also the species.[94] Quoting Aristotle's *Ethics*, he insists that "pleasure confirms the act and makes it better and more lasting." But for Evrart none of this philosophically grounded justification applies to those who "abuse the aforementioned act."[95] Setting homosexual interaction in front of this clearly delineated procreative backdrop, Evrart, unlike Pietro, mobilizes a full range of derogatory language. Even those born with anatomical defects are possessed of an inclination that is "very crazy and detestable"; they have a "perverse nature, effeminate and imperfect"; and they seek enjoyment by a "route displeasing to nature and to all reason."[96]

Evrart's commentary not only reflects the general atmosphere of stigmatization conveyed in other texts and in marginalia but also brings to the foreground the problem of the natural and the unnatural implicit in the genre of the *problema* and, in Pietro's phrase, "their monstrous nature": How are we to understand the natural causes of natures that are contrary to nature? Loyally but reluctantly Evart reproduces the explanations of the unnatural nature: "It seems that Aristotle wants to attribute this very crazy and detestable will to a natural cause."[97] And he even summarizes and cites the authorities for the position that habit can be converted into something like nature. In terms of his moral position, however, such arguments are of no consequence, since natures of this sort, whether innate or acquired, must be overcome by Reason, who governs "the powers of the senses and even habit and nature" insofar as these things lie in human hands.[98] Reason is the ally of universal nature, and together they are stern governors. Individuals' deviations from the teleological order of nature are not only wrong but also "crazy." For Evrart, acts are either agreeable or abominable to nature; nature desires, demands, and ordains certain things; and it is Nature who orders the excommunication of these "unnatural [*desnatures*]" men. More than any other author or reader, Evrart sustained a tone of condemnation, and more than any other he provided a persistent basis for his harsh judgment: the rule of nature. A physician, he incorporated Pietro's quotation from Avicenna but, unlike Pietro, he concurred that the condition described in *problema* IV.26 is not a medical one at all. It is "mental, not natural." It is thus as a natural philosopher that he spoke in this case—neither to mock theologians, as Burley had, nor to circumscribe moral accountability, as Pietro had—but to appeal to the authority of nature, reason, and Aristotle for the enforcement of norms and the stigmatization of men whose desires do not obey them. In a sense, he too is patrolling the boundaries of natural philosophy. His aim is not, however, to acquire more territory for it but to protect it from incursions by material that might pollute it.

In part because of its length, Evrart's work provides more systematic underpinnings to his reactions—from distaste to condemnation—than do the briefer utterances of others. But his repeated use of words like "bad," "depraved," and "abominable" is not simply the logical consequence of his philosophical position on the order of nature or his interpretation of the *Ethics*. The more diffuse sources of his revulsion are manifest in what he says about the problem of speech, which he likewise develops more fully than other scholars. He was among the readers of the *Problemata*, some of them undoubtedly physicians or students of natural philosophy, others perhaps clerics or students of theology, who not only regarded certain inclinations, acts, body parts, and individuals as "monstrous," "corrupted," and "unnatural" but also saw them as *nephandus*, literally "unspeakable," to cite Pietro's term. They passed over the subject in silence or announced their refusal "to write many things because of the shamefulness." That the shame and filth adhered to the speech itself is clear even in the margins.[99] Nevertheless, most copyists carried on without protest, and most authors discussed the text, whether highlighted with a label like "sodomites" or partially obscured behind the screen of "women and their equivalents."

Evrart gave voice to both the impulse to speak and the impulse to remain silent. More than a quarter of his discussion of *problema* IV.26 is devoted to this predicament. On the one hand, he does not want to "pause for long"; on the other hand, he does not want to be accused of "negligence" as a commentator.[100] He makes clear from the start how unpleasant it is to deal with a subject that is despised even by nature, and he announces that, when he gets to that specific topic, he will "touch on it a little . . . , in general, and rather lightly."[101] At various points in his exposition of the *Problemata*, Evrart had noted the difficulty of rendering an already difficult Latin work into the vernacular, and this complaint surfaces at IV.26 as well. In addition, he may have had his broader audience in mind in minimizing his engagement with the less *gracieuse* aspects of the text and in modulating his diction. His language is delicate—unlike Pietro, he does not mention any specific body parts or actions—but neither is it especially evasive. And when Evrart, unlike Burley and others, finally does address Aristotle's two questions, he does not demur by dealing primarily with the first, about the sources of sexual pleasure in general. Rather, his assertions to the contrary notwithstanding, he devotes the bulk of his commentary to those men who are "configured in their reproductive members in such a way that the [spermatic] material cannot pass and flow quickly to the proper place as nature desires and commands it." And he adds a less expansive discussion of those who acquire their condition from a "bad habit."[102]

To justify his undertaking, Evrart mounts a defense of natural philosophical research that echoes in many respects the one offered by Burley for the entire group of questions about sex. And he draws on some of the same authoritative sources and adds others. At the most general level, he invokes the rationale for the work as a whole. With its jumble of miscellaneous questions on subjects ranging from the pedestrian to the outlandish, what could be the purpose, the final cause, of such a book? One answer Evart had given in the prologue to his translation and commentary was that knowledge of the causes of natural phenomena fulfills the intellect, that is (in Aristotelian terms well known to his audience), it actualizes the potential of a human being.[103] At *problema* IV.26 he recalled that same general principle: "to understand the above-mentioned causes of things in nature [produces] great fortune and great perfection of the understanding, as was stated at the beginning of this book."[104] But in this case there were apparently strong reasons for excluding homosexual pleasures from the group of questions about sex, even though the subjects of others in the group were hardly respectable. A more specific defense of IV.26 was apparently called for, one that would explain why Aristotle (and, by implication, Evrart himself) could not "forget or pass over the general subject of Book IV without making any mention of it."[105] The answer lay in the nature of the inquiry: "Philosophy is called the science of truth, as Aristotle says elsewhere. And thus, from this perspective, nothing is ugly, no matter how much the thing is not considered beautiful in itself."[106] The authority of Aristotle and the discipline of natural philosophy have provided Evrart with the permission, perhaps the imperative, to speak of the unspeakable, that is, to explain "the act of generation however it may be carried out," rather than to ignore or suppress it.[107]

Whereas Burley had loudly defended his philosophical voice at the beginning of the questions about sex, only to lower it when he arrived at the one about men engaged in anal stimulation, Evrart (who had not felt called upon to mount a defense at the opening of Book IV) followed through on his stated determination to speak about both nature and Aristotle at *problema* 26. His willingness and ability to do so were supported, not undermined, by the delicacy of his language, the conservatism of his moral position, and his expressions of outrage.

However great their differences, both authors illustrate the array of forces operating upon those who dealt with the widely stigmatized pleasures and people. Avicenna, a medical authority, had placed men with the condition called "halubnathi" beyond the reach of medicine; Aristotle, a philosophical authority, had placed "those with whom intercourse is had" beyond the boundaries of vice; church doctrine had categorized their sin as against nature and thus against its Creator, wherefore Dante had placed them deep in hell; and social codes on manliness and public order imposed regulatory regimes upon them. No reader of the Latin *Problemata* was insulated from these social and cultural forces, and their impact is often visible in the texts and margins of manuscripts circulating in the fourteenth and fifteenth centuries—though it is often difficult to distinguish exactly which is at work. Yet, all who have left traces were subjected to countervailing influences as well. The case for scholarly treatment of any subject, "no matter how vile," gathered its own strength from the prominence of natural philosophy in universities' curricula and the subculture of students in the arts faculties; from the status accorded by Aristotle's authorship of the *Problemata*; and from the social realities of towns and courts. Neither matter-of-fact explanations nor laughter nor condemnation nor silence arose from a single source, nor did any one of them entirely prevail in late medieval discussions of sex. Yet the *problema* and the problem of "men with whom intercourse is had," of the "soft and effeminate," of the "sodomites" and "those who have been sodomized" occupied a special place in the conflict between speech and silence.

Epilogue

> Tell me where is fancy bred,
> Or in the heart or in the head?
> How begot, how nourished?
> Reply, reply.
>
> —William Shakespeare, *Merchant of Venice* 3.2

Given the Aristotelian dictum (often cited in the Middle Ages) that "All people desire to know," the impulse to explain desire is not surprising. From Plato, whose mythical account has us searching for our long-lost other halves, to Goethe, whose characters attract each other like chemical elements, the history of Western thought is strewn with imaginative attempts to answer the question posed by Shakespeare on the occasion of the betrothal of that famous cross-dresser, Portia. Scientists, along with poets, philosophers, and divines have tried their hand at the game, and so it is that we find the readers of the *Problemata* struggling with questions about the sources and mechanisms of desire and pleasure in a particular group of individuals. They accepted as fact the regularly occurring existence of men whose sexual impulses differed not only from what was usual but also from what was teleologically determined, and they undertook to explain the phenomenon. That is, at the invitation of a brief passage in an authoritative work, they set out to produce a science of sodomy.

Neither science nor the sodomite is a singular, discrete object, as the life history of *Problemata*, Book IV, *problema* 26, has shown. In the fourteenth and fifteenth centuries natural philosophers struggled to understand and establish the parameters of their discipline. They disagreed among themselves and with outsiders about what constituted appropriate subjects of investigation and consequently about what constituted decent discourse. The commitment of many of them to providing natural explanations for desires and pleasures they regarded as unnatural led them to examine and refine the very foundations of their enterprise, as they adjudicated the relationships between universal and individual nature and between natures arising from the body and those arising from habit. The *Problemata*-related texts exhibit little consensus about the taxonomy of the men in question. They might be sorted by their anatomies and excretions, by the causes of their dispositions, by their age, by their position on an active-passive scale, or by their moral accountability. They were a heterogeneous group, but they were a group. And although one author mentioned that some such men occasionally also have intercourse with women, and a few manuscripts retain a reference to masturbation, what they are doing and enjoying involves stimulation around the anus of one man by another.

If, in the late Middle Ages, not only the intellectual enterprise of providing natural explanations but also the kind of men placed under scrutiny for their sexual passions turn out to have been complicated, then their historical fate will be complicated as well. Precisely because there was diversity and disagreement in the Middle Ages, it makes no sense to debate whether a direct line of descent connects natural philosophy with science or the men of *problema* IV.26 with homosexuals. Looked at chronologically, the lines of filiation between medieval natural philosophy and modern science are thickened and thinned, tangled and broken, but the earnest attempt by early commentators and readers to identify existing phenomena and to search for explanations based on the known substances and processes of nature is not entirely foreign to the modern eye. Similarly, the sodomites and others featured in the commentaries and the margins of manuscripts are not the direct ancestors of gay men. The most significant discontinuity has to do with self-identification. There were undoubtedly some readers who recognized themselves in the Aristotelian text, and some of them in turn may even have intervened in the discussion as authors or as annotators. (How often have I wished I could know who wrote "Nota," and why!) But the discourse was defined and controlled by others. Yet, even so, just as there are aspects of medieval natural philosophy that make it recognizable to modern scientists, so there are elements of the people and pleasures described in these texts that command a sense of kinship.

One chapter of a huge ancient compilation of puzzling phenomena has provided not only diverse perspectives on the history of science and the history of homosexuality but also a case study of how investigations of natural causes subsist in a historical environment. Embedded within the formal scholastic considerations of habit that becomes nature and the typically scholastic intertextual dialogue among the *Problemata*, the *Ethics*, and the *Canon of Medicine* were beliefs and attitudes that can be identified as leitmotifs in late medieval society and culture. These have been represented here as they emerged from and shaped the comparison of certain men to women. The imagery and associations prompted by that analogy infused the terse text of *problema* IV.26 with a plethora of meanings that lent force to the stigmatization of men with homosexual desires, even if it did not lend clarity. Conversely, the special status of homosexual acts and pleasures within the already suspect category of nonteleological sex manifested itself in conflicts about speech and silence and in limitations on sexual humor. Just as there was disagreement within the sober considerations of scholastic exposition, so there was diversity within the less structured expression of opinions. If women had their detractors like the Pseudo-Albertus Magnus, they had their defenders like Christine de Pizan; and if one reader regarded the subject of male-male sex acts as extremely disgusting to speak of, another ticked off without comment the main points in the explanation he was reading. We know little about the named and identified authors like Pietro d'Abano; we know less about the anonymous ones like the "Prague" commentator, labeled only by circumstantial evidence; and we know almost nothing about redactors or scribes who skipped *problema* IV.26. But we do know that they had different views and sensibilities that affected their posture toward the subject of the Aristotelian text.

This inextricability of scientific inquiry, coexisting discourses, webs of beliefs, and competing values is not peculiar to the Middle Ages. It too has a history, elements of which are still recognizable, even though their content and conditions change. Medieval and modern efforts to understand homosexual desire in natural terms are radically different in a number of ways, not least the cultural and institutional weakness of medieval natural philosophy as compared to the power of modern science. But as with the separate strands of sodomy/homosexuality and natural philosophy/science, so with the implications of their mutual encounter, comparable issues surface. Indeed, the underlying question of what science can and should explain is once again in play. Modern theories that postulate a somatic basis for homosexual desire by emphasizing hormones, neurons, or genes can be both affirming and limiting, as can medieval theories centered on superfluities, anatomical blockages, and astrological influences. And, now as well as then, theories that center on psychological processes involving experience, habit, preference, or choice allow more room for agency and flexibility and less room for a sense of an inherent way of being. Both sets of explanatory principles reduce a complex array of experiences and self-perceptions to a few components that hardly encompass the constituent parts of individuals' fancy or desires, much less of their loves. In addition, the medieval associations cascaded outward, spilling over into categories such as sodomy and womanliness, and raising issues such as blame, just as modern associations radiate outward, intersecting with discourses of rights and freedoms but still bearing the burdens of social prejudices and constraints.

Yet these possible logics were not and are not immutable; the effects and implications of scientific approaches are not only historically contingent but also malleable. In the fourteenth century, Aristotle could be used as the basis for reinterpreting Aristotle, as the protean phrase "beyond the boundaries of vice" has shown; and in the twentieth century, the medically sanctioned marginalization embedded in AIDS research could be turned on its head by advocacy organizations like Act Up. The struggle to define and own the "natural," visible in medieval philosophers' ambivalence about "monstrous natures," resurfaces again and again in settings where nature can serve as an instrument either of social regulation or of personal liberation. The presiding judge in a recent trial to determine the fate of a California ban against same-sex marriage took note of the way in which the "natural" can be deployed. According to the opinion he issued in the case, "recognizing that [the law] must advance a secular purpose to be constitutional," defenders of the prohibition had (newly) adopted a number of nonreligious arguments, among them that the ban promotes "naturally procreative sexual relationships" and the "natural and mutually beneficial bond between children and their biological parents." In the course of rejecting that argument, the judge held that the law could not in fact have the effect of propelling gay men and lesbians into heterosexual marriages, citing the testimony of one of the plaintiffs, who had described himself as "a natural-born gay."[1] No matter how many times you throw a stone up, medieval Aristotelians might have said, it will not learn to go upward.

Whether or not it is allowed to stand, Judge Walker's ruling is a judicial landmark not only for lesbians and gay men but for everyone's fundamental rights and their equal protection under the law. It will not, however, settle the issues about nature and natures that permeated the discussions prompted by *Problemata* IV.26, nor about the relationship of what is natural (as distinguished from chosen or created) and what is right. As my intellectual ancestors said, a *problema* is a difficult question, subject to doubt and disputation, that battles in the mind.

Appendix

Pietro d'Abano, *Expositio Problematum Aristotelis*, IV.26

A Text

To offer a fixed text of Pietro d'Abano's *Problemata* commentary is to contravene the very premise of this book, based as it is on the malleability inherent in manuscript culture. But failure to provide the reader with the means to make judgments and other uses of the material would be churlish. A critical edition of the whole commentary by Pieter De Leemans et al. is forthcoming. A translation by Faith Wallis, based on the 1475 Mantua edition, is available in Kenneth Borris, ed., *Same-Sex Desire in the English Renaissance: Sourcebook of Texts, 1470–1650* (New York: Routledge, 2004), 132–40.

The following is a transcription of BNF lat. 6540. That manuscript is a copy of Pietro's second version, but the scribe who prepared it also made use of a copy of Pietro's first version, which allowed him to verify and emend readings. Occasional comparisons to BAV lat. 2175, 42vb–54va (= VL), and Cesena, BCM D.XXIV.2, 36rb–38va (= CM), both also of Pietro's second version, provide supplementary readings. These are selected to offer alternatives to terms and grammatical forms, and to fill in possible omissions. Variants that register a very different understanding of the text from that of the scribe of BNF lat. 6540 are not included here but are discussed in the body of this book and recorded in the notes.

Following the medieval convention, Pietro's quotations from the text of Bartholomeus of Messina's translation of the *Problemata* are underlined. Quotations from other works are in quotation marks. The lemmas used by Pietro to divide the text appear here in bold. They also serve as tag words for the

English passages from Barthalomeus's Latin in the Introduction and chapters 1 and 2. References to Pietro's sources are to the editions listed in the Works Cited. "Avicenna" refers to Avicenna, *Liber canonis medicine.*

[Particula IV, problema 26]
63rb

¶**[Propter quid quidam]**

Querit Aristoteles in hoc capitulo duas questiones. Et prima est quare est quod gaudent sive delectantur illi cum quibus actus coitus peragitur ut mulieres aut eis proportionati? [63va] Vel cum quibus, id est a quibus. Hoc est, quare gaudent coeuntes? [VL and CM: Et ut ad unum sit dicere propter quid gaudent coeuntes] sive se habeant in ratione agentis sive patientis. Quod videtur statim in secunda questione cum querit quare est quod quidam agunt et patiuntur quidam vero non sed patiuntur.

63va cont'd.

¶Deinde **Aut quia** solvit. Et primo ad questionem primam, secundo ad secundam. Secunda: quibus utique. Prima pars in tres. Primo enim assignat causam prime dubitationis, dicens ideo aliqui delectantur in coitu et cum gaudent [VL and CM: congaudent] adinvecem. Quia unicuique in quo fit digestio ex qua recipit iuvamentum inest aliquis locus et receptaculam [VL and CM: receptaculum] in quem superfluitas illius iuvativi naturaliter est apta nata segregari et evacuari a natura, aliquo tempore retenta ipsa ventositate stimulata. Et hoc est quod dicit quod talis sequestratio causatur propter laborem et motum excitantem ventositatem que egrediens inducit primitus inflationem in locum per quem transit.

¶Deinde motu mediante et labore sequestrat superfluitatem a parte in qua erat et extra depellit sicut apparet in egerentibus. Unde Avicenna primo *Canonis* fen prima [bk. I, fen 1, doctr. 6, ch. 3, 24ra] ventositas iuvat ad expulsionem quoniam "prohibet dissolutionem que ad expellendum iuvat." Et hoc est quod dicit Aristoteles eam inflamationem [VL and CM: inflationem] facere et auxiliatur in grossitie eius quod debet expelli. Et quia: "vilum secundum in [CM and Avicenna omit: in] latitudinem" expulsivum "contrahit et inspissat" [ibid.].

¶Et hoc declarat illic **ut urina** appropriando unicuique superfluitati suum receptaculum, dicens quod nostra [VL and CM: urina] que est superfluitas secunde digestionis evacuatur in vesicam. Ipsa enim vesica ut dictum est in prima particula [I.40] est receptaculum humidi indigesti.

¶Superfluitas autem alimenti que est scilicet prime digestionis stercus dicta segregatur in intestina et retinetur in tribus inferioribus precipue scilicet in monoculo, colon, et longanone [VL: longaone].

¶Alia autem tria superiora ut duodenum, ieiunum et gracile retinent magis ad hoc purum et iuvativi [VL: iuvativum] alimenti. Lacrima vero in oculum que eius est superfluitas. Exprimitur enim ex ipso [VL adds: ad cerebrum]. Mucillagines quidam que sunt superfluitates primi et secundi ventriculi cerebri magis per foramen dictum collatorium sive alchasim sub primo existens ventriculo evacuantur in nares. Similiter et sanguis segregatur in venas et cetera.

¶Dubitaret aliquis quomodo numeret sanguinem inter superfluitatem [VL: superfluitates] cum ipse sit humorum naturalissimus et nature filius cuius naturalis locus sunt vene.

¶Dicendum quod per sanguinem hic debet licet [VL: omits] intelligi sanguis superfluus melancholicus multis superfluitatibus aliorum humorum admixtus non autem naturalis. Talis autem est sanguis menstruus et per ymoroydas emissus. Per venas autem est intelligere extremitates venarum que ad fossulas terminantur matricis et extremitates venarum 5 circa anni [*sic*] circulum terminatarum. Per evacuationem etiam huiusmodi superfluitatis possumus intelligere evacuationem omnem superfluitatis tertie digestionis ut sudoris transpirationis insensibilis ut superfluitas hic tangatur omnis digestionis.

¶Deinde **similiter utique**, applicat ad propositum hoc ad superfluitatem spermaticam de qua est principalis intentio. Et dicit quod eodem modo genitura sive sperma segregatur et [VL and CM add: in virgam] evacuatur testiculis mediantibus. Ipsi enim ut dictum est vehementer ad hanc faciendam segregationem operantur. Sic ergo causa delectationis et gaudii est quia superfluum expellendum et expellitur ad locum in quem naturaliter est natum expelli, ita quod in hoc conservatur nostre [63vb] compositionis armonia etc.

63vb cont'd.

¶[CM omits ¶] Notandum est quod in coitu est maxima delectatio respectu aliorum [VL: aliarum] evacuationum, eo quod virtus in eo maxime viget animalis sensitiva cum et materia ipsius super partes transeat sensibilissimas. Quod et opus existit eo quod talis actus evacuationis non sic ordinatur ad conservationem individui ut alii, sed potius ad conservationem speciei. In aliis autem est econtra, non [VL: et ideo] propter magnam necessitatem ipsius et utilitatem necessarium fuit delectationem maximam adesse et precipue ne propter fastidium quod in ipso est relinquatur. Unde propter delectationem perceptam in ipso animalia magis moventur ad ipsum melius quia magis perficiunt, quia "delectatio propria confirmat operationes et diuturniores et meliores facit," sicut dicitur 4° *Ethicorum* [X.5, 1175b15]. Huic quoque [VL adds: genitalium delectationi in emittendo proportionatur] delectatio meri [VL: minus] in suscipiendo qui naturali sensu et a [VL: omits] animato [VL adds: solum] preter alia membrorum cum grandi appetit delectatione sicut dicit Galenus in 4° *De iuvamentis membrorum* [cf. *De usu partium*, XIV] et Aly in *Tegni* de signis [*Pantegni*, Theorica, X(.12?)]. Hoc autem factum est quia non propter se appetit solum sed propter cetera membra sicut coitus non fit gratia individui sed gratia speciei magis.

¶Deinde **quibus utique**. Solvit ad secundam questionem et primo premittit quandam divisionem bimembrem, dicens quod illi [VL: illis] quibus pori sive meatus non sunt naturaliter plasmati ut [VL and CM add: directe terminentur in virgam potest dupliciter contingere. Aut propter id quod] hii meatus qui terminantur ad virgam extendantur [VL: extenduntur; CM: excecantur], materia opilativa interventiente, sicut oculi excecantur cum humor viscosus cadit in nervos opticos sicut est videre in eunuchis et pro, id est in, effimineatis [*sic*] vel in eunuchis qui carent testiculis aut parvos omnino et debiles habent [cf. *Problemata* IV.3, 54rb–va] et in illis quorum complexiones infrigidate sunt et humectate in tantum quod redacti sunt in naturam et mores muliebres. Aut potest aliomodo contingere utpote quia pori tendentes in virgam exterius non solum sint excecati et obstructi ut prius sed quod sint omnino excecati versus virgam in exterius et sint aperti aliqualiter versus radicem virge et anum.

¶In talibus quidem humiditas spermatica concurrit in anum et dispersa inde. Sic tandem mota et subtiliata per confricationem graditur [VL and CM: egreditur] foras etc.

¶Deinde **signum autem**. Prosequitur declarandum membra divisionis predicte ut inde appareat solutio. Et primo agit de illis in quibus hec humiditas non egreditur naturaliter et libere cum hic [VL and CM: hoc] fuerit eis a nativitate. Secundo quando talis dispositio sive passio eis propter pravam consuetudinem contingerit. Ibi secunda, Et quibus quidem in anum. Item primo manifestat secundum membrum divisionis pertacte; secundo primum. Ibi secunda, Qui autem natura. Dividitur iterum prima in tres. Primo enim ostendit quod oportet in hiis humiditatem spermaticam ad anum decurrere, dicens quod signum humidi [VL: huiusmodi] non possunt emittere humidum spermaticum ad virgam sed ad anum est quia quando tales coeunt fit collectio et coangustatio loci existentis circa virgam superius et dissolutio fit et apertio illarum partium que sunt circa anum. Ita quod recte advenit contrarium eius quod oportet esse in coitu naturali sicut apparuit in secundo problemate [IV.2]. Propter quod sicut in coitu naturali humidum spermaticum "velud a manu expressum" circa anum existens cogitur pervenire ad virgam. Ita in his cum fiat omnino contrarium oportet humidum id ad anum emitti. Et ideo si aliquis superfluat in materia coitus aut forte usus fuerit alias evacuatione inde materia spermatica ad hunc locum pervenit circa anum existentem. Propter quod quando appetitus coeundi supervenit appetit per fricationem et motum fortem expellere huiusmodi materiam spermaticam et quia nata est coligi circa annum ad locum ad quem nata est coligi secundum naturam [64ra] eius monstruosam. Et quia nata est coligi circa annum desiderat quod inde expellatur et ideo appetit motum confricationis ibidem fieri etc.

64ra cont'd.

¶Deinde **desiderium**. Quia fecerat mentionem de desiderio ostendit hic secundo unde causetur desiderium istius expulsionis dicens quod talis motus desiderii causatur a duobus, a cibis et potibus inflativis [VL: inflammativis] sicut rapis, pestinacis, carnibus [VL adds: caprinis], columbinis, paserinis, vino novo et talibus. Ex hiis enim generatur humor ventosus ex quo fit spermatis materia stimulativa ad coitum propter eius instabilitatem.

¶Interdum vero contingit ab intelligentia practica vel ymaginatione eo quod ymaginetur delectationem quam fortasse alias [VL: anima] in hoc actu habuit materia etiam spermatica ventosa non etiam ad hoc ipsum stimulante.

¶Interdum vero propter utrumque potest contingere cum itaque desiderium movetur ab altero istorum aut ab utroque. Spiritus ventosus per confricationem et motum excitatus concurrit ad partes circa radicem virge existentes et ex inde spermatica humiditas effluit in locum sibi secundum naturam istam orbatum [CM: orbatam] destinatum. Et hoc sive fuerit talis humiditas subtilis aquea ut in illis quorum meatus essent latiores sive sit solum ventosa que tamen subtiliata per fricationem egreditur in illis etiam quorum meatus sunt angustissimi etc.

¶Notandum sicut dicitur in 3° *De anima* [III.10, 433a10–14] duo esse motus principia, scilicet appetitus et intellectus. Et hoc siquis ponat intellectum velud fantasiam vel ymaginacionem. Non enim fit motus huiusmodi semper propter intelligentiam sed propter fantasiam in multis nostrum [CM: nostram] et in aliis animalibus omnibus in quibus non est intelligentia neque ratio sed fantasia. Quod idem dicit in libro *De causa motus animalium* [*De motu animalium*, VI, 700b17–19]: "Videmus quidem animalia moventia esse intellectum, fantasiam, electionem, voluntatem et concupiscentiam que tamen omnia reducuntur in intellectum et appetitum" et hec duo ulterius in appetibili collocantur.

¶Eorum autem que circa hec principia motuum existunt "quoddam est movens, quoddam est quo fit motus, aliud autem est motum, etc. Movens autem est duplex. Quoddam est omnino immobile" [*De anima*, III.10, 433b13–14] et per se et per accidens ut accidentale, bonum semper et "non aliquando sic et aliquando non," quod est quidem divinum, sempiternum et honorabile; quoddam est motum ab huiusmodi bono divino et sempiterno et est movens alia sub ratione boni vel apparentis boni sicut appetibile et intellectuale quod finis est et terminus opositionis [CM: operationis] practice. Illud autem quo fit motus est organum ut membrum. Tertium autem quod movetur est ipsum animal quorum motuum hec duo principia tetigit, cum dixit desiderium, etc. In libro vero *De causa motus animalium* [VIII, 701b17–19] ponit talem moventium ordinem ut quod "passiones ydonee moveant partes organicas et appetitus passiones et appetitum fantasia [VL: fantasie; CM: fantasiam] que quidem intellectu aut sensu movetur." Quod et Avicenna in fine primi *6ⁱ naturalium* [I.5] pulcerime ordinavit.

¶Deinde **Hoc exeunte**. Hic tertio comparat sic dispositos ad quosdam impotentes ad hoc sperma in coitu emittere, ostendendo quid eis accidat humiditate ista egrediente, dicens quod tali humiditate spermatica egrediente in eis fiunt distensiones et expansiones alarum et oris cum aliquali delectatione ut incurrant alices, ossitationes et pruritus oripilationes per totum propter materiam ventosam confricatione calefactam transeunte [VL and CM: transeuntem] super membra eorum sensibilia [VL adds: vel pudendorum extensiones]. Iste quidem distensiones eveniunt pueris qui non sunt adhuc in tempore in quo possint coire [64rb] ad hoc [VL: adhuc] ex quibus fit eis delectari.

64rb cont'd.

Item tales distentiones delectabiles contingunt illi [VL: illis; CM: eis] qui iam pervenerunt ad tempus in quo deberent coire qui quidem cessant a coitu nullo humido corpulento ab eis egrediente sed spiritu solum ventoso quod accidit quoniam humidum eorum consumptum est. Quod potest esse propter multotiens coire. Ita quod propter inflationem [VL inflammationem] corporis ex illo totum humidum conversum fit [VL and CM: sit] in spiritum ventosum autem [VL and CM: aut] propter egritudinem sicut in ethicis et ptisicis aut fortasse [VL: fortassis] accidit propter complexionem eius natura superflue calidam et siccam, humidum resolventem in spiritus ventosos sicut fit in passeribus. Et licet tales forte non habeant sperma tempore iuventutis habebunt tamen circa etatem consistentie et ultra sicut dicunt Aristoteles, Galenus et Avicenna in libris *De generatione animalium* [Aristotle, I.20, 728a10–14] quibusdam contingere. Nunc autem dicit Aristoteles si aliter [VL and CM: alicui] accidat aliquid predictorum sive puero inperfecto sive viro ut quod patiatur vel propter alterum horum scilicet propter emissionem humiditatis subtilis aut ventose et hoc sive motus ab intellectu aut alimentis appetit ire et evacuari a materia huiusmodi potenta [VL and CM: contenta] subtili vel ventosa quoniam evacuata [VL adds: contenta] potest quietari et gaudere.

¶Notandum est quod sermo de delectatione puerorum coire adhuc impotentium et quorundam virorum hic breviter insinuatus exprimitur in primo *De generatione animalium* [I.20, 728a10–14] sub hiis verbis: "Delectatio accidit in coitu non solum propter emissionem spermatis sed propter emissionem spiritus ex quo spiritu constituto et condensato fit sperma. Et hoc est manifestum in pueris illis qui nondum possunt spermatizare quando fuerint prope tempus in quo debet [VL and CM: debent] spermatizare. Et similiter est manifestum in viris sterilibus. In hiis enim omnibus fit delectatio propter decisionem spiritus" et confricationem.

¶Deinde **qui autem**. Prosequitur primum membrum divisionis predicte. Et primo ostendit quomodo in talibus humiditas spermatica segregetur et per consequens quomodo delectentur dicens quod illi qui naturaliter a primis componentibus effeminati sunt vel eunuchi sic habuerunt poros constitutos et plasmatos ut taliter excecatos quod in eos nulla humiditas evacuetur aut modica valde expellitur ex eis in locum in quem expellitur in illis qui naturaliter secundum naturam habent hos poros.

Vel aliter expellitur modica tamen ipsis humiditatibus [CM: habentibus] ubi, id est, locum secundum naturam suam monstruosam quia expellitur in hunc locum monstruose plasmatum. Et hoc ideo quia non fuerunt constituta [VL and CM: constituti] secundum naturam recte operantem. Qui autem causam et modum humidi [VL and CM: huiusmodi] monstruositatis [CM and VL add: et] aliarum considerare voluerit 9^{m} [VL and CM: quartum] *De generatione animalium* [IV.3, 769b11–4, 773a31] diligenter inspiciat [VL adds: et videat contra].

¶Deinde **quare necesse est**. Hic secundo concludendo ex predictis ostendit quomodo tales se habeant ad curam istius male plasmationis pororum quibus sperma in virgam defertur. Et primo ostendit quod non possint curari per cyrurgiam secando, dicens, si igitur tales pori sunt sic male plasmati cum sint egri oportet eos curare, quod per sectionem necesse est fieri et divisionem subtillem cum ferro cum potissime peccatum plasmationis sectione curetur.

Sed hoc non est faciendum quoniam aliquando talis seccatio vel sectio inducit corruptionem pororum omnia [VL: omninam; CM: omnimodam] et mortem plerumque; aliquando inducit perversionem et mutationem in deterius [64va] in partibus pororum sicut accidit non parum secantibus plagas et ut cicatrices rectificent, propter quod non est facienda.

64va cont'd.

¶Item si fieret qui vir erat primo mutaretur in mulierem et sic mutaretur in peius. Unde bene dictum est ex ea perversionem fieri. Per talem enim seccationem partes circa testiculos existentes corrumpentur ita quod rederetur talis eunuchus et effeminatus cum prius non sic esset. Non tamen intelligendum quod [VL and CM add: vir] essentialiter mutetur in mulierem sed potius accidentaliter acquiriendo actus et mores muliebres.

¶Deinde **necesse est igitur**. Docet hic qualiter tales debeant curari dicens primo quod oportet illud quod inest in istis de spermatica humiditate expurgatura divertere ab ista parte et movere ad aliam partem quod fit cum cibis et medicinis exsiccantibus sperma et materiam eius ad alias regiones divertentibus ut ad urinas et sudores et cetera. Unde Avicenna 3° *Canonis*, fen 20^{a} [bk. 3, fen 20, tr. 2, ch. 42, fol. 358ra] loquens de huiusmodi dispositione et priori magis dicit quod "stulti sunt homines qui volunt eos curare quoniam vitium egritudinis eorum est medicamentum non naturale. Sed si cura confert eis tunc est illud quod frangit desiderium eorum ex tristitia, fame, vigiliis, carcere et percussione."

¶Deinde philosophus infert unum corelarium dicens quod sic sunt dispositi ideo redduntur insatiabiles et insani semper appetentes fricari ut humiditas illa seminalis expurgetur. Sicut et mulieres sunt insatiabiles et maxime iuvencule quarum meatus adhuc sunt angusti eo quod modica est humiditas que cogitur expelli mota et liquata per fricationem que quidem [VL and CM: quia] modica est cito infrigidatur non potens expelli. Propter quod iterum appetunt fricari ut expellatur. Sic enim contingit ipsis sicut corpori existenti in continua effluxione sudoris de qua [VL and CM: quo] dictum est in 22° capitulo particula secunda [VL and CM: 2^{e} particule] [*Problemata* II.22].

¶Notandum est quod mulieres insatiabiles sunt precipue iuvencule circa annum 14^{m} existentes de quibus fere est sermo in 9 *De ystoriis animalium* [VII.1, 581b11–21] et cetera. Iuvencule "maxime incitantur ad venereorum usum" cum incipiunt et "venerea passe incontinentes fiunt." Cuius quidem causa est quia tempore illo propter humoris spermatici et menstrui descesum [VL and CM: descensum] incipiunt inspissari genitalia et fissura vulve claudi et labia eius mollificari etiam ingrossari, pilosa fieri. Et incipit tunc ad coitum insanire. Non tamen emittit in desiderio et quanto plus coit aut manu se fricaverit tanto plus appetit eo quod per talem fricationem humor attrahitur sed non emittitur. Et cum humore attrahitur calor et ventositas consurgit sed cum corpus frigidum sit et clausum in poris non cito semen emittit.

¶**Et quibus [VL adds: quidem]**. Agit de illis quibus predicta propter pravam consuetudinem et obscenam adveniunt quales sunt sodomite. Et dividitur in partes 2. Primo enim repetit illud quod dictum est de secundo membro divisionis predicte. In quibus humiditas confluit ad anum ad hoc ut convenientius condescendat ad illos qui ad hoc operantur prevaricatione turpissima, dicens quod inter illos in quibus humiditas spermata [VL and CM: spermatica] decurrit ad anum quidam sunt qui appetunt pati et fricari vehementer circa anum ut hec expellatur humiditas, sed quibus fit descensus [CM: decursus] istius humiditatis partim in anum et partim in virgam. Hii appetunt fricare et fricari secundum utramque istarum partium. Appetunt enim virga alios fricare et cum [64vb] virgis aliorum in ano fricari ab eis. Inter istos qui sic appetunt agere et pati secundum utramque partem quidam magis desiderant agere seu fricari [VL and CM: fricare] quidam vero magis pati secundum quod humiditas stimulativa magis est nata confluere in virgam vel in anum.

64vb cont'd.

¶Notandum sicut dicit Avicenna 3º *Canonis* fen 20ª [bk. 3, fen 20, tr. 2, ch. 42, 358ra] de huiusmodi passione quam vocat haluebathi. "Dixerunt [VL adds: quidam] causam halubnathy esse eo quod nervus sensibilitatis qui venit ad virgam ramificatur in illis in duos ramos quorum subtilis radici virge continuantur [VL: conterminantur] et grossus declinat ad capud virge, quare indigent illo qui est tenuis fricatione vehementi donec ssentiat [*sic*], quare" prosternit se ad iacendum super hominem [Avicenna: "spargitur super ipsum hominem" for "posternit . . . hominem"]. Et tunc occurrit eis complimentum [VL: complementum] libidinis quod videtur hic plus [VL omits: hic plus] Philosophus hic dicere. Ipsi tamen Avicenne non fuit hoc visum. Ymo dicit quod "hec res sit sicut illud quod remotum est a veritate." Quare autem hoc sit falsum Avicenna non adducit rationem, qui etiam contrarius videtur sententie Philosophi. Non video quid impediat a nativitate talem inordinationem et monstruositatem fieri cum idem statim Avicenna faciant [VL and CM: faciat] sermonem de hermostrodicis [VL: hermofroditis] qui magis appent [VL and CM: apparent] apud naturam monstruosi.

¶ Deinde **quibusdam autem**. Prosequitur hic secundo agendo de illis qui scelerato vicio sodomitico sunt damnati [VL and CM: depravati]. Et primo ostendit quomodo talis passio et profana operatio propter consuetudinem accidat et in naturam convertatur, dicens quod in aliquibus preter [VL: propter] consuetudinem, que potius coruptela est dicenda, predicta contingit passio que predictis inerat a nativitate. Qui moventur ad hunc actum propter delectationem quoniam quecumque accidit facere gaudere et delectare illis [VL and CM: illos] qui emittunt genituram recte [VL: ratione] coitum exercendo. Ita accidit illis quibus talia fiunt. Propter hunc enim actum profanum appetunt exercere ad hoc ut gaudeant et delectentur. Vel aliter quia quecumque et cetera, id est tota eorum intentio non est nisi emittere genituram id est humiditatem illam seminalem ut inde eos accidat gaudere et delectari. Et hoc est quod dicit quod ita hoc modo desiderant illi qui hec operantur nephanda quos Arisoteles 7º *Ethicorum* [VII.7, 1149a1] ait fore "extra terminos malicie." In quorum extirpationem lex inclamat: "Insurgant leges armentur vita [VL and CM: iura] ultore gladio feriantur" [*Codex Theodosii* 9.7.3]. Unde Avicenna 3º *Canonis*, fen 20ª: "Coitus cum infantibus fetidus [CM: fedus] est apud multitudinem gentium et prohibitus in lege." Hunc autem actum sepe consuetudine exercendo habituata consuetudine fit transitus in naturam, sicut apparebit in primo capitulo 28 particule [*Problemata* XXVIII.1]. Unde in libro *Memoria et reminiscentia* [II, 452a27–29] "consuetudo est tanquam natura." Quomodo autem consuetudo in naturam convertatur pulchre determinat Ypocras in *Libro de aere et aqua* [XIV] inducendo quosdam quibus ex consuetudine matrum capita manibus post natum [VL and CM: nativitatem] elongando consuetudine mutatio facta est in naturam.

¶Dubitatur quoniam non videtur consuetudo converti in naturam ita quod natura prior suum ordinem reliquat quoniam "nullum natura existentium aliter" et aliter "asuescit, puta si lapis decies milies" prohiciatur in sursum non asuescit, sicut dicitur in secundo *Ethicorum* [II.1, 1103a20–24].

¶Dicendum quod naturalium duplex est motus et inclinatio. Unus quidem est qui sequitur [VL: consequitur] commixtionem primam elementorum dominantium in mixto et formam datam a generante. Et secundum hoc nullum acquirit consuetudinem in contrarium scilicet dominantis elementi in ipso cum secundum hoc omnia sint inanimata.

¶Alius est motus et inclinatio qui sequitur formam superiorem resultatam ex mixtione elementorum aut deforis advenientem. Etiam [VL: Et sic] aliquid potest habere [65ra] inclinationem et motum ad id quod fuerit ei primo contrarium sicut ex multotiens agere ex quo sicut dicitur 2º *Ethicorum* [I.1; cf. VII.14, 1152a29–30] acquiritur consuetudo que tandem multotiens convertitur in naturam. Et isto modo intelligitur hic Aristotelis sermo et non primo modo secundum quem procesit abiectio [VL: obiectio] et cetera.

65ra cont'd.

¶Deinde **propter hoc**. Hic secundo infert corolarium [*sic*]. Et primo ostendit quomodo se habeant ad hunc actum incipientes ipsum exercere tempore pubertatis dicens, et ideo quicumque non ante pubertatem et tempus in quo possunt emittere sperma sed circa hanc consueverunt suponi et fricari circa anum eo quod in consuetudine talis eis relinquitur mero [VL and CM: memoria]. Ex multis enim partibus [VL: particularibus] sensationibus et delectationibus usu perceptis causatur in istis memoria. Iste enim est modus agnationis [VL: adgenerationis; CM: generationis] memorie sicut dicitur in fine *Posteriorum* [II.19, 100a2–14], propter quod longa interveniente consuetudine ita desiderant ac [VL: hac] si natura ad hoc esset [VL and CM: essent] producti. Et hoc est possibile nam frequens consuetudo et longa fit sicut ipsa inesset natis a nativitate eorum natura et cetera. Unde Tulius in prohemio *Rethorice veteris* [I.2, § 3]: "Consuetudo propter vetustatem obtinet vim nature" et cetera.

¶Notandum est quod istud nephandum opus sodomiticum quidam exercent puerorum inter coxas confricatione quod in [CM omits; VL: et] plurimi agunt hec [VL and CM: hodie]; alii autem fricationem faciendo circa anum et virgam in ipsum imponendo sicut in vulvam imponitur et de talibus videtur esse sermo Aristotelis magis.

¶Deinde **si autem**. Ostendit quomodo se habeant ad hoc illi qui ante pubertatem supponuntur dicens quod si huiusmodi turpis luxuria accidat tempore quo mollis et tenera est natura istorum qui supponuntur ut quando non possunt sperma emittere, omnia predictorum istis cito adveniunt ut delectatio inmensa et emissio seminis ad anum et cetera. Et consueta convertuntur in naturam et magis quam illis qui [VL adds: non] ante pubertatem incipiunt. Et de ista quidem materia est sermo Avicenne 3° *Canonis* fen 20ª [bk. 3, fen 20, tr. 2, ch. 42, fol. 358ra] in capitulo allegato prius de halubuachi.

¶Causam vero istius actus nephandi dicit Aristoteles in 7° *Ethicorum* [VII.7, 1149a1] esse "bestialitatem" omnia [VL: omnina; CM: omniam] et perversionem anime. Propter quod non inmerito vult eos "extra terminos" omnis "malicie" existere. Ptholomeus vero in 8. [VL: 80°] verbo *Centiloquii* [80] et in 4° *Quadripartiti* [IV.4] assignat causam ex coniunctione vel aspectu malorum cum Venere in signo malorum et maxime in domo 7ª et 6ª Iovis coniunctione vel aspectu ab hiis remoto et cetera.

Abbreviations

AL	= Lacombe et al., *Aristoteles Latinus*
Arsenal	= Site Arsenal, Bibliothèque Nationale de France [Paris]
BAV	= The Vatican, Biblioteca Apostolica Vaticana
BCM	= Biblioteca Comunale Maletestiana [Cesena]
BJ	= Biblioteka Jagiellońska [Krakow]
BL	= London, British Library
BM	= Aristotle, *Problemata*, translated by Bartholomeus de Messina. All references are to Book IV, *problema* 26, unless otherwise indicated.
BNF	= Paris, Bibliothèque Nationale de France
BN Marciana	= Venice, Biblioteca Nazionale di San Marco
Bodl.	= Oxford, Bodleian Library
CA	= Universitätsbibliothek Erfurt/Gotha, Collectio Amploniana [Erfurt]
CLM	= Munich, Bayerische Staatsbibliothek, Codex Latinus Monacensis
CU	= Cambridge, Cambridge University Library
EC	= Evrart de Conty, *Les Problemes d'Aristote*. All references are to Book IV, *problema* 26, unless otherwise indicated.
Escorial	= San Lorenzo de El Escorial, Real Biblioteca del Monasterio de San Lorenzo
GT	= Aristotle, *Problemata*, translated by George of Trebizond. All references are to Book IV, *problema* 26, unless otherwise indicated.
JJ	= Pietro d'Abano, *Expositio Problematum Aristotelis*, redacted by Jean de Jandun. All references are to Book IV, *problema* 26, unless otherwise indicated.
KB	= Koninklijke Bibliotheek [The Hague]
NK	= Národní Knihovna České Republiky [Prague]

PA = Pietro d'Abano, *Expositio Problematum Aristotelis*. All references are to Book IV, *problema* 26, unless otherwise indicated.

Sorbonne = Paris, Bibliothèque Universitaire de la Sorbonne

StB = Stadtbibliothek [Nuremberg]

TG = Aristotle, *Problemata*, translated by Theodore Gaza. All references are to Book IV, *problema* 26, unless otherwise indicated.

UB = Universitätsbibliotek or Universiteitsbiblioteek [Leipzig, Ghent]

WB = Walter Burley, *Epitome Problematum Aristotelis*. All references are to Book IV, *problema* 26, unless otherwise indicated.

Notes

INTRODUCTION

1. Walter Burley, *Problemata* [hereafter referred to as WB; references are to IV.26 unless otherwise indicated], IV.26, BLd 153, 110r: "Nullum naturale turpe sed omnia munda mundo existunt. Excusatio lectori propter inhonestatem materie. Non tamen est spurnendum vilia contemplari quia in quolibet quantocumque vili aliquid divinum et mirabile continetur secundum Philosophum .13. *De animalibus*. Non vilia contemplando sed operando vel volendo viles [Oxford, St. John's 113, BL Royal: viciosi] efficimur. *Ethicorum*." Aristotle, *De animalibus: Michael Scot's Arabic-Latin Translation*, part 2, *Books XI–XIV: Parts of Animals*, ed. Aafke M. I. Van Oppenraaij, Aristoteles Semetico-Latinus, 5 (Leiden: Brill, 1998), XI, 4–5 [*De partibus animalium*, I.5, 645a4–17], p. 27; *Ethica nicomachea: Translatio Roberti Grosseteste Lincolniensis sive "Liber Ethicorum," A. recensio pura*, fasc. 3 of *Ethica nicomachea*, ed. René Antoine Gauthier, 5 fascicules, Aristoteles Latinus, 26, 1–3, Corpus Philosophorum Medii Aevi, Union Académique Internationale (Leiden: Brill, 1972–74), II.2, 03b26–30 [hereafter referred to as *Ethica nicomachea*, AL]; cf. Walter Burley, *Expositio Gaulteri Burlei super decem libros Ethicorum Aristotelis* (Venice: Andrea Torresani Asula, 1500), bk. 2, tract 1, ch. 2, 25rb–va. Gijs Coucke points out that Walter is drawing upon an older medieval tradition: " 'Non adeo est honesta ut delectabilis': Sexual Pleasure in Medieval Medicine: The Case of Peter of Abano," in *Piacere e dolore: Materiali per una storia delle passioni nel Medioevo,* ed. Carla Casagrande and Silvana Vecchio, Micrologus 29 (Florence: SISMEL Edizioni del Galluzzo, 2009), 126–27 and n. 27.

2. Table to Pietro d'Abano, *Expositio Problematum Aristotelis*, Gottweig 138, 238v, "Sodomiticum peccatum quidam faciunt natura moti."

3. Aristotle, *Les Problèmes*, ed. and trans. Pierre Louis, Collection des Universités de France (Paris: Les Belles Lettres, 1991), 3 vols., vol. 1, IV.26, 879a36–880a5. Subsequent references to the Greek are to this edition.

4. For a recent review, from the perspective of medieval history, of the issues surrounding the term, see Helmut Puff, *Sodomy in Reformation Germany and Switzerland, 1400–1600*, Chicago Series on History, Sexuality, and Society (Chicago: University of Chicago Press, 2003), 1–14.

5. "Prague" commentary, Prague, NK VIII.A.19, 79va; Evrart de Conty, *Problemes* [hereafter referred to as EC; all references to IV.26 unless otherwise indicated], IV.26, BNF fr. 24281, 116v.

6. With its limited source base, realm of discourse, social locus, geographical extent, and chronological range, this book will not settle any of the larger questions about what it means to do the history of homosexuality. It aspires to provide material for those engaged in the discussion, in which several (among many) milestones and recent contributions to the literature on pre-modern Europe include: John Boswell, *Christianity, Social Tolerance, and Homosexuality: Gay People in Western Europe from the Beginning of the Christian Era to the Beginning of the Fourteenth Century* (Chicago: University of Chicago Press, 1980); Carolyn Dinshaw, *Getting Medieval: Sexualities and Communities, Pre- and Postmodern* (Durham, N.C.: Duke University Press, 1999); essays in Glenn Burger and Steven F. Kruger, eds., *Queering the Middle Ages*, Medieval Cultures 27 (Minneapolis: University of Minnesota Press, 2001).

7. David Piché with Claude Lafleur, ed., *La condemnation parisienne de 1277*, Sic et Non (Paris: J. Vrin, 1999), p. 72 and § 166, p. 128: "Nonnulli Parisius studentes . . . quosdam manifestos et execrabiles errores . . . tractare et disputare presumunt. . . . Quod peccatum contra naturam, utpote abusus in coitu, licet sit contra naturam speciei, non tamen est contra naturam individui."

8. Aristotle, *Problemata*, translated into Latin by Bartholomeus de Messina [hereafter referred to as BM; references to IV.26 unless otherwise indicated], BNF lat. 15081, 12vb: "Quibusdam autem fit et ex consuetudine passio hec."

9. Pietro de Abano, *Expositio Problematum Aristotelis* [hereafter referred to as PA; all references to IV.26 unless otherwise indicated], BAV 2175, 52rb.

10. Aristotle, *Ethica nicomachea*, fasc. 3, VII.7, 48b29–30.

11. Medieval Latin copies do not all agree on the number and division of the *particule*. The first *particula* has a clear division within it and was sometimes taken to be two *particule*, and there was some confusion about the short groups at the end. The confusion was recognized in the Middle Ages as attested in a manuscript of Evrart de Conty's commentary, The Hague, KB 133 A 3, vol. I, 1r. (The remark is absent from the autograph, BNF fr. 24281). I have followed the standard numbering of the Greek text to which most of the Latin versions correspond. In cases where they do not, I cite the standard *particula* number in square brackets.

12. The textual history is reviewed in the Introduction to Aristotle, *Problèmes*, ed. Louis, vii–xxxv. Subsequent references to the Greek text are to this edition. See also *Problems*, ed. and trans. W. S. Hett, rev. ed., 2 vols., Loeb Classical Library (Cambridge, MA: Harvard University Press, 1970), Robert Mayhew, ed., 2 vols., Loeb Classical (Cambridge, MA: Harvard University Press, 2011), and *Problemata physica*, trans. Hellmut Flashar, Aristoteles Werke in deutscher Übersetzung 19 (Berlin: Akademie, 1975). On the genres and Latin texts of problem literature, see Brian Lawn, *I Quesiti salernitani: Introduzione alla storia della letteratura problematica medica e scientifica nel Medio Evo e nel Rinascimento*, trans. Alessandro Spagnuolo ([Salerno]: Di Mauro, 1969 (a revised version of *Salernitan Questions: An Introduction to the History of Medieval and Renaissance Problem Literature* [Oxford: Clarendon Press, 1963]); Ann Blair, "The 'Problemata' as a Natural Philosophical Genre," in *Natural Particulars: Nature and the Disciplines in Renaissance Europe*, ed. Anthony Grafton and Nancy Siraisi, Dibner Institute Studies in the History of Science and Technology (Cambridge, MA: MIT Press,

1999), 171–204; Paolo Cherchi, "Il quotidiano, i 'Problemata' e la meraviglia: Ministoria di un microgenere," *Intersezioni* 21 (2001): 243–75; Iolanda Ventura, "Aristotelis fuit causa efficiens huius libri: On the Reception of Pseudo Aristotle's 'Problemata' in Late Medieval Encyclopedic Culture," in *Aristotle's "Problemata" in Different Times and Tongues*, ed. Pieter De Leemans and Michèle Goyens, Medievalia Lovaniensia, ser. 1, Studia 39 (Leuven: Leuven University Press, 2006), 113–44, and "*Quaestiones* and Encyclopedias: Some Aspects of the Late Medieval Reception of the Pseudo-Aristotelian 'Problemata' in Encyclopedic and Scientific Culture," in *Schooling and Society: The Ordering and Reordering of Knowledge in the Western Middle Ages*, ed. Alasdair A. MacDonald and Michael W. Twomey, Groningen Studies in Cultural Change (Leuven: Peeters, 2004), 23–42.

13. WB, Oxford, St. John's 113, 1ra: "Causa materialis sive subiectum libri *Problematum* Aristotelis secundum Petrum Badulanensem [*sic*] est scibile commune sub dubitatione prolatum." Cf. Oxford, BL Digby 153, 102r: "est stabile quoddam sub dubitatione prolatum," followed shortly by "quodlibet scibile"; and PA, Prologue, BNF 6540, 1ra: "Materia sive subiectum huius libri vel scientie est scibile pene commune sub dubitatione prolatum."

14. Aristotle, *Problemata*, V.5. Of the single-answer questions, a few are posed in more than one place and given different answers, e.g., XI.16, 34, and 62 on the voices of women, children, and eunuchs.

15. PA, Prologue, BNF lat. 6540, 1rb: "Problema quidem est grecum latine probationem inportans. Est enim questio difficilis aliquid continens quod disputatione solvendum est quod et voraginem videtur [CM XXIV.D.2 omits] videtur denotare ut ea [CM: enim] voret et decipiat propositum est problema." Cf. Aristotle, *Topics*, I.11, 104b1–17. The opinion of Alexander Aphrodisius to the contrary went unheard in the Middle Ages. It was revived by humanists, but Pietro knew of an earlier medieval definition. See Blair, "*Problemata* as a Natural Philosophical Genre," 175–6.

16. "Erfurt" commentary, Erfurt, CA F. 263, 35vb: "est speculacio contendens ad mentem." See Maaike Van der Lugt, "Genèse et postérité du commentaire de Pietro d'Abano sur les 'Problèmes' d'Aristote: Le succès d'un hapax," in *Médecine, astrologie et magie au Moyen Âge: Autour de Pietro d'Abano*, ed. J.-P. Boudet, F. Collard, and N. Weill-Parot, Micrologus (Tavarnuzze: SISMEL, Edizioni del Galluzzo, forthcoming), 166–69.

17. Rudolf Seligsohn, *Die Übersetzung der pseudo-aristotelischen "Problemata" durch Bartholomaeus von Messina: Text und textkritische Untersuchungen zum ersten Buch* (Berlin: Ebering, 1934); Gerardo Marenghi, "Un capitolo dell'Aristotele medievale: Bartolomeo da Messina traduttore dei 'Problemata Physica,'" *Aevum* 36 (1962), 268–83; Michèle Goyens and Pieter De Leemans, "Traduire du grec au latin et du latin au français: Un défi à la fidélité," in *Pratiques de traduction au Moyen Âge: Actes du colloque de l'Université de Copenhague 25 et 26 octobre 2002*, ed. Peter Anderson (Copenhagen: Museum Tusculanum Press, University of Copenhagen, 2004), 204–13. An earlier translation of forty-three *problemata* was not in circulation in the period under consideration here. It contained three questions on eunuchs, one of which is of some interest for its terminology: "Quare inquid differit spaton ab eunuchus. R. quia eunuchis ab infantia auferuntur testiculi spaton autem iam dure castratur." Valentine Rose, ed. "Problemata" [translatio vetustissima], *Aristoteles pseudepigraphus* (Leipzig: B. G. Teubner, 1863; repr. Hildesheim: Georg Olms, 1971), § 3, p. 667.

18. Prefatory to BM, Vat. 2112, 1r: "Iste liber non obstante dicto aliquorum in oppositum compilatus fuit ab Aristotele, cum de ipso in 2° *Metheorum* ac in plerisque locis *Parvorum naturalium* et *Libris animalium* faciat multotiens mentionem." In a similar vein, PA, Prologue, BNF 6540, 1rb. See Joan Cadden, "Preliminary Observations on the Place of the 'Problemata' in Medieval Learning," in *Aristotle's "Problemata" in Different Times and Tongues*, ed. Pieter De Leemans and Michèle Goyens, Mediaevalia Lovaniensia, ser. 1, Studia 39 (Leuven: Leuven University Press, 2006), 8–9, and, on the continuing debate over authorship and its significance, Ann Blair, "Authorship in the Popular 'Problemata Aristotelis,'" *Early Science and Medicine*, 4 (1999): 189–227.

19. John Monfasani, ed., *Collectanea Trapezuntiana: Texts, Documents, and Bibliographies of George of Trebizond*, Medieval and Renaissance Texts and Studies 25; Renaissance Society of America, Renaissance Texts and Studies 8 (Binghamton, NY: Medieval and Renaissance Texts and Studies, 1984), § XIX, 131–32.

20. PA, Prologue, Ghent 72, 1r. On the relationship of the work to the Aristotelian canon and the dissemination and audiences for the *Problemata*, see Joan Cadden, "Preliminary Observations on the Place of the 'Problemata' in Medieval Learning," 1–19; on this image, figure 3 and 17–19.

21. Johannes de Janduno, *Quaestiones super 8 libros Physicorum Aristotelis* (Venice: Iuntas, 1551; repr Frankfurt: Minerva, 1969), Preface, [xi]rb: "Liber autem de *Problematibus*, quo ad magnam eius partem congregatus est ex multis accidentibus naturalibus manifestis ad sensum, quorum tamen causę sunt difficiles et occultae, pertinent ad diversas partes scientiae praenominatas. . . . Et scias quod liber ille de *Problematibus* communiter invenitur corruptus et incorrectus, et non est multum expositus ab aliquo noto aut famoso, et ideo pauci student in eo, et pauciores intelligunt eum sufficienter. Quia multa et pulcherrima [*sic*] teoremata mirabilis delectationis sunt in eo congregata." The date of this work is uncertain. Schmugge dates Jean's writings to the period from 1315–1324, but this seems late if Jean knew of Pietro's work. He was already a famous master by 1310 and may well have lectured on the *Physics* before then. Ludwig Schmugge, *Johannes von Jandun, 1285/89–1328*, Pariser Historische Studien, 5 (Stuttgart: Anton Hiersemann, 1966), 14–20.

22. A group centered at the University of Leuven is preparing an edition of Pietro's commentary under the direction of Pieter De Leemans. The commentary is available in many early printed editions, of which the first is *Expositio Problematum Aristotelis*, ed. Stephanus Illarius (Mantua: Paulus Johannis de Puzpach, 1475). For the most recent scholarship and bibliography on Pietro and that work, see Pieter De Leemans, ed., *Philosophy Between Text and Tradition: Essays on Peter of Abano's "Expositio Problematum Aristotelis"* (forthcoming). In addition, for biographical and bibliographical information, see Leo Norpoth, "Zur Bio- und Wissenschaftslehre des Pietro d'Abano, Mediziners, Philosophen und Astronomen in Padua," *Kyklos: Jahrbuch für Geschichte und Philosophie der Medizin*, 3 (1930): 292–353; Eugenia Paschetto, *Pietro d'Abano, medico e filosofo* (Florence: Nuovedizioni Vallechi, 1984). For the most recent authoritative account of Pietro's life and his *Expositio Problematum Aristotelis* I rely on Van der Lugt, "Genèse et postérité du commentaire de Pietro d'Abano sur les 'Problèmes' d'Aristote."

23. Gijs Coucke points out that, unlike some other medical writers of his time, Pietro

makes scant reference to experiences derived from medical practice: "From Abstinence to Promiscuity: Man, Beasts and Eunuchs in Peter of Abano's 'Expositio Problematum,'" in "Edition of Peter of Abano, 'Exp.' IV, with Synthesis, Annotations and Essays," vol. 2 of *Philosophy Between Text and Tradition: The Reception of Aristotle's "Problemata" in the Middle Ages: Peter of Abano's "Expositio Problematum,"* Ph.D. dissertation, Katholieke Universiteit Leuven, 2008, 209–12.

24. On the relationship between natural philosophy and medicine in his work, see Nancy G. Siraisi, "Pietro d'Abano and Taddeo Alderotti: Two Models of Medical Culture," *Medioevo* 11 (1985): 139–62, and Coucke, "From Abstinence to Promiscuity," in "Edition of Peter of Abano, 'Exp.' IV," 207–38.

25. Paschetto, *Pietro d'Abano*, 8, suggests as much, contrasting Pietro's approach with systematic Aristotelianism. On the character of Pietro's naturalism and rationalism, see Marie-Thérèse d'Alverny, "Pietro d'Abano et les 'naturalistes' à l'époque de Dante," in *Dante e la cultura veneta: Atti del convegno de studi organizzato della Fondazione "Giorgio Cini" in collaborazione con l'Istituto Universitario di Venezia, l'Università di Padova, il Centro di Studi Danteschi, e i comuni di Venezia, Padova, Verona: Venezia, Padova, Verona, 30 marzo–5 aprile 1966*, ed. Vittorio Branca and Giorgio Padoan (Florence: Leo S. Olschki, 1966), 107–19; Graziella Federici Vescovini "Peter of Abano and Astrology," in *Astrology, Science, and Society: Historical Essays*, ed. Patrick Curry (Woodbridge, U.K.: Boydell, 1987), 19–39, and "L'antropologia naturale de Pietro d'Abano," *Paradigmi* 15 (1997): 525–41.

26. Pieter De Leemans, "Was Peter of Abano the Translator of Pseudo-Aristotle's 'Problemata physica'?" *Bulletin de Philosophie Médiévale* 49 (2007): 103–18. He made other translations as well: Marie-Thérèse d'Alverny, "Pietro d'Abano traducteur de Galien," *Medioevo* 11 (1985): 19–64.

27. Prompted by his comment "in linguam iam latinam transduxi" (PA, Prologue, BNF 6540, 1rb), scholars have searched for signs that Pietro translated the Aristotelian work itself, but this was not the case. See Luigi Olivieri, *Pietro d'Abano e il pensiero neolatino: Filosofia, scienza e ricerca dell'Aristotele greco tra i secoli XIII e XIV*, Centro per la Storia della Tradizione Aristotelica nel Veneto, Saggi e Testi 23 (Padua: Antenore, 1988), 75–81; Graziella Federici Vescovini, "'L'Expositio succincta Problemata Aristotelis' de Pierre d'Abano," in De Leemans and Goyens, *Aristotle's "Problemata" in Different Times and Tongues*, 58–65; De Leemans, "Was Peter of Abano the Translator of Pseudo-Aristotle's 'Problemata physica'?" Pietro's source here is *Abubecri Rasis ad regem Mansorem de re medicina Liber II translatus ex arabico in latinum a Gerardo Cremonensi,* ed. Richard Foerster, in *Scriptores physionomonici graeci et latini,* 2 vols., Bibliotheca Scriptorum Graecorum et Latinorum Teubneriana (Leipzig: B. G. Teubner, 1893), ch. 57, vol. 2, 187, to which Pietro refers in the chapter on eunuchs of his *Compilatio phisonomie,* pt. 2, ch. 10 [bis], decisio 22, [36]r.

28. There have been many speculations about the circumstances of and reasons for his having been censured. Lynn Thorndike, *A History of Magic and Experimental Science During the First Thirteen Centuries of Our Era*, vol. 2, rev. ed. (New York: Macmillan, 1929), appendix 8, 938–47, and "Peter of Abano and the Inquisition," *Speculum* 11 (1936): 132–33; Bruno Nardi, "Intorno alle dottrine di Pietro d'Abano," in *Saggi sull'aristotelismo padovano dal secolo XIV al XVI*, Studi sulla Tradizione Aristotelica nel Veneto, 1 (Florence: G.

C. Sansoni, 1958); Paolo Marangon, *Il pensiero ereticale nella Marca Tevigiana e a Venezia dal 1200 al 1350* (Abano: Francisci, 1984), 66–104. Graziella Federici Vescovini reviews the evidence and opinions (including those associating Pietro with necromancy or Averroism) in the introduction to Pietro de Abano, *Trattati di astronomia: Lucidator dubitabilium astronomiae, De motu octavae sphaerae e altre opere*, ed. Graziella Federici Vescovini, 2nd ed., Il Mito e la Storia, 3 (Padua: Programma, 1992), 21–30 and "Pietro d'Abano tra biografia e fortuna: Due 'ritratti' quadrocenteschi," *Medioevo* 16 (1990): 293–321. On his non-Averroistic views, see also her "Pietro d'Abano e Taddeo da Parma sull'imaginazione," in *Intellect et imagination dans la philosophie médiévale/Intellect and Imagination in Medieval Philosophy/Intelecto e imaginação na filosofia medieval, Actes du XIe Congrès International de la Société Internationale pour l'Étude de la Philosophie Médiévale (SIEPM)*, ed. Maria Cândida Pacheco and José F. Meirinhos, 3 vols. Société Internationale pour l'Étude de la Philosophie Médiévale, Rencontres de Philosophie Médiévale 11, 1–3 ([Turnhout]: Brepols, 2006), vol. 3, 1783–95. At a later date, the same Dominicans acquired a copy of Pietro's *Problemata* commentary, now Paris, Bibliothèque Mazarine 3520. See Sante Ferrari, "Per la biografia e per gli scritti di Pietro d'Abano," *Atti della R. Academia dei Lincei: Memorie della classe di scienze morali, storiche e filologiche*, ser. 5, vol. 15 (1915), 680.

29. Pietro d'Abano, *Conciliator controversiarum quae inter philosophos et medicos versantur Petro Abano Patavino* (Venice: Iuntas, 1565), facsimile in *Opere de Pietro d'Abano*, ed. G. I. Ludwig (Padua: Il Glifo, 1982), diff. 9, 15vb.

30. Van der Lugt, "Genèse et postérité du commentaire de Pietro d'Abano sur les 'Problèmes,'" 165, citing *Conciliator*, diff. 48, and *Expositio Problematum Aristotelis*, X.13 (PA, BNF lat. 6540, 98va).

31. Tiziana Pesenti, "Per la tradizione del testimento di Pietro d'Abano," *Medioevo* 6 (1980): 533–42.

32. Pieter De Leemans, "Peter of Abano, Commentator and Translator," paper delivered at the workshop "Philosophy Between Text and Tradition: Petrus de Abano and the Reception of Aristotle's 'Problemata' in the Middle Ages," Freiburg im Briesgau, December 13, 2007, forthcoming in De Leemans, ed., *Philosophy Between Text and Tradition*. Luigi Olivieri discusses some evidence of Pietro at work on the project (*Pietro d'Abano e il pensiero neolatino*, 65–133) which suggests the existence of a first version that has not survived. Van der Lugt, "Genèse et postérité du commentaire de Pietro d'Abano sur les 'Problèmes,'" 163.

33. PA, Prologue, BNF lat. 6540, 1rb: "Liber hic non potest plene intelligi nisi ab illo qui philosophiam secundum omnem partem eius inspexerit."

34. Jean de Jandun's redaction of Pietro d'Abano's *Expositio Problematum Aristotelis* [hereafter referred to as JJ], Prologue, BNF lat. 6542, 1ra. For Jean's exposition of *Problemata* IV.26 I have relied in the first instance on BAV Ottob. 1764 (see Coucke, "Edition of Peter of Abano, 'Exp.,' IV," introduction, xxxi–xxxiv), and all references are to IV.26 unless otherwise indicated. That manuscript, however, lacks what Kuksewicz judges to have been the original form of the Prologue, for which I use BNF lat. 6542. Zdzislaw Kuksewicz, "Les 'Problemata' de Pietro d'Abano et leur 'rédaction' par Jean de Jandun," *Medioevo* 11 (1985): 119–120. On Jean's life and thought, see Schmugge, *Johannes von Jandun*, and Stuart MacClintock, *Perversity and Error: Studies on the "Averroist" John of Jandun*, Indiana University

Publications, Humanities Series 37 (Bloomington: Indiana University Press, 1956). On the relevant aspects of Marsilius's career, see William J. Courtenay, "University Masters and Political Power: The Parisian Years of Marsilius of Padua," in *Politische Reflexion in der Welt des späten Mittelalters/Political Thought in the Age of Scholasticism: Essays in Honour of Jürgen Miethke*, ed. Martin Kaufhold (Leiden: Brill, 2004), 208–23, on Marsilius's career at Paris; and Maurice de Wulf, *Histoire de la philosophie médiévale*, vol. 3, *Après le treizième siècle*, 6th ed. (Paris: Librairie Philosophique J. Vrin, 1947), § 423, 125–28.

35. JJ, Prologue, BNF lat. 6542, 1ra: "diligendissimus [*sic*] vir et doctor excellentissimus." Jean's prologue has been published by Tine Swaenepoel in "Les deux versions des 'Problemata' de Pierre d'Abano," *Filologia Mediolatina* 16 (2009), 243–45.

36. The date of Jean's intervention has been the subject of speculation. Maurice de Wulf, *Histoire de la philosophie médiévale*, vol. 3, § 423, 126, suggested 1315 based on what is known about Marsilius of Padua's travels; Schmugge, *Johannes von Jandun*, 27, n. 156, agreed without further support. The colophon of ms. BAV. Ottob. 1764, 226rb, reads: "Anno dominum M°CCC° 18° 6ª die augusti completa fuit scriptura huius expositionis per manum magistri Johanis de Genduno Parisius in domo scolarum de Navarra. Deo gratias." While the colophon itself may have been copied from an earlier manuscript, it is the best evidence available. At a *dubium* that Jean has appended to Pietro's treatment of IV.12 (45ra), that manuscript also contains, in the copyist's hand, the unique annotation, "Addidi." The *dubium* is edited by Coucke in "Edition of Peter of Abano, 'Exp.,' IV," as Appendix II. The scribe of JJ in Erfurt, CA f. 15 (47ra) has copied the "addidi" but the colophon mentions only PA and gives the established date for his second version, 1310; in CU Peterhouse 79, [29]ra there is a stray "a" in the margin where the word may have appeared in the copyist's model.

37. Kuksewicz, "Les 'Problemata' de Pietro d'Abano," minimizes the extent of Jean's revision; Swaenepoel, "Les deux versions des 'Problemata,'" argues that his adaptation of Pietro's text was greater than Kuksewicz allowed.

38. In an attack on the views of the Franciscan theologian Landolfo Caraccioli, Jean compared fashions in academic positions to fashions in dress. Charles Joseph Ermatinger, "John of Jandun in His Relations with Arts Masters and Theologians," in *Arts libéraux et philosophie au moyen âge: Actes du quatrième Congrès International de Philosophie Médiévale, Université de Montréal, . . . 27 août–2 septembre, 1967* (Paris: Librairie Philosophique J. Vrin, 1969), 1181, n. 30. See also Schmugge, *Johannes von Jandun*, 20, n. 118.

39. For example, de Wulf calls him the "leader" of his generation of Paris Averroists: *Histoire de la philosophie médiévale*, vol. 3, § 423, 126. See also MacClintock, *Perversity and Error*. The term connotes a general commitment to Aristotelianism and the dignity (even primacy) of philosophy, as well as adherence to certain specific opinions about such subjects as the individual soul that were contained in the interpretations of the Spanish Muslim philosopher Averroes (Ibn Rushd), whose commentaries on Aristotle were widely relied on at Western universities. The concept has fallen into disuse, as historians have found it difficult to locate significant concentrations of such beliefs in the works of individuals, much less to identify a movement. Jean did, however, insist on the autonomy of philosophy. See Zdzislaw Kuksewicz, "Jean de Jandun et sa conception de la philosophie," in *Was ist Philosophie*

im Mittelalter? . . . Akten des X. Internationalen Kongresses für mittelalterliche Philosophie der Société Internationale pour l'Étude de la Philosophie Médiévale, 25. bis 30. August 1997 in Erfurt, ed. Jan A. Aertsen and Andreas Speer, Miscellanea Medievalia, Veröffentlichungen des Thomas-Instituts der Universität zu Köln (Berlin: Walter de Gruyter, 1998), 428–34.

40. Schmugge, *Johannes von Jandun*, 23–24.

41. Only testimony from after his condemnation for other reasons characterizes his university teachings as erroneous. Schmugge, *Johannes von Jandun*, 19.

42. De Leemans, "Peter of Abano, Commentator and Translator," and Coucke, "Edition of Peter of Abano, 'Exp.,' IV."

43. With respect to the text of Book IV: Jean de Jandun's redaction is found in CU Peterhouse 79; Ghent, UB 72; Erfurt, CA f.15; BNF lat. 6542 and 15454; Sorbonne 122; Paris, Arsenal 723; Paris, Mazarine, 3520; Seville, Biblioteca Capitular y Colombina 7.7.9; BAV lat. 2176 and Ottob. 1764. Pietro d'Abano's first version is represented by Nuremberg, StB Cent.III.38; BNF lat. 6541; and BAV lat. 2174; his second version is represented by Cesena, BCM D.XXIV.2 and S.VI.2–3; BAV lat. 2175; BN Marciana 2465; BNF lat. 6540 and 6541A. References to PA are based on BNF lat. 6540 unless variants call for attention. BN Marciana 1547 (263) and BL Rawlinson C. 116 contain partial texts. On Pietro's two versions, see De Leemans' article in De Leemans, ed. *Philosophy Between Text and Tradition.*

44. Coucke, "Expositio: Edition of Peter of Abano 'Exp.' IV," lxxvii–lxxviii. Like Coucke and for similar reasons, I have used BNF lat. 6540 as a base when variants are not relevant to a specific point. I have noted a small number of instances in which my readings of various manuscripts diverge from his.

45. Johannes de Janduno, *Quaestiones super 8 libros Physicorum*: "Et scias quod liber ille *De Problematibus* communiter invenitur corruptus et incorrectus." Whether Jean's accusation was aimed at the translation itself or at the copyists, he clearly expected Aristotle's works to be "corrected." Goyens and De Leemans, "Traduire du grec au latin et du latin au français," 207.

46. Coucke details the patterns of "contamination" among the manuscripts of Pietro's *Expositio*: "Edition of Peter of Abano 'Exp.' IV," lxxvii–lxxix. There is circumstantial evidence that the medieval practice of comparing and collating manuscripts was involved. For example, the versions PA and JJ crossed paths at least twice: BNF lat. 6541A (fifteenth century, Italian) contains the first half of Pietro's version and the second half of Jean's; Erfurt, CA F 15 (fourteenth century, Italian) contains the first half of Jean's and the second half of Pietro's, suggesting that manuscripts of Pietro's and Jean's texts had been in each other's presence for collating and copying. See Kuksewicz, "Les 'Problemata' de Pietro d'Abano," 132–37; Władysław Seńko, *Repertorium commentariorum medii aevi in Aristotelem latinorum quae in bibliothecis publicis parisiis asservantur (Bibliothèque Nationale, Arsenal, Mazarine, Sorbonne, Ste Geneviève),* Opera Philosophorum Medii Aevi, Textus et Studia, vol. 5, 2 fascicules (Warsaw: Akademia Teologii Katolickiej, 1982), fasc. 1, 91; Miecislaus Markowski, *Repertorium commentariorum medii aevi in Aristotelem Latinorum quae in Bibliotheca Amploniana Erffordiae asservantur* (Wrocław: Zakład Narodowy im. Ossolińskich for the Polska Akademia Nauk Instytut Filozofi I Sociologii, 1987), 167. Evrart de Conty mentions variant readings. See Françoise Guichard-Tesson, "Le métier de traducteur et de commentateur au

XIVe siècle d'après Evrart de Conty," *Le Moyen Français* 24–25 (1990), 161. And a copyist of his work commented on discrepancies among *Problemata* texts: The Hague, KB 133A3.1, IVa.

47. The alphabetical version survives in Gray's Inn, ms. 2 (fourteenth century) and Oxford, Magdalen College, ms. 65 (fifteenth century).

48. On Burley's life and works, see Conor Martin, "Walter Burley," in *Oxford Studies Presented to Daniel Callus*, Oxford Historical Society, n.s., 16 (Oxford: Clarendon Press for the Oxford Historical Society, 1964), 194–230.

49. Jacques Chiffoleau, "*Contra naturam*: Pour une approche casuistique et procédurale de la nature médiévale," *Micrologus* 4 (1996), 276–80.

50. Katharine Park, "Albert's Influence on Late Medieval Psychology," in *Albertus Magnus and the Sciences: Commemorative Essays, 1980*, ed. James A. Weisheipl, Studies and Texts 49 (Toronto: Pontifical Institute of Mediaeval Studies, 1980), 511–16.

51. On the work, its attribution to Burley, and the extant manuscripts, see Maaike Van der Lugt, "Aristotle's 'Problems' in the West: A Contribution to the Study of the Medieval Latin Tradition," in Pieter De Leemans and Michèle Goyens, eds., *Aristotle's "Problemata" in Different Times and Tongues*, Mediaevalia Lovaniensia, ser. 1, Studia 39 (Leuven: Leuven University Press, 2006), 71–111 on 77–92, which corrects some errors in Cadden, "'Nothing Natural Is Shameful': Vestiges of a Debate About Sex and Science in a Group of Late Medieval Manuscripts," *Speculum* 76 (2001): 78–85. See also Lynn Thorndike, "Peter of Abano and Another Commentary on the 'Problems' of Aristotle," *Bulletin of the History of Medicine* 29 (1955): 517–23; Lawn, *Salernitan Questions*, 95; and James A. Weisheipl, "Repertorium Mertonlense," *Mediaeval Studies* 31 (1969): 174–224 at #31, 200. The association of almost all extant manuscripts of his abridgment with Oxford is one reason for dating it after his return to England. Only one, Bodl. Digby 206, has any links to the other side of the Channel. Since it does not contain a dedication to Richard de Bury or to the pope, which some works from his post-Paris do, it may predate that phase of his life, in which case its composition might be dated to the early 1330s.

52. WB, Prologue, Bodl. Digby 153, 102r. The reference is to *De regimine principium*, sometimes known in vernaculars as *The Mirror of Princes*. On academic, especially Aristotelian, prologues, see A. J. Minnis, *Medieval Theory of Authorship: Scholastic Literary Attitudes in the Later Middle Ages*, 2nd ed. (Philadelphia: University of Pennsylvania Press, 1988), ch. 1, esp. 28–29.

53. An edition of Evrart's *Problemes* to be published by Champion is under way. See Françoise Guichard-Tesson and Michèle Goyens, "Comment éditer l'autographe d'une traduction de traduction," *Scriptorium* 63 (2009): 173–205. Until its appearance, for biographical and bibliographical details, see the apparatus of Evrart de Conty, *Le livre des eschez amoureux moralisés*, ed. Françoise Guichard-Tesson and Bruno Roy, Bibliothèque du Moyen Français 2 (Montreal: Ceres, 1993); Guichard-Tesson, "Evrart de Conty, poète, traducteur et commentateur," in De Leemans and Goyens, eds., *Aristotle's "Problemata" in Different Times and Tongues*, 145–173, and the bibliography in that volume, 310–15; and Caroline Boucher with Laurent Brun, "Evrart de Conty," *Archives de Littérature du Moyen Âge*, last modified March 25, 2011, accessed January 9, 2013, http://www.arlima.net/eh/evrart_de_conty.html#pro (http://www.arlima.net/no/58).

54. Guichard-Tesson and Roy, eds., *Le livre des eschez amoureux moralisés*, liii–lvi; Danielle Jacquart, *La médecine médiévale dans le cadre parisien, XIe–XVe siècle*, Penser la Médecine ([N.p.]: Fayard, 1998), 153–60, esp. 160.

55. See, for example, Léopold Delisle, *Recherches sur la Librairie de Charles V*, 2 vols. and atlas of plates (Paris: H. Champion, 1907), and Claire Richter Sherman, *Imaging Aristotle: Verbal and Visual Representation in Fourteenth-Century France* (Berkeley: University of California Press, 1995). Christine de Pizan celebrated Charles's patronage of learning in *Le livre des fais et bonnes meurs du sage roy Charles V*, ed. S. Solente, Société de l'Histoire de France, Série antérieur à 1789, 2 vols. (Paris: Librairie Ancienne Honoré Champion, 1936–40).

56. On Evrart's *Problemes*, see Guichard-Tesson, "Le métier de traducteur et de commentateur au XIVe siècle d'après Evrart de Conty" and "Le souci de la langue et du style au XIVe siècle: L'autographe des 'Problèmes' d'Evrart de Conty," *Le moyen français* 33 (1993): 57–84; Michèle Goyens, "Évrart de Conty: Traducteur, adapteur et commentateur des 'Problèmes' d'Aristote," in *"Pour acquerir honneur et pris": Mélanges de moyen français offerts à Guiseppi Di Stefano*, ed. Maria Colombo Timelli and Claudio Galderisi (Montreal: CERES, 2004), 123–35; and Michèle Goyens and Pieter De Leemans, "'Et samble qu'il woeille dire . . .': Evrart de Conty traducteur de Pierre d'Abano," in *The Medieval Translator: Traduire au Moyen Âge*, ed. Jacqueline Jenkins and Olivier Bertrand, The Medieval Translator: Traduire au Moyen Âge 10 (Turnhout: Brepols, 2007), 285–302. Evrart may have had a manuscript containing both Bartholomeus's text and Pietro's commentary, but there were also many independent copies of the Aristotelian text available in Paris, and he refers to textual variants (Guichard-Tesson, "Le métier de traducteur et de commentateur," 161). Since he speaks of "expositeurs" in the plural (EC, BNF fr. 24281, 116v), he may have read or heard the expositions by additional scholars. His reference (Prologue, 1r) to Virgil's *Georgics* 2.490 suggests a direct or indirect relationship to Burley, whose epitome opens with that line of verse.

57. Michèle Goyens, "Le développement du lexique scientifique français et la traduction des 'Problèmes' d'Aristote par Evrart de Conty (c. 1380)," *Thélème: Revista Complutense de Estudios Franceses*, special issue (2003): 189–207; and Joëlle Ducos, "Traduction et lexique scientifique: Le cas des *Problèmes* d'Aristote traduits par Evrart de Conty," in *Traduction et adaptation en France à la fin du Moyen Âge et à la Renaissance: Actes du colloque organisé par l'Université de Nancy II, 23–25 mars 1995*, ed. Charles Brucker (Paris: Honoré Champion, 1997), 237–47. On French scientific translations, see Serge Lusignan, *"Parler vulgairement": Les intellectuels et la langue française aux XIIIe et XIVe siècles* (Paris: J. Vrin 1986), and, with reference to Charles V's court, Lusignan, "Nicole Oresme traducteur et la pensée de la langue française savante," in P. Souffrin and A. Ph. Segonds, eds., *Nicole Oresme: Tradition et innovation chez un intellectuel du XIVe siècle*, Science et Humanisme (Paris: Les Belles Lettres, 1988), 93–104.

58. Guichard-Tesson, "Le métier de traducteur et de commentateur," 138–39, 146; Jacquart, *La médecine médiévale dans le cadre parisien*, 281–82.

59. Guichard-Tesson, "Le métier de traducteur et de commentateur," 144–46; Caroline Boucher, "Des problèmes pour exercer l'entendement des lecteurs: Evrart de Conty,

Nicole Oresme et la recherche de la nouveauté," in De Leemans and Goyens, *Aristotle's "Problemata" in Different Times and Tongues*, 174–97.

60. Evrart de Conty, *Le livre des eschez amoureux moralisés.*

61. At least one scribe who made a copy of Bartholomeus's translation regarded the whole work as "useful for many things." BM, BAV lat. 2112, colophon, 34vb: "Explicit iste liber utilis ad multa." I am grateful to Pieter De Leemans for calling this to my attention.

62. EC, *Problemes*, Prologue, BNF fr. 24281, Prologue, 1r: "Li princes des phylosophes Aristotes voelt estraire et eslire de pluseurs materes et de diverses sciences comme esmerveillables et delitables entre les questions qui se pocent [Hague, Koninklijke Bibliotheek 133A3: peuent] faire des choses que on voit en nature. . . . Et de ce s'ensieut il que cils livres qui est ainsi inquisitis des causes de pluseurs merveilles entre les autres est appetables et amables pource que li entendemens humains desire naturalment a savoir et a cognoistre les causes des merveilles de nature. Et en ce se delite souverainement et a la verite ceste congnissance des causes fait moult a la perfection et a la felicite humaine." See Aristotle, *Metaphysics*, I.1; Virgil, *Georgics* (2.490).

63. EC, IV.1, BNF fr. 24281, 102r: "Pource qu'il y ha en ceste matere moult de doubtes et moult de grans merveilles."

64. Now BNF fr. 24281 and 24282, the two-volume autograph was in the huge library of the monastery of St. Victor before 1443, where it was classified with works of philosophy and logic. Gilbert Ouy, *Les manuscrits de l'abbaye de Saint-Victor: Catalogue établi sur la base du répertoire de Claude de Grandrue (1514)*, 2 vols., Bibliotheca Victorina, 10 ([Turnhout]: Brepols, 1999), vol. 1, *Introduction, Concordances, Index*, 15–17, 54; vol. 2, *Texte*, 603–4.

65. Soon after its completion, professionally prepared, beautifully decorated copies on parchment or vellum are documented as having been in the libraries of Charles VI of France and other members of the royal family, and by the end of the fifteenth century two copies were also in the possession of ruling families in the Burgundian cultural orbit. Not all the entries in medieval catalogues can be identified with extant copies. We know that Louis, duc d'Orléans, received a copy in 1398; Jean, duc de Berry, received one in 1405; and Charles VI received one from the duc de Guienne in 1410. See Le Roux de Lincy, *La bibliothèque de Charles d'Orléans à son château de Blois en 1427*, Extrait de la Bibliothèque de l'École des Chartes 5 (Paris: Firmin Didot Frères, 1843), p. 44, #39; and Léopold Delisle, *Le cabinet des manuscrits de la Bibliothèque Nationale: Étude sur la formation de ce dépot comprenant les éléments d'une histoire de la calligraphie, de la miniature, de la reliure, et du commerce des livres à Paris avant l'invention de l'imprimerie*, Histoire Générale de Paris (Paris: Imprimerie Impériale and Imprimerie Nationale, 1868–81; repr. Amsterdam: Gérard Th. van Heusden, 1969), vol. 3 (1881), p. 137, #472; p. 183, #153. A manuscript from around 1400 (Jena, Gall. f. 18), the planned illuminations of which were not completed, bears the coat of arms of the dukes of Cleves: Walther Dexel, *Untersuchungen über die französischen illuminierten Handschriften der Jenaer Universitätsbibliothek vom Ende des 14. bis zur Mitte des 15. Jahrhunderts*, Zur Kunstgeschichte des Ausland, 115 (Strasbourg: J. H. Heitz [Heitz und Mündel], 1917), 19–23 and plate 4. The Hague, KB 133, produced in the southern Netherlands in the second half of the fifteenth century, was later in the library of the count of Nassau: Anne S. Korteweg, *Boeken van Oranje-Nassau: De bibliotheek van de graven van Nassau*

en prinsen van Oranje in de vijftiende en zestiende eeuw, Exhibition, Museum Meermanno Westreenianum (The Hague: Koninklijke Bibliotheek, Den Haag, 1998), pl. 28. I am grateful to Dr. Korteweg and to Professor Gregory Clark for information about this manuscript.

66. Aleksander Birkenmajer, Georges Lacombe, and Lorenzo Minio-Paluello, *Aristoteles Latinus: Codices*, 3 vols.: vol. 1, revised edition, Union Académique Internationale, Corpus Philosophorum Medii Aevi (Bruges: Desclée de Brouwer, 1957); vol. 2 (Cambridge: Academia, 1955); [vol. 3] *Supplementa altera* (Bruges: Desclée de Brouwer, 1961).

67. EC, BNF fr. 24281, 116v: "I'en toucherai un petit ci apres en general et assez legierement. Et qui le volra veir plus a plain si lise le latin et le texte de Aristote et des expositeurs."

68. Ownership, for example, by Petrus Roselli and the monastery of St. Victor, provides corroborating evidence on the Paris scene, as does documentation of a Flemish notary lecturing on the work to about sixty Paris students (Jacquart, *La médecine médiévale dans le cadre parisien*, 175–76). In addition, references to the *Problemata* are found in various works of natural philosophy and medicine.

69. For a full account of the texts and manuscripts, see Van der Lugt, "Aristotle's *Problems* in the West"; also Cadden, "Preliminary Observations on the Place of the *Problemata*."

70. Brugge, Openbare Biliotheek 481, 1ra–104ra; The Hague, KB 70.E.9, 1ra–82rb; Naples, Biblioteca Nazionale VII.AA.39, 47vb–116ra, all copies of "Flemish commentary"; BNF lat. 14728, 277r–313va, "St. Victor commentary"; BAV lat. 901, 122v–35v, "Vatican 901 commentary."

71. "Bavarian," Ghent, UB 178, 91ra–157vb; CLM 12021, 118ra–79rb; Krakow, BJ 2095, 227ra–44bis r.; "Aachen," Erfurt, CA F. 263, 16ra–29va; 35vb–39vb; 40ra–55va; "Erfurt," CLM 4710, 262ra–293rb; Krakow, BJ 654, 1r–25vb; Erfurt, CA Q16, 1ra–40vb; Leipzig, UB 1435, 125ra–53vb; "Vatican," BAV lat. 2481, 89ra–96rb.

72. The "Vatican," "St. Victor," "Bavarian," and "Aachen" texts are clearly derived from Jean de Jandun's edition of Pietro's commentary; the "Flemish" and "Erfurt" texts are derived from some version of Pietro's work.

73. Prague, NK VIII.A.19, 79rb: "Propter quid aliqui sunt viri qui volunt tantum fricari et quidam tantum fricare et quidam volunt fricari et etiam fricare in coytu." A second copy of this work is found in Prague, NK I.C.25.

74. BM, Escorial, f.I.11, 7ra.

75. Aristotle, *The "Problemata physica" Attributed to Aristotle: The Arabic Version of Hunain ibn Ishaq and the Hebrew Version of Moses ibn Tibbon*, ed. Lou S. Filius, Aristoteles Semitico-Latinus 11 (Leiden: Brill, 1999).

76. Avicenna, *Liber canonis*. Nancy G. Siraisi, *Avicenna in Renaissance Italy: The "Canon" and Medical Teaching in Italian Universities After 1500* (Princeton: Princeton University Press, 1987), 43–76; Danielle Jacquart, "La réception du 'Canon' d'Avicenne: Comparaison entre Montpellier et Paris aux XIIIe et XIVe siècles," in *Histoire de l'École médicale de Montpellier: Actes du 110e Congrès national des sociétés savantes (Montpellier, 1985): Section d'histoire des sciences et des techniques* (Paris: Comité des Travaux Historiques et Scientifiques, 1985), vol. 2, 69–77.

77. For example, the bibliophile-physician Giovanni di Marco owned Books I, II, and IV of the *Canon*: Anna Manfron, *La biblioteca di un medico del quattrocento: I codici di Giovanni di Marco da Rimini nella Biblioteca Malatestiana*, Archivi di Bibliofilia (Turin: Umberto Allemandi for the Instituzione Biblioteca Malatestiana, Cesena, 1998), D.XXV.3, D.XXV.5, 202–4. The rare commentary of Jacques Despars on the relevant chapter of Book III is discussed in my Chapter 2.

78. Avicenna, *Liber canonis medicine*, trans. Gerard of Cremona (Venice: Paganinis, 1507; reprinted Hildesheim: Georg Olms, 1964), bk. 3, fen 20, tr. 1, ch. 11, 352vb–53ra, and ch. 42, 358ra. Translations of part of the former and all of the latter from the twelfth-century Latin version are provided by Faith Wallis in Kenneth Borris, ed., *Same-Sex Desire in the English Renaissance: A Sourcebook of Texts, 1470–1650* (New York: Routledge, 2004), 128–30. See Mark D. Jordan, *The Invention of Sodomy in Christian Theology* (Chicago: University of Chicago Press, 1997), 115–23.

79. For a helpful list of Arabic terms relating to masculinity and male sexuality (including not *ubna* but *ḥulāq*) see Everett K. Rowson, "Gender Irregularity as Entertainment: Institutionalized Transvestism at the Caliphal Court in Medieval Baghdad," in *Gender and Difference in the Middle Ages*, ed. Sharon Farmer and Carol Braun Pasternack, Medieval Cultures 32 (Minneapolis: University of Minnesota Press, 2003), 66.

80. JJ, BNF lat. 15454, margin, 54rb: "De alubnati secundum Avicennam et est actus sodomiticus."

81. Both the Latin and the Arabic are a little unclear, and it is possible Avicenna is indicating that one of the two people is the man himself. I am grateful to Everett Rowson for providing me with a translation from the Arabic. A translation is also contained in Bassem Nathan, "Medieval Arabic Medical Views on Male Homosexuality," *Journal of Homosexuality* 26 (1994): 38–39.

82. Avicenna, *Liber canonis medicine*, bk. 3, fen 20, tr. 1, ch. 42, 358ra: "Et sunt in veritate prostrate anime et maligne nature et male consuetudinis et complexionis muliebris. . . . Et stulti homines sunt qui volunt eos curare. Nam initium egritudinis eorum meditativum est non naturale. Si vero confert cura eius tunc est illud quod frangit desiderium eorum ex tristicia et fame et vigiliis et carcere et percussione."

83. On lovesickness, Hans Hinrich Biesterfeldt and Dimitri Gutas, "The Malady of Love," in *Studies in Islam and the Ancient Near East Dedicated to Franz Rosenthal*, ed. Jeanette Wakin, *Journal of the American Oriental Society* 104 (1984), 23. On sexuality and homosexuality in Muslim societies more generally: Norman Roth, "A Research Note on Sexuality and Muslim Society," in Vern L. Bullough and James A. Brundage, eds., *Handbook of Medieval Sexuality*, Reference Library of the Humanities 1696 (New York: Garland, 1996), 319–27; Sabine Schmidtke, "Homoeroticism and Homosexuality in Islam: A Review Article," *Bulletin of the School of Oriental and African Studies* 62 (1999): 260–66; Basim F. Musallam, *Sex and Society in Islam: Birth Control Before the Nineteenth Century*, Studies in Islamic Civilization (Cambridge: Cambridge University Press, 1983); Franz Rosenthal, "ar-Rāzī on the Hidden Illness," *Bulletin of the History of Medicine* 52 (1978): 45–60, reprinted in *Science and Medicine in Islam: A Collection of Essays*, Collected Studies 330 (Aldershot, U.K.: Variorum, 1990), item 9; Khaled El-Rouayheb,

Before Homosexuality in the Arab-Islamic World, 1500–1800 (Chicago: University of Chicago Press, 2005).

84. The Latin ("quare spargitur super ipsum hominem") is unclear, in part because "spargere" means "to spray" or "to spread" and in part because it is in the passive or impersonal here ("it is lain/sprayed upon"). In any case, the Latin translation is faulty. In the Arabic, according to Rowson (see above, note 81), violent friction activates the sensation and enables the man to engage in the active role in sex.

85. Gentilis Fulginatis, *Expositiones cum textu Avicenne* (Venice: Baptista de Tortis, 1492; repr. film, *Italian Books Before 1601*, Cambridge, Mass: General Microfilm, n.d., roll 307), bk. 3, fen 20, ch. 42, [ccc5]ra–b. On Gentile: Roger K. French, *Canonical Medicine: Gentile da Foligno and Scholasticism* (Leiden: Brill, 2001).

86. For summaries of some other treatments in Arabic medical works, including a chapter on women by as-Samau'al ibn Yahyâ, see Danielle Jacquart and Claude Thomasset, *Sexuality and Medicine in the Middle Ages*, trans. Matthew Adamson (Princeton: Princeton University Press, 1988), 122–26.

87. In the case of Aristotle's works they typically had an older Latin translation made from an Arabic translation and a newer one made from the original Greek. Such a comparison was not possible in the case of the *Problemata*, since the Arabic version was never translated.

88. For a few details about Suriano, see Joan Cadden, "On a Daughter's Death, Venice, 1477," *Medieval Feminist Newsletter* 24 (1997): 15–17.

89. BM; George of Trebizond's and Theodore Gaza's translations of Aristotle's *Problemata* [hereafter referred to as GT and TG, respectively; all references to IV.26 unless otherwise indicated], BL add. 21978, 2ra–110ra. Suriano's substantive interest is attested by his elaborate coded system of annotation.

90. John Monfasani, "George of Trebizond's Critique of Theodore Gaza's Translation of the Aristotelian 'Problemata,'" in De Leemans and Goyens, *Aristotle's "Problemata" in Different Times and Tongues*, 275. See also Monfasani, "The Pseudo-Aristotelian 'Problemata' and Aristotle's 'De animalibus' in the Renaissance," in Grafton and Siraisi, eds., *Natural Particulars*, 205–47. Gaza's translation was printed a number of times: in 1473, 1501, 1562, and so on. The last of these is available in reprint in *Extra ordinem naturalium libri quibus nonnulli etiam additi sunt Aristoteli ascripti. Alexandri Problematum libri duo*, vol. 7 (1562) of *Aristotelis opera cum Averrois commentariis* (Venice: Junctas, 1562–74; repr. Frankfurt am Main: Minerva, 1962), 1ra–98rb.

91. George of Trebizond, "Scholia to the Aristotelian 'Problemata,'" in Monfasani, ed., *Collectanea Trapezuntiana*, § 157, 640–66.

92. On Gaza, see John Monfasani, "Theodore Gaza as a Philosopher: A Preliminary Survey," in *Manuele Crisolora e il ritorno del greco in occidente: Atti del Convegno Internazionale (Napoli, 26–29 giugno 1997)*, ed. Riccardo Maisano and Antonio Rollo (Naples: Istituto Universitario Orientale, Dipartimento di Studi dell'Europa Orientale and Dipartimento di Studi del Mondo Classico e del Mediterraneo Antico, 2002), 269–81.

93. BM, BNF lat. 15081, 12va: "Propter quid quidam cum quibus coitur gaudent"; GT, BL add. 21978, 18va: "Quamobrem nonnulli venerem patientes delectant"; TG in

Extra ordinem naturalium libri quibus nonnulli etiam additi sunt Aristoteli ascripti, IV."27," 18vb: "Cur nonnulli rem veneream cum voluptate patiantur." In the Greek: "Διὰ τί ἔνιοι ἀφροδισιαζόμενοι χαίρουσι, καὶ οἱ μὲν ἅμα δρῶντες, οἱ δ'οὔ."

94. *Compilatio phisonomie* (Padua: Petrus Maufer, 1474), reproduced in Manuscripta: Microfilms of Rare and Out-of-Print Books, list 14 (Saint Louis: Pius XII Memorial Library, Saint Louis University, n.d.). On the nature of that work, see Eugenia Paschetto, "La fisiognomica nell'enciclopedia delle scienzie di Pietro d'Abano," *Medioevo* 11 (1985): 97–111.

95. On "cycles of salience," see Valerie Traub, "The Present Future of Lesbian Historiography," in *A Companion to Lesbian, Gay, Bisexual, Transgender, and Queer Studies*, ed. George Haggerty and Molly McGarry, Companions to Cultural Studies (Oxford: Blackwell, 2007), 124–45.

96. In addition to the present study, see Faith Wallis, "Guilio Guastavini's Commentary on Pseudo-Aristotle's Account of Male Same-Sex Coitus, *Problemata* 4.26," in *The Sciences of Homosexuality in Early Modern Europe*, ed. Kenneth Borris and George Rousseau (London: Routledge, 2008), 57–73.

97. Jordan, *The Invention of Sodomy*. See also William Burgwinkle, *Sodomy, Masculinity, and Law in Medieval Literature: France and England, 1050–1230*, Cambridge Studies in Medieval Literature (Cambridge: Cambridge University Press, 2004).

98. John Boswell, *Same-Sex Unions in Premodern Europe* (New York: Villard, 1994), 10.

99. Adrienne Rich, "Compulsory Heterosexuality and Lesbian Existence," *Signs* 5 (1980): 631 60; Michel Foucault, *The History of Sexuality*, tr. Robert Hurley, 3 vols. (New York: Vintage Books, 1988–90).

100. PA, BNF lat. 6540, 65ra.

101. See the account of the literature and debates, as well as a peace proposal, in Traub, "The Present Future of Lesbian Historiography," to which may be added the critique of gay pride's influence in David M. Halperin and Valerie Traub, eds. *Gay Shame* (Chicago: University of Chicago Press, 2009). For a compact review of the literature on medieval sexuality more generally, see Ruth Mazo Karras, *Sexuality in Medieval Europe: Doing unto Others* (New York: Routledge, 2005), 160–81.

102. Traub, "The Present Future of Lesbian Historiography," 133–34. Her list includes twenty-six general themes plus ten more specifically relevant to women, and she invites the addition of others. The proposal could be applied as a practical approach to the framework articulated in Judith M. Bennett's "'Lesbian-Like' and the Social History of Lesbianisms," *Journal of the History of Sexuality* 9 (2000): 1–24.

103. The science of physiognomy offers some relevant material. Joseph Ziegler, "Sexuality and the Sexual Organs in Latin Physiognomy 1200–1500," in *Sexuality and Culture in Medieval and Renaissance Europe*, ed. Philip M. Soergel, Studies in Medieval and Renaissance History, 3rd ser., 2 (New York: AMS Press, 2005), 86 and 103, n. 11.

104. Avicenna, *Liber canonis medicine*, bk. 3, fen 21, tr. 4, ch. 22, 377va: "De carne addita. . . . Et quandoque advenit ei ut faciat cum mulieribus simile quod fit eis cum quibus coitur." Cf. Fedwa Malti-Douglas, "Tribadism/Lesbianism and the Sexualized Body in Medieval Arabo-Islamic Narratives," in *Same Sex Love and Desire Among Women in the*

Middle Ages, ed. Francesca Canadé Sautman and Pamela Sheingorn (Basingstoke, U.K.: Palgrave, 2001), 123–41, based on a source in which "medical lore mixes with anecdotes and poetry" (127).

105. See Helen Rodnite Lemay, "William of Saliceto on Human Sexuality," *Viator* 12 (1981): 165–81 on 179.

106. The occasional pro forma inclusion of women in the context of categories derived from discourse on men is seldom illuminating. See, e.g., George of Trebizond, *Claudii Ptolemaei Centiloquium sive Aphorismi a Georgio Trapezuntio ex graeco in latinum versi et commentariis illustrati*, in Guido Bonatus, *De astronomia tractatus X, Universum quod iudiciariam rationem nativitatum, aëris, tempestatum, attinet, comprehendentes. Adiectus est Cl. Ptolemaei Liber fructus cum comentariis utilissimus Georgii Trapezuntii* (Basel: Nicolaus Pruknerus, 1550), aphorism 80, col. 49: "Hi vero sunt qui ad podicem exeunt, et sic natura non culpa effoeminantur, masculi et foeminae depravantur, quae res rarissime, sicut omnia monstra, solet accidere."

107. Louis Compton, "The Myth of Lesbian Impunity," in *The Gay Past: A Collection of Historical Essays*, ed. Salvatore J. Licata and Robert P. Petersen (New York: Harrington Park Press, 1985), 11–25; Carol Lansing, "Donna con Donna? A 1295 Inquest into Female Sodomy," in *Sexuality and Culture in Medieval and Renaissance Europe*, ed. Philip M. Soergel, Studies in Medieval and Renaissance History, 3rd ser., 2 (New York: AMS Press, 2005), 109–22. See also Helmut Puff, "Female Sexuality: The Trial of Katherina Hetzeldorfer (1477)," *Journal of Medieval and Early Modern Studies* 30 (2000): 41–61, and Joan Cadden, *Meanings of Sex Difference in the Middle Ages: Medicine, Science, and Culture*, History of Medicine (Cambridge: Cambridge University Press, 1993), 223–24.

108. Jacquart and Thomasset, *Sexuality and Medicine in the Middle Ages*, 45–47.

109. Katharine Park, "The Rediscovery of the Clitoris: French Medicine and the Tribade, 1570–1620," in *The Body in Parts: Fantasies of Corporeality in Early Modern Europe*, ed. David Hillman and Carla Mazzio (New York: Routledge, 1997), 170–93; and Valerie Traub, "The Psychomorphology of the Clitoris, or, The Reemergence of the Tribade in English Culture," in *Generation and Degeneration: Tropes of Reproduction in Literature and History from Antiquity Through Early Modern Europe*, ed. Valeria Finucci and Kevin Brownlee (Durham, NC: Duke University Press, 2001), 153–86.

110. Cadden, *Meanings of Sex Difference*, 201–18; Lorraine Daston and Katharine Park, "The Hermaphrodite and the Orders of Nature: Sexual Ambiguity in Early Modern France," in *Premodern Sexualities in Europe*, ed. Louise Fradenburg and Carla Freccero, *GLQ: A Journal of Lesbian and Gay Studies* 1 (1995): 419–37; repr. in *Premodern Sexualities in Europe*, ed. Louise Fradenburg and Carla Freccero with Kathy Lavezzo (New York: Routledge, 1996), 117–36. The hermaphrodite was a topic of discussion among certain theologians and canonists: Maaike Van der Lugt, "L'humanité des monstres et leur accès aux sacrements dans la pensée médiévale," HAL: halshs-00175497, version 1, consulted at http://halshs.archives-ouvertes.fr./ December 15, 2011, pp. 17–21 (also published, without references in Latin, in *Monstre et imaginaire social: Approches historiques*, ed. A. Caiozzo and A.-E. Demartini [Paris: Créaphis, 2008], 135–61).

111. A few of the many studies, theoretical interventions, and resources: Compton,

"The Myth of Lesbian Impunity"; Judith Brown, "Lesbian Sexuality in Medieval and Early Modern Europe," in *Hidden from History: Reclaiming the Gay and Lesbian Past*, ed. Martin Bauml Duberman, Martha Vicinus, and George Chauncey Jr. (New York: New American Library, 1989), 67–75; Jacqueline Murray, "Twice Marginal and Twice Invisible: Lesbians in the Middle Ages," in *Handbook of Medieval Sexuality*, ed. Bullough and Brundage, 191–222; Karma Lochrie, *Covert Operations: The Medieval Uses of Secrecy*, The Middle Ages (Philadelphia: University of Pennsylvania Press, 1999), "Presumptive Sodomy and Its Silences," *Textual Practice* 13 (1999): 295–310, and *Heterosyncrasies: Female Sexuality When Normal Wasn't* (Minneapolis: University of Minnesota Press, 2005); Bennett, "'Lesbian-Like' and the Social History of Lesbianisms"; Traub, "The Present Future of Lesbian Historiography."

112. Specifically historical resources and treatments include: Jacquart and Thomasset, *Sexuality and Medicine in the Middle Ages*; Joyce E. Salisbury, *Medieval Sexuality: A Research Guide*, Medieval Bibliographies 5 (New York: Garland, 1990); Bullough and Brundage, eds., *Handbook of Medieval Sexuality*; James A. Brundage, *Law, Sex, and Christian Society in Medieval Europe* (Chicago: University of Chicago Press, 1987); Bernd-Ulrich Hergemöller, *Sodom und Gomorrha: Zur Altagswirklichkeit und Verfolgung Homosexueller im Mittelalter* (Hamburg: MännerschwarmSkript, 1998); Ruth Mazo Karras, *Sexuality in Medieval Europe*, including "Further Reading"; Puff, *Sodomy in Reformation Germany and Switzerland.*

113. Robert Mills, *Pain, Pleasure and Punishment in Medieval Culture* (London: Reaktion, 2005), 83–105.

114. For example, Ruth Mazo Karras and David Lorenzo Boyd, "The Interrogation of a Male Transvestite Prostitute in Fourteenth-Century London," *GLQ: A Journal of Lesbian and Gay Studies* 1 (1995): 459–65, and "'Ut cum muliere': A Male Transvestite Prostitute in Fourteenth-Century London," in Fradenburg and Freccero, *Premodern Sexualities in Europe*, 101–16; Helmut Puff, "Localizing Sodomy: The 'Priest and Sodomite' in Pre-Reformation Germany and Switzerland," *Journal of the History of Sexuality* 8 (1997): 165–95; Michael Goodich, ed., *Other Middle Ages: Witnesses at the Margins of Medieval Society*, The Middle Ages (Philadelphia: University of Pennsylvania Press, 1998), 104–49.

115. On Poggio, see below, Chapter 2

116. In Coucke's edition ("Edition of Peter of Abano, 'Exp.' IV"), the text of the Bartholomeus translation is on pp. 136–37, that of Pietro's exposition (with lemmas in italics) on pp. 138–56, and a paraphrase of the *problema* on pp. 331–36.

117. BM, BNF lat. 15081, 12va: "Propter quid quidam cum quibus coitur gaudent. Et hii quidem simul agentes, hii autem non." Here and at subsequent lemmas, I have provided a text based on the manuscript cited here with supplementary readings from BNF lat. 6540 and BAV Chisiani G.V.131.

118. BM in BN Marciana 2488, margin at "quidam cum quibus," 204rb: "scilicet passive ad modum femellarum"; at "quidam agentes," "ut mares."

119. "Prague" *Problemata*, Prague, NK VIII.A.19, 79rb.

120. JJ, BAV Ottob. 1764, 58va–b: "illi cum quibus actus coytus peragitur scilicet vir et mulier aut eis proportionati"; likewise "Vatican" *Problemata*, BAV lat. 2481, 93rb.

121. PA, BNF lat. 6540, 63rb: "Hoc est, quare gaudent coeuntes sive se habeant in ratione agentis sive patientis."

122. CU Peterhouse 220 in WB, [57]r.

123. Pietro often announced such omissions and sent readers to the earlier treatment. For example, at IV.16 on the beneficial effects of coitus for phlegmatic diseases, he refers the reader to his discussion of the same topic at *problema* I.50.

124. This gloss on "cum quibus" is clear in the manuscripts of Pietro's version A and Jean's edition; Pietro's version B is less consistent. Cf. Coucke, "Edition of Peter of Abano, 'Exp.,' IV," 138 and apparatus at line 44.

125. WB, *Problemata* IV.15. On Burley's treatment, see Cadden, " 'Nothing Natural Is Shameful,' " 81.

126. PA, BNF lat., 6541, 73ra; BAV lat. 2174, 30ra, and Nuremberg, StB III.38, 31va.

127. JJ, BAV Ottob. 1764, 58vb: "quidam vero alterum solum exercent ut apparebit circa finem sermonis." BNF lat. 6541 and BAV lat. 2175 (PA) have "seorsum," as does BNF lat. 6542, the corrected version of Jean's redaction; all the other manuscripts of the latter read "solum." Nuremberg StB Cent.III.38 (PA) is ambiguous. Pietro also mentions masturbation "with the hand" as an alternative to intercourse for adolescent girls, but this occurs less than halfway through the text and is therefore not what this phrase refers to.

CHAPTER 1. MOVED BY NATURE

1. George of Trebizond, *Claudii Ptolemaei Centiloquium sive Aphorismi*, aphorism 80, col. 49. "Si ergo Saturnus coniungitur et dignitatem habet in domo septima, transfert propter Saturnum qui sordibus praeest coitum nati ad immunda, facietque ut contra universalem naturam monstrosa quadam et particulari natura quasi foemina patiatur." He completed the translation and commentary in the mid-1450s: John Monfasani, *George of Trebizond: A Biography and a Study of His Rhetoric and Logic*, Columbia Studies in the Classical Tradition 1 (Leiden: Brill, 1976), 118–19.

2. George of Trebizond, *Claudii Ptolemaei Centiloquium*, aphorism 80, col. 49: "Necesse est ut quod a natura ad genitales partes sperma depellitur cum nullos ibi canales, nullos transitus inveniat, ad propinquos meatus derivetur."

3. Table to PA, Göttweig, Benediktanerstiftsbibliothek 138, s.v. "S," 238v: "Sodomiticum peccatum quidam faciunt natura moti." This is the table's only reference to IV.26. The table is distinct from those in BNF lat. 6543 and BN Marciana 2672.

4. See Bert Hansen, *Nicole Oresme and the Marvels of Nature: A Study of His "De causis mirabilium" with Critical Edition, Translation, and Commentary*, Studies and Texts 68 (Toronto: Pontifical Institute of Mediaeval Studies, 1985), and Lorraine Daston and Katharine Park, *Wonders and the Order of Nature, 1150–1750* (New York: Zone, 1998).

5. The theme is present in Alain de Lille's influential "De planctu naturae," ed. Nikolaus M. Häring, *Studi medievali*, ser. 3, 19/2 (1978): 797–879. For both Latin and vernacular examples of unregimented nature, see Joan Cadden, "Trouble in the Earthly Paradise: The Regime of Nature in Late Medieval Christian Culture," in *The Moral Authority of Nature*, ed. Lorraine Daston and Fernando Vidal (Chicago: University of Chicago Press, 2004), 207–31; Katharine Park, "Nature in Person: Medieval and Renaissance Allegories

and Emblems," in Daston and Vidal, *Moral Authority of Nature*, 50–73; George Economou, *The Goddess Natura in Medieval Literature* (Cambridge, MA.: Harvard University Press, 1972); Albert Zimmermann and Andreas Speer, eds., *Mensch und Natur im Mittelalter*, 2 vols., Miscellanea Medievalia, Veröffentichungen des Thomas-Instituts der Universität zu Köln, 21/1–2 (Berlin: Walter de Gruyter, 1991); and Gary P. Cestaro, "Queering Nature, Queering Gender: Dante and Sodomy," in *Dante for the New Millennium*, ed. Teodolinda Barolini and H. Wayne Storey, Fordham Series in Medieval Studies 2 (New York: Fordham University Press, 2003), 90–103.

6. Johannes de Janduno, *Quaestiones super 8 libros Physicorum Aristotelis*, bk. II, q. 14, 39ra: "An monstra, quae peccata naturae dicuntur, ab ipsa natura intenta sint," on Aristotle's *Physics*: *Physica: Translatio vetus*, ed. Fernand Bossier and Jozef Brams, 2 fascicules, fasc. 2, Aristoteles Latinus 7.1, Corpus Philosophorum Medii Aevi Academiarum Consociatarum Auspiciis et Concilio Editum, Union Académique Internationale (Leiden: Brill, 1990), II.8, 199b3–5, fasc. 2, 88.

7. Walter Burley, *In Physicam Aristotelis expositio et quaestiones* (Venice: Simone de Luere, Andree Torresani de Asula, 1501; repr. Hildesheim: Georg Olms, 1972), 57vb: "Accipiendo monstrum pro individuo substantie cui accidit dispositio disconveniens sic natura simpliciter intendit monstrum. Sed si monstrum accipiatur pro toto aggregato sic dico quod natura secundum quod intendit tale monstrum et illud quod sic intendit a natura fit a casu." See 57va for his discussion of *natura simpliciter* and *secundum quod*. See also Johannes de Janduno, *Quaestiones super 8 libros Physicorum*, bk. II. q. 14, 39ra. Their interpretation is not entirely typical of *Physics* commentators.

8. The phrase "against nature [*contra naturam*]," common in other medieval contexts appears only sporadically in texts related to the *Problemata*. On the history of concept, see Jacques Chiffoleau, "'Contra naturam': Pour une approche casuistique et procédurale de la nature médiévale," *Micrologus* 4 (1996): 265–312; on its less consistent history in premodern legal settings, see Helmut Puff, "Nature on Trial: Acts 'Against Nature' in the Law Courts of Early Modern Germany and Switzerland," in Daston and Vidal, *The Moral Authority of Nature*, 232–53. George of Trebizond completed his translation of the *Centiloquium* in 1454: Monfasani, *Collectanea Trapezuntiana*, 750–51.

9. BM, BNF lat. 15081, 12va: "¶Aut quia est unicuique superfluitati locus in quem apta nata est secundum naturam segregari et labore facto ventositas exiens inflare facit et segregare per ipsam. ¶Ut urina quidem in vesicam, superfluitas cibi in intestina, lacrima autem in oculum, musillagiens [BNF lat. 6540: mucillagines] autem in nasum, sanguis autem in venas."

10. BNF lat. 6541A reads "inflamationem" twice; Cesena, BCM D.XXIV.2 and BAV Ottob. 1764 (JJ), "inflationem." Avicenna, *Liber canonis medicine*, bk. 1, fen 1, doctr. 6, ch. 3, 24ra, in a discussion of the natural faculties.

11. Given the prestige of Avicenna, Pietro's use of a transliterated Arabic term, *alchasim*, in this section on the processing of superfluities may have added the weight of authority for some readers, though perhaps not for others. Copyists tripped over it, and Jean de Jandun dropped it. Variants include *albasius* and *chaiasim*. Among the manuscripts of Jean's redaction, only BN 6542 (JJ with apparent corrections based on PA) includes it.

12. "Erfurt" *Problemata*, IV.26, CLM 4710, [268]rb.

13. PA, BNF lat. 6540, 63va: "humorum naturalissimus." JJ and some PA manuscripts (Cesena, BCM D.XXIV.2 and S.VI.2, and BNF lat. 6541A) lack the sentence in which "naturalissimus" occurs.

14. PA, BNF lat. 6540, 63va, "intelligi sanguis superfluus melancholicus multis superfluitatibus aliorum humorum admixtus non autem naturalis."

15. BM, BNF lat. 15081, 12va: "Similiter utique hiis et genitura in testiculis et virga."

16. PA, BNF lat. 6540, 63va–b: "Sic ergo causa delectationis et gaudii est quia superfluum expellendum et expellitur ad locum in quem naturaliter est natum expelli, ita quod in hoc conservatur nostre compositionis armonia etc."

17. PA, BNF lat. 6540, 63vb: "Notandum est quod in coitu est maxima delectatio respectu aliorum [BAV lat. 2175: aliarum] evacuationum."

18. It is omitted in some PA mss. (Cesena D.XXIV.2 and S.VI.2, and BNF lat. 6541A); JJ retains it. The "Bavarian," "Aachen," and "Erfurt" texts make no reference to this passage. There is no analogue at this point in the more independent "Prague" *Problemata*.

19. EC, BNF fr. 24281,116v, 117r: "ordenance de nature"; "que nature ordena"; "ordené de nature." In other respects Evrart followed Pietro closely on the reasons why coitus provides the greatest pleasure.

20. Coucke offers an outline and summary of *problema* IV.26 in "Edition of Peter of Abano's 'Exp.,'" 331–36.

21. BM, BNF lat. 15081, 12va: "'Quibus utique' pori non secundum naturam se habent. Si autem propter id quod excedant [BAV Chisiani G.V.131: excecantur] ut [BNF lat. 6540 omits] hii qui sunt in virga ut accidit eunuchis et effeminatis. Aut et aliter in anum confluit humiditas huiusmodi et enim egreditur ita."

22. CLM 4710, [268]rb and Krakow, BJ 654, 5vb. Or, in one manuscript, the "males and veins of the anus"(*masculos* instead of *musculos*): ms. Leipzig, UB 1435, 130va.

23. The reading "exsecantur," etc., is found in manuscripts of BM (e.g., Angers, Bibliothèque Municipale 450; BNF lat. 3121A and 6540), and JJ (Sorbonne 122; Erfurt, CA F. 15; BNF lat. 15454; Ghent, UB 72). It also occurs in one manuscript of the "Bavarian" *Problemata* (Ghent, UB 178). In medieval Latin, the two words were homophones.

24. "Erfurt" *Problemata*, IV.26, CLM 4710, [268]rb: "Quando autem precinduntur loca talia vel clauduntur ad loca vacua proxima expelluntur talia vi nature."

25. BM at "excecantur," BNF lat. 6327, gloss, 21va: "cirurigi incidunt."

26. PA, IV.3, BNF lat. 6540, 54rb: "Eunchus [*sic*] est homo inperfectus, muliebris, lenis, non operosus, malorum morum, in actum coytus non potens aut debiliter valde." Here he also mentions a third sense of "eunuch": a person with a religious commitment to abstinence. On the varied connotations and ambiguities of "eunuch," see Mathew Kuefler, *The Manly Eunuch: Masculinity, Gender Ambiguity, and Christian Ideology in Late Antiquity* (Chicago: University of Chicago Press, 2001), 31–36. Some of these have been entertained in connection with Chaucer's "Pardoner's Tale." See, for example, Alastair Minnis, "Chaucer and Queering the Eunuch," *New Medieval Literature 6* (2003): 107–28.

27. PA, BNF lat. 6540, 63vb: "eunuchis qui carent testiculis aut parvos omnino et debiles habent." Cf. Pietro's *Compilatio phisonomie*, pt. 2, ch. 10 bis, decision 20, [36]v.

28. Glosses on BM at "ut accidit eunuchis et effeminatis": BN Marciana 2488, 204rb: "scilicet occasionatus"; BNF lat. 6327, 21va: "occasionatis."

29. PA, BNF lat. 6440, 63vb: "Aut potest aliomodo contingere utpote quia pori tendentes in virgam exterius non solum sint excecati et obstructi ut prius sed quod sint omnino excecati versus virgam in exterius et sint aperti aliqualiter versus radicem virge et anum. In talibus quidem humiditas spermatica concurrit in anum et dispersa inde. Sic tandem mota et subtiliata per confricationem graditur [Cesena, BCM D.XXIV.2: egreditur] foras." Except for BAV lat. 2175, the manuscripts of both versions of PA (but not of JJ) read "non solum non sint [or: sunt] excecati et obstructi ut prius." I understand this simply to mean that the pores are obstructed in a different way.

30. BM, BAV lat. 10452, 75vb.

31. BM, BNF lat. 15081, 12va: "Signum autem in coitu collectio huius loci et dissolutio eorum que sunt circa anum. Si igitur superhabundat aliquis coitu hic [BAV Chisiani G.V.131: hiis] huc convenit quare quando desiderium superveniat hoc desiderat fricatione ad quod colligitur."

32. Pietro points out that the structure of the Aristotelian text requires him to treat the groups with wholly blocked and partly blocked vessels in reverse order from the way they were first introduced. At the lemmas Signum autem, Desiderium, and Hoc exeunte he discusses those who have some sort of anal emission; then, at the lemmas Qui autem and Quare necesse est, he considers those who have none (or, as he later concedes, very little).

33. Aristotle, *De generatione animalium: Translatio Guillelmi de Moerbeka*, ed. H. J. Drossaart Lulofs, Aristoteles Latinus, 17/2.v, Union Académique Internationale, Corpus Philosophorum Medii Aevi (Bruges: Desclée de Brouwer, 1966). The reference is to the discussion of such phenomena as polydactyly and hermaphroditism treated in Book IV, especially chapters 3 and 4.

34. "Erfurt" *Problemata*, CLM 4710, [268]rb: "experimentaliter comprobatur."

35. Aristotle, *Problemata*, IV.26. For this interpretation of the Greek, see ed. and trans. Pierre Louis, 3 vols., Collection des Universités de France (Paris: Les Belles Lettres, 1991), vol. 1, 89, n. 57.

36. PA, BNF lat. 6540, 63vb, margin: "Nota quod aliquis apetit in ano suponi."

37. PA, IV.2, BNF lat. 6540, 53vb: "Non enim naturaliter contingit emitti sperma nisi plene coaduventur [Cesena, BCM D.XXIV.2: coadiuventur] partes predicte circa anum existentes."

38. Pietro is quoting *Problemata* IV.2: "velud a manu expressum."

39. PA, BNF lat. 6540, 63vb: "Ita quod recte advenit contrarium eius quod oportet esse in coitu naturali."

40. "Bavarian" *Problemata*, Krakow, BJ 2095, 227va: "'contrarium quod oportet esse in coitu naturali,' scilicet ut materia spermatica exprimatur ab ano in virgam cum hic a virga in anum exprimatur."

41. PA, BNF lat. 6540, 64rb, in the lemma Qui autem.

42. PA, BNF lat. 6540, 63vb–64ra: "Propter quod quando appetitus coeundi supervenit appetit per fricationem et motum fortem expellere huiusmodi materiam spermaticam

et 'quia nata est colligi' circa annum ad locum ad quem nata est colligi secundum naturam eius monstruosam."

43. BM, BNF lat. 15081, 12va: "Si igitur superhabundat aliquis coitu"; PA first version (Nuremberg, StB Cent. III.19, BNF lat. 6541, BAV lat. 2174) and JJ, BAV Ottob. 1764, "Et ideo si aliquis superfluat in coitu et materia eius" (and the like); PA second version (BNF lat. 6540 and 6541A, Cesena, BCM D.XXIV.2 and S.VI.2, BAV lat. 2175, BN Marciana 2465): "Et ideo si aliquis superfluat in materia coitus aut forte fuerit usus" (and the like).

44. BM, BNF lat. 15081, 12va-b: "Desiderium autem et a cibis et ab intellectu fit. Quando enim movebitur hoc a quocumque ventositas concurrit et talis superfluitas confluit ubi apta nata est sive subtilis fuerit aut ventosa."

45. The exception among the JJ manuscripts is BNF lat. 6542, which lacks a number of variants of JJ and includes elements omitted by JJ, suggesting that the scribe had access to PA as well.

46. PA, BNF lat. 6540, 64ra. BAV lat. 2174, 30rb and 2175, 51rb read "cibis et potibus inflammativis": presumably thinking of the heat that is a physiological component of seed and desire.

47. EC, BNF fr. 24281, 117r: "Et porce deffent Ovides en son livre tels viandes a ciaulz qui se voelent retraire de fole amour."

48. Aristotle, *De anima*, III.10, 433a10–14, and *De motu animalium*, VI, 700b17–19; Avicenna, *Sextus naturalium*, I.5 in *Liber de anima seu Sextus de Naturalibus: Édition critique de la traduction latine médiévale*, ed. S. van Riet, 2 vols., Avicenna Latinus (Louvain: Peeters, 1972), 99–101.

49. Cf. Joannes de Janduno, *Super libros Aristotelis de anima* (Venice: Hieronymus Scotus, 1587; reprinted Frankfurt: Minerva, 1966), bk. III, q. 39, col. 433: "Intellectus et appetitus communiter sumpti, sunt principia essentia motum [*sic*] in cunctis animalibus."

50. Aristotle, *De motu animalium*, VIII, 701b17–19.

51. Federici Vescovini, "Pietro d'Abano e Taddeo da Parma sull'imaginazione."

52. Aristotle, *De motu animalium*, 6, 700b 24–30.

53. Following Aristotle's own allusions in *De motu animalium*, 4, 699b32, and 6, 700b9.

54. The word *confricatio* (as distinguished from *fricatio*, employed elsewhere) might suggest that two people are involved, but later Pietro uses the term in connection with both intercourse and masturbation.

55. PA, BNF lat. 6540, 64ra: "Spiritus ventosus per confricationem et motum excitatus concurrit ad partes circa radicem virge existentes et ex inde spermatica humiditas effluit in locum sibi secundum naturam istam orbatum destinatum." The final phrase was (and is) hard to sort out. BNF lat. 6540 and 6541A, and BAV lat. 2174 and 2175 all read "orbatum destinatum." But variants occur in both PA and JJ, including: Nuremberg, StB Cent. III.19 and BNF lat. 6541, "orbatam destinatam"; and Cesena, BCM D.XXIV.2 and S.VI.2, "orbatam destinatum." Cf. PA, IV.10, BNF lat. 6540, 56vb.

56. BM, BNF lat. 15081, 12vb: "Hoc exeunte quemadmodum distensiones pueris et hiis qui in etate quandoque nullo humido exeunte quiescunt, quandoque extinguitur humidum. Si autem neutrum horum patiatur desiderat quousque aliquid horum accidit."

57. PA, BNF lat. 6540, 64ra: "Et hoc sive fuerit talis humiditas subtilis aquea ut in illis quorum meatus essent latiores sive sit solum ventosa que tamen subtiliata per fricationem egreditur in illis etiam quorum meatus sunt angustissimi."

58. PA, IV.9, BNF lat. 6540, 56vb: "'tertius est meatus per quem transit humiditas subtilis dicta alguedi que emittiur maxime cum quis tangens mulierem in copulam eius languescit.'" Avicenna, *Liber canonis medicine*, bk. III, fen 20, tr. 1, ch.1, 352ra. See also Avicenna, glossary, 553vd: "Alguadi est humor qui egredit de virga quando quis tangit mulierum."

59. PA, BAV lat. 2175, 52va and vb. No other surviving manuscript contains this variant, but it is present in the 1485 and 1482 editions of Pietro's exposition.

60. Avicenna, *Liber canonis medicine*, bk. I, fen 1, doctr. 3, ch. 3, 3vb: "Deinde est etas algulemati vel alguadi et rehac, id est cum ab eo egreditur sperma que est quousque barbescant." Both the association with youth and the idea of a fluid elicited by touch (or sight) are present in Aristotle's *Generation of Animals*, with which Pietro was familiar. Avicenna, *De animalibus*, trans. Michael Scot, in *Logica, Sufficientia, De celo et mundo, De anima, De animalibus, De intelligentiis, Alfarabius De intelligentiis, Philosophia prima*, fols. 29ra–64rb (Venice: [Putzbach,] 1508; repr. Frankfurt-am-Main: Minerva, 1961), bk. 17, 62rb: "Et quando etiam iste olent masculum vel audiunt ipsum appetunt luxuriari, sicut etiam iuvenibus hominibus accidit qui possunt spermatizare propter solum tactum vel visum pulchre mulieris." Cf. Aristotle, *De generatione animalium*, III.1, 751a14–19. Avicenna has omitted a reference to male quadrupeds who react similarly, and he has added that the humans are "young men." Some Greek manuscripts and the Latin translation used by Pietro present the analogy as an explanation for the quickness of the phenomenon (*velocitatis autem causa*): *De generatione animalium*, ed. Drossaart Lulofs, 89. William of Moerbeke's translation leaves open the possibility that copulation ensues (*permota fuerint corpora ad coitum*), but the context (unfertilized eggs) rules out that reading.

61. "Erfurt" *Problemata*, CLM 4710, [268]rb: "exalationibus insensibilibus."

62. PA, BNF lat. 6540, 64ra–b: "propter materiam ventosam confricatione calefactam transeunte [Cesena, BCM D.XXIV.2: transeuntem] super membra eorum sensibilia.

63. BNF lat. 6540, 63vb, margin: "Sperma potest egredi per anum mediante fricatione."

64. "Bavarian" *Problemata*, Krakow, BJ 2095, 227va: "Hiis accidit in tali actu delectatio quedam sicud pueris non habentibus materiam spermatis adhuc sed solum quemdam ventositatem erigentem virgam."

65. PA, BNF lat. 6540, 64rb: "Exprimitur in primo *De generatione animalium* sub hiis verbis: 'Delectatio accidit in coitu non solum propter emissionem spermatis sed propter emissionem spiritus ex quo spiritu constituto et condensato fit sperma. Et hoc est manifestum in pueris illis qui nondum possunt spermatizare quando fuerint prope tempus in quo debet [Cesena, BCM D.XXIV.2: debent] spermatizare. Et similiter est manifestum in viris sterilibus. In hiis enim omnibus fit delectatio propter decisionem spiritus' et confricationem." Aristotle, *De generatione animalium translatio Guillelmi de Moerbeka*, I.20, 728a10–14.

66. "Erfurt" *Problemata*, CLM 4710, [268]rb: "exalationibus insensibilibus." *Problemata* XXX.1 (perhaps the *problema* best known in the Middle Ages and the Renaissance)

confirms the ability of boys approaching puberty to experience pleasure from the passage of spirit, as they do in masturbation.

67. PA, BNF lat. 6540, 64rb.

68. PA, BNF lat. 6540, 64rb: "Fortasse accidit propter complexionem eius natura superflue calidam et siccam."

69. Men who fail to produce semen in their youth (*in tempore iuventutis*), may be able to do so later (*circa etatem consistentie*). The divisions between life stages or "ages" are impossible to fix. Pietro d'Abano (in a question on whether *iuvenes* or *pueri* are *temperatior*) includes a discussion of the nonsense (*garrulitatem*) about various ways the life stages are divided: "Similiter consistentia est duplex: una quidem iuventutem praecedens, in qua pausat augmentum, apparens *De regimine sanitatis*; altera vero subsequens illam, in qua corpus et virtus conservantur a casu saltem apparenti iam praetacta." *Conciliator controversiarum quae inter philosophos et medicos versantur Petro Abano Patavino* (Venice: Iuntas, 1565), diff. 26, facsimile in *Opere de Pietro d'Abano*, ed. G. I. Ludwig (Padua: Il Glifo, 1982), 39rb. Cf. Aristotle, *Generation of Animals*, V.7, 787b29–30, and I.20, 728a10–14; Avicenna, *De animalibus*, bk. 9, ch. 1, 40va–b; and Avicenna, *Liber canonis medicine*, bk. I, fen 1, doctr. 3, ch. 3, 3va, which, however, does not fit well: "etas consistendi" or "pulchritudinis" is between 30 and 35 or 40. This is also where "alguadi" turns up as linked to puberty (3vb). I have not located a relevant passage in Galen. See Elizabeth Sears, *The Ages of Man: Medieval Interpretations of the Life Cycle* (Princeton: Princeton University Press, 1986), 25–31.

70. "Erfurt" *Problemata*, CLM 4710, [268]rb: "Sicud istis pruritus et delectatio fit in ano."

71. PA, BNF lat. 6540, 64rb: "Appetit ire et evacuari a materia huiusmodi potenta [Cesena, BCM D.XXIV.2: contenta] subtili vel ventosa quoniam evacuata potest quietari et gaudere."

72. PA, BNF lat. 6540, 64rb: "propter complexionem eius natura superflue calidam et siccam, humidum resolventem in spiritus ventosos sicut fit in passeribus."

73. Beryl Rowland, *Birds with Human Souls: A Guide to Bird Symbolism* (Knoxville: University of Tennessee Press, 1978), 157–58.

74. Aristotle, *Historia animalium,* V.2, 539b29–31.

75. Albertus Magnus, *De animalibus libri XXVI*, ed. Hermann Stadler, 2 vols., Beiträge zur Geschichte der Philosophie des Mittelalters 15–16 (Münster: Aschendorff, 1916–21), bk. V, tract 1, ch. 2, § 10, p. 411: "sicut patet in passeribus qui velocissimi sunt coitus [*sic*] et iterant coitum multotiens una vice post aliam"; bk. XXIII, tract 1, ch. 24, § 98, p. 1509: "Multi est valde coitus et generationis ita quod in una hora forte vicesies coit; et caro eius ad libidinem incendit et constipationem inducit, calida enim est et sicca et ideo parum passer impinguatur, et est avis mobilis propter humidi consumptionem superflui." Pietro may well have been familiar with some of Albertus's work, but a common source or parallel extrapolations seems more likely. Avicenna, the most likely influence on both, mentions but does not elaborate on the Aristotelian facts concerning the longevity of male sparrows (*De animalibus,* bk. 8, ch. 7, 38va–b).

76. PA, BNF lat. 6540, 63vb–64ra: "Propter quod quando appetitus coeundi supervenit appetit per fricationem et motum fortem expellere huiusmodi materiam spermaticam

et quia nata est colligi circa annum ad locum ad quem nata est coligi secundum naturam eius monstruosam."

77. BM, BNF lat. 15081, 12va: "Qui autem natura sunt effeminati ita constituerunt quod ibi quidem non segregatur aut modica ubi habentibus secundum naturam expellitur in locum autem hunc quia extra naturam constituerunt. Mares enim existentes sic dispositi sunt."

78. "Prague" *Problemata*, Prague, NK VIII.A.19, 79rb: "Et in illis fluit materia spermatica versus culum et emittunt sperma in coytu. Et illi volunt fricari per impositionem virge virilis ad oculos [*sic*] eorum." The modern reader is invited to interpret the scribal insertion of "eyes" into the picture.

79. PA, BNF lat. 6540, 64rb: "illi qui naturaliter a primis componentibus 'effeminati sunt' vel eunuchi."

80. The assertion that eunuchs, not just effeminates, are constituted in this way from the start (*a primis componentibus*) is a little puzzling, especially since they are not mentioned at this point in the ancient text. Perhaps for this reason, the manuscripts of the Jean de Jandun redaction as well as those of Evrart de Conty omit eunuchs here.

81. PA, BNF lat. 6540, 64rb: "Illi qui naturaliter a primis componentibus 'effeminati sunt' vel eunuchi sic habuerunt poros constitutos et plasmatos ut taliter excecatos quod in eos nulla humiditas evacuetur 'aut modica' valde expellitur ex eis in locum in quem expellitur in illis qui naturaliter secundum naturam habent hos poros."

82. PA, BNF lat. 6540, 64rb: "locum 'secundum naturam' suam monstruosam quia 'expellitur in hunc locum' monstruose plasmatum. Et hoc ideo quia non fuerunt constituta [Cesena, BCM D.XXIV.2: constituti] secundum naturam recte operantem."

83. Aristotle, *De generatione animalium*, IV.3–7.

84. *Problema* IV.13 asks under what conditions the offspring of human parents is to be considered nonhuman. On varieties of monstrosity and the questions they raised, see Adam Davidson, "The Horror of Monsters," in James J. Sheehan and Morton Sosna, eds., *The Boundaries of Humanity: Humans, Animals, Machines* (Berkeley: University of California Press, 1991), 36–67; Daston and Park, *Wonders and the Order of Nature*, 173–214; Maaike Van der Lugt, *Le ver, le démon et la vièrge: Les théories médiévales de la génération extraordinaire: Une étude sur les rapports entre théologie, philosophie naturelle et médecine*, L'Âne d'Or, 16 (Paris: Les Belles Lettres, 2004), and "L'humanité des monstres et leur accès aux sacrements dans la pensée médiévale," HAL: halshs-00175497, version 1, consulted at http://halshs.archives-ouvertes.fr./ December 15, 2011, published without references in Latin in *Monstre et imaginaire social: Approches historiques*, ed. A. Caiozzo and A.-E. Demartini (Paris: Créaphis, 2008), 135–61.

85. EC, BNF fr. 24281, 117r: "li malvais regart des infortunes du ciel"; "la malvaise disposition qui lor est donne de nature des lor premiere generation."

86. BM, BNF lat. 15081, 12vb: "Quare necesse est locum hunc secari ipsorum. Secatio aut hec quidem omnino facit corruptionem, hec aut perversionem. Illa quidem igitur non est. Mulier enim utique fieret."

87. "Aachen" *Problemata*, Erfurt, CA F. 263, 23rb: "Et tunc prosequit de illis de cura ipsorum qui habent istomodo poros et meatus obstructos quod non debent per cirurgiam

secari vel scindi cum aliquo ere vel alico alio instrumento inscisivo quia tunc fieret forte totalis extrasecatio et obstructio vel incurreunt mortem vel perversionem mentis vel totaliter fierent sicud mulier. Ita quod fierent effeminati et acquireunt mores mulierum."

88. The translations of Louis, Hett, Mayhew and Flashar are in agreement on this reading. George of Trebizond, helpfully, offers: "Obstructio autem omnis in toto quidem corruptionem facit in parte vero perversionem. Illa igitur non est. Esset enim utique femina. Necesse igitur est perversionem esse et alio seminalem expulsionem perverti" (BL add. 21978, 18va).

89. GT renders πήρωσις as *obstructio* (BL add. 21978, 18vb); TG as *laesio* (in *Extra ordinem naturalium libri quibus nonnulli etiam additi sunt Aristoteli ascripti*, 19ra).

90. According to Liddell and Scott, πήρωσις is "a maiming or being maimed in the limbs or senses"; related words convey disability, incapacity, weakness. Louis, ed., *Problèmes*, vol. 1, 194, n. 4, observes that in Aristotle the word usually refers to "un développement incomplet d'une partie du corps," citing *Historia animalium*, II.1, 498a32; and *De generatione animalium*, II.3, 737a28, IV.1, 766a26, and V.3, 784a10, though it can also refer to a "mutilation" (*Historia animalium*, VII.1, 581b22). *Seco* and related terms indicate cutting or dividing. It can refer specifically to surgical actions, including castration, though it can also convey less intentional senses, such as to wound.

91. One BM manuscript reads "blinded" or "obstructed" (*cecatio*) rather than "cut" (*secatio*), not only recalling the earlier description of the condition but also suggesting that the connection with surgery was not inevitable. For "cecari" and "cecatio," BAV Chisiani G.V.131, 75rb; for "secari" and "secatio," BNF lat. 15081, 12va; Cesena, BCM D.XXIV.2, 36rb; Escorial f.I.11, 7ra. The phrase is omitted in BL add. 21978. In some geographical areas, the substitution might be easy to make if the scribe were writing from dictation, but the fact that the copyist of the BAV Chisiani ms. was apparently Italian makes the variant more interesting.

92. BM, BNF lat. 6327, 21va, gloss: "possibile."

93. Cf. Coucke, "From Abstinence to Promiscuity" for the idea that castration was a putative cure. On signs, Ziegler, "Sexuality and the Sexual Organs in Latin Physiognomy 1200–1500."

94. PA, BNF lat. 6540, 64rb: "Hoc 'non est' faciendum." A glossator of BM in Escorial, f.I.11 (7ra) produced the same reading, but there are signs elsewhere that he was following Pietro's exposition.

95. PA, BNF lat. 6540, 64rb–va: "Sectio inducit 'corruptionem' pororum omnia [Cesena, BCM D.XXIV.2: omnimodam] et mortem plerumque. . . . Per talem enim seccationem partes circa testiculos existentes corrumpentur ita quod rederetur talis eunuchus et effeminatus cum prius non sic esset."

96. BM, BNF lat.15081, 12vb: "Necesse est igitur perverti et alibi moveri quidem seminalis expurgationis. Propter quod insatiabiles quemadmodum mulieres. Modica enim humiditas non cogitur exire et infrigidatur cito."

97. "Bavarian" *Problemata*, Krakow, BJ 2095, 227va–b: "Interim autem quod nec emittitur spiritus ventosus nec materia aquea subtilis ab hiis non quiescunt sed desiderant fricari quousque aliud horum accidat."

98. "Aachen" *Problemata*, Erfurt, CA F. 263, 23rb: "Quanto magis fricantur tanto magis attrahitur . . . quasi nichil emittitur et sic magis intenduntur in furia ipsarum."

99. Aristotle, *Problemata*, II.22.

100. PA, BNF lat. 6540, 64va: "Igitur docet hic qualiter tales debeant curari dicens primo quod oportet illud quod inest in istis de spermatica humiditate expurgatura divertere ab ista parte et movere ad aliam partem quod fit cum cibis et medicinis exsiccantibus sperma et materiam eius ad alias regiones divertentibus ut ad urinas et sudores et cetera."

101. Avicenna, *Liber canonis medicine*, bk. 3, fen 20, tr. 1, ch. 42, 358ra. In printed editions, the heading is "De aluminati" (Padua, 1476; Venice, 1483, 1486, 1490, 1500; etc.). The readings contained in the manuscripts of Pietro's *Problemata* commentary vary, but more closely approximate Nuremberg, StB Cent. III.19, 152rb, "De alubnati."

102. PA, BNF lat. 6540, 64va: "Unde Avicenna 3° *Canonis*, fen 20ª [tr. 2, ch. 42, 358ra] loquens de huiusmodi dispositione et priori magis dicit quod 'stulti sunt homines qui volunt eos curare quoniam vitium [Cesena, BCM D.XXIV.2: initium] egritudinis eorum est medicamentum [Cesena, BCM D.XXIV.2: meditativum] non naturale. Sed si cura confert eis tunc est illud quod frangit desiderium eorum ex tristitia, fame, vigiliis, carcere et percussione.'"

103. "Mental" is found in BCM XXVI.D.6, 37ra ("initium egritudinis eorum est meditativum non naturale") and similarly in BNF lat. 6541A, both representing Pietro's second version, and in BNF lat. 15454 (JJ).

104. For example, the manuscript Nuremberg, StB, Cent. III.19 of Avicenna's *Canon* has "meditativum" as does the Venice 1482–83 edition; the Venice 1492 edition gives "medicamen."

105. Avicenna, *Liber canonis medicine*, bk. 3, fen 20, tr. 1, ch. 42, 358ra. A careful study of the Latin manuscripts would be required to attribute the interpolation to Gerard of Cremona with certainty. The sentence does not occur in four Arabic texts of or derived from Ibn Sīnā examined by Everett K. Rowson, to whom I am indebted for this information (personal communication, June 20, 2003).

106. Derek Neal, "Disorder of Body, Mind, or Soul: Male Sexual Deviance in Jacques Despars's Commentary on Avicenna," in Borris and Rousseau, *The Sciences of Homosexuality in Early Modern Europe*, 51.

107. EC, BNF fr. 24281, 117r: "Car le commencement de lor maladie n'est pas naturel ains est engendrés par meditation et par forte pensee. . . . Car nulle medicine ne porroit transmuer ceste fole pensee."

108. Table to PA, Gottweig 138, 238v: "Sodomiticum peccatum."

109. Glosses on BM in BN Marciana 2488, 204rb, and BNF lat. 6327, 21va: "occasionatus"; "Prague" *Problemata*, Prague, NK VIII.A.19, 79rb: "fricare," "fricari," and so on.

110. PA, BNF lat. 6540, 63vb, margin: "Sperma potest egredi per anum mediante fricatione"; "Erfurt" *Problemata*, CLM 4710, [268]rb: "Et hoc etiam fit aliquando quando habundat tanta materia spermatis ut . . . non potest expurgari sufficienter per coitum."

111. WB, general prologue, Bodl. Digby 153, 102r: "Felix qui poterit causas cognoscere rerum." Virgil, *Georgics*, 2.490. Two manuscripts, CU EE.I.22 and BL Royal 12.E.16, contain the canonical reading of the quotation, "Felix qui poterit rerum cognoscere causas."

CHAPTER 2. HABIT IS A KIND OF NATURE

1. BM, BNF lat. 15081, 12vb: "Quibusdam autem fit et ex consuetudine passio hec."

2. "Aachen" *Problemata*, Erfurt, CA F. 263, 23rb: "Qui circa iuventute [*sic*] talia consuevunt facere iam consuetudo est eis natura quod semper [quasi?] fricant illas partes propter delectationem vel volunt fricari ab aliis et maxime isti sodomite. Et id quod ipsi faciunt in iuventute est eis in memoriam. Et quia tunc sunt molles luxuria eis tunc vel postea frequenter accidit."

3. PA, BNF lat. 6540, 64va: "Et primo enim repetit illud quod dictum est de secundo membro divisionis predicte. In quibus humiditas confluit ad anum. . . ."

4. BM, BNF lat. 15081, 12vb: "Et quibus quidem in anum hii pati desiderant; quibuscumque autem in utraque hii et agere et pati. In utraque [BNF lat. 6540 & BAV Chisiani G.V.131: utrum] autem plus hoc magis desiderant."

5. The Prague commentator dispenses with those who enjoy both rubbing and being rubbed before moving on to a discussion of habits; the Bavarian commentator omits any mention of those with two sources of pleasure.

6. PA, BNF lat. 6540, 64va–b: "'pati' et fricari vehementer circa anum ut hec expellatur humiditas, . . . Appetunt enim virga alios fricare et cum virgis aliorum in ano fricari ab eis."

7. This digression and a later one appear in both versions of Pietro's commentaries, as well as in all the manuscripts of Jean's redaction, but a they have been omitted in three related copies of PA version B, the two Cesena manuscripts, and BN lat. 6541A.

8. Avicenna, *Liber canonis medicine,* bk. 3, fen 20, tr. 1, ch. 42, 358ra: "Quare indiget ille qui tenuis est fricatione vehementi donec sentiat, quare spargitur super ipsum hominem. Et tunc advenit ei explementum libidinis." Pietro ignores the difference between the "pores" of the *Problemata* and the "nerves" of the *Canon*, perhaps because he was aware of the ambiguity associated with various vessels in anatomical texts and translations. Cf. Neal, "Disorder of Body, Mind, or Soul," 50, who regards the distinction as fatal to a connection between the two works on this point.

9. PA, BNF lat. 6540, 64vb: "Avicenna non adducit rationem, qui etiam contrarius videtur sententie Philosophi. Non video quid impediat a nativitate talem inordinationem et monstruositatem fieri cum idem statim Avicenna faciant [Cesena, BCM D.XXIV.2: faciat] sermonem de hermostrodicis [Cesena, BCM D.XXIV.2: hermafroditis] qui magis appent [Cesena, BCM D.XXIV.2: apparent] apud naturam monstruosi." On the triad of monster/hermaphrodite/sodomite: Maaike van der Lugt, "Pourquoi Dieu a-t-il créé la femme? Différence sexuelle et la théologie médiévale," in *Ève et Pandora: La création de la première femme,* ed. Jean-Claude Schmitt, Le Temps des Images ([Paris]: Gallimard, 2001), 95–101.

10. See Danielle Jacquart and Gérard Troupeau, "Traduction de l'arabe et vocabulaire médical latin: Quelques exemples," in *La lexicographie du latin médiéval et ses rapports avec les recherches actuelles sur la civilisation du moyen âge, Paris 18–21 octobre 1978,* Colloques Internationaux du Centre National de la Recherche Scientifique, 589 (Paris: Éditions du Centre National de la Recherche Scientifique, 1981), 367–76.

11. Jordan, *Invention of Sodomy*, 87, speaking of Alain de Lille.

12. JJ, BNF lat. 15454, 54rb (margin): "De alubnati secundum Avicenna et est actus sodomiticus." The manuscript contains the names of two owners, Petrus Rosellus, *medicus* (Ernest Wickersheimer, *Dictionnaire biographique des médecins en France au Moyen Âge*, 2 vols. (Paris: Droz, 1936, reprinted Hautes Études Médiévales et Modernes 34, 1–2, Geneva: Droz, 1979), 242), and Franciscus Guillebon, doctor of theology.

13. Many medieval Latin spellings would retain its basic core, "alubnati": an initial "h" or one added after the "t" is of no consequence, and the final "i" might be written "y," but other variants betray uncertainly. For example, the careful copyist of BNF lat. 6540 tripped over the first instance of the word. He first wrote "halubachi" but corrected himself in the conventional manner by putting dots under the error, and then wrote "haluebathi." When the term recurred just a few words later, he settled on "halubnathy" (BNF lat. 6540, 64vb). Among the other attempts to reproduce the term were "alumbnati," "halbubuathi," "habibuathi," and "albuachi." The fact that in some medieval hands "c" and "t" are written identically and the "u" and "n" are often very difficult to distinguish contributed to the confusion.

14. At least one manuscript (Nuremberg, StB, Cent. III.19) reads "alubnati." All the editions I have seen (Padua 1476, Venice 1482–83, 1486, 1490, 1507, etc.) give "De aluminati." The Venice 1523 edition with commentary (reproduced in Neal, "Disorder of Body, Mind, or Soul," 42) uses "aluminati" in the title and "alubuati" within the chapter. Avicenna, *Liber canonis medicine*, glossary, 554vb: "Alumen iameni, idest scissum." In that edition, bk. III, fen 20, tr. 1, ch. 42 is "De aluminati."

15. See, for example, Jordan, *Invention of Sodomy*, 27. Gregory Hutchenson points out, however, that for Spain, at least, much of this supposed association is an artifact of modern Spanish historiography: "The Sodomitic Moor: Queerness in the Narrative of the *Reconquista*," in *Queering the Middle Ages*, ed. Glenn Burger and Steven F. Kruger, Medieval Cultures 27 (Minneapolis: University of Minnesota Press, 2001), 99–122.

16. BM, BNF lat.15081, 12vb: "Quibusdam autem fit et ex consuetudine passio hec. Quecumque enim utique facere accidit ipsis gaudere et emittere genituram. Ita desiderant autem facere quibus hec fiunt et magis consuetudo quemadmodum natura fit."

17. One scribe even substituted the word "nature" for the word "habit" at the beginning of this passage, and two others followed his lead. The two Cesena manuscripts and BNF lat. 6541A contain the variant.

18. Aristotle, *De memoria et reminiscentia* II, 452a27–29. See also *Ethica nicomachea*, fasc. 3, VII.14, 1152a30–32.

19. Hippocrates, *Airs, Waters, Places*, 14. The Latin translations did not circulate as widely as those of the Hippocratic works included in medical curricula. Pearl Kibre, *Hippocrates Latinus: Repertorium of Hippocratic Writings in the Middle Ages*, rev. ed. (New York: Fordham University Press, 1985), 25–28.

20. George of Trebizond, "Scholia to the Aristotelian 'Problemata,'" in *Collectanea Trapezuntiana*, ed. Monfasani, text 157, 4.26, p. 647: "Consuetudo in omnibus quasi altera natura est, maxime vero in rebus venereis." The passage even holds that the change becomes hereditary, though PA's readers may not have known it.

21. Aristotle, *Ethica nicomachea*, fasc. 3, II.1, 1103a20–21.

22. PA, BNF lat. 6540, 64vb: "Alius est motus et inclinatio qui sequitur formam superiorem resultatam ex mixtione elementorum aut de foris advenientem."

23. This is a central argument of the *Nicomachean Ethics*. See, for example, II.1, 1103a30–b16.

24. It does not appear in this context in the Latin translations of the *Ethics*, where forms of *dispono* appear.

25. *Ethica nicomachea*, fasc. 3, II.5, 1106a5–6: "Ad hec autem secundum passiones quidem, moveri dicimur; secundum virtutes et malicias, non moveri, set disponi qualiter."

26. The objection and response as a whole is missing in the group of manuscripts of Pietro's second version prone to block omissions (Cesena, BCM, D.XXIV.2 and S.VI.2, and BNF lat. 6541A).

27. BM, BNF lat.15081, 12vb: "Propter hoc utique quicumque non ante pubertatem sed circa pubertatem consueverunt supponi propter hoc quod fit ipsis in usu memoria. Simul autem memoria delectatio. Propter consuetudinem autem sicut apti nati desiderant pati. Multa quidem et consuetudo sicut et in natis fit."

28. Here middle aorist infinitive: ἀφροδισιάζεσθαι. At the very beginning of the *problema*, Bartholomeus renders the passive participle ἀφροδισιαζόμενοι with the impersonal *coitur*.

29. PA, BNF lat. 6540, 64vb: "Qui moventur ad hunc actum propter delectationem quoniam 'quecumque accidit facere gaudere' et delectare illis [BAV lat. 2175 and BCM D.XXIV.2: illos] qui 'emittunt genituram' recte [BAV lat. 2175: ratione] coitum exercendo. 'Ita' accidit illis quibus talia fiunt. Propter hunc enim actum profanum appetunt exercere ad hoc ut gaudeant et delectentur."

30. Aristotle, *Analytica posteriora: Translationes Iacobi, anonymi sive "Ioannis," Gerardi et Guillelmi de Moerbeka*, ed. Laurentiius Minio-Paluello and Bernardus G. Dod, Aristoteles Latinus IV, 1–4, 2 and 3 rev. ed., Corpus Philosophorum Medii Aevi, Union Académique Internationale (Bruges: Desclée de Brouwer, 1968), trans. Guillelmus, ed. Dod, fasc.4, II.19, 100a4–14.

31. Cicero, *De inventione*, I.2, § 3.

32. PA, BNF lat. 6540, 65ra: "Et hoc est possibile nam frequens consuetudo et longa fit sicut ipsa inesset natis a nativitate eorum natura et cetera."

33. JJ, Seville, Biblioteca Capitular y Colombina 7-7-9, 53rb (margin): "Frequens consuetudo et longa fit sicut ipsa inesset a nativitate."

34. PA, BNF lat. 6540, 64vb. The word does not appear in the margins of the manuscripts of the Bartholomeus translations that I have examined.

35. Aristotle, *Ethica nicomachea*, VII.7, 1149a1.

36. Most readers of the Jean de Jandun redaction would not have encountered that explication, and so might have seen what was acquired by habit in a different light. In all manuscripts of the PA version, Signum autem contains the gloss "passio sive dispositio."

37. BM, BNF lat. 6327, glosses, 21va.

38. PA, BNF lat. 6540, 64va–b: "Hii appetunt fricare et fricari secundum utrumque istarum partium. Appetunt enim virga alios fricare et cum virgis aliorum in ano fricari ab eis. Inter istos qui sic appetunt agere et pati secundum utrumque partem quidam magis

desiderant agere seu fricari [Cesena, BCM D.XXIV.2: fricare] quidam vero magis pati [Cesena, BCM D.XXIV.2 adds: et fricari] secundum quod humiditas stimulativa magis est nata confluere in virgam vel in anum."

39. "Aachen" *Problemata*, Erfurt, CA F. 263, 23rb: "Et tunc ostendit qui per pravam consuetudinem patiuntur talem fricationem. Dicit quod quidam sunt in quibus talis humiditas est circa anum et quidam in quibus una pars est circa anum et alia circa virgam, modo in quibus est circa anum illa humiditas illi desiderant fricari et pati sed in quibus est pars circa anum et pars circa virgam desiderant agere et pati, fricare et fricari. Et hoc secundum plus et minus secundum quod aliqui magis et minus sunt consueti et stimulati." Cf. PA, BNF lat. 6540, 64vb: "Inter istos qui sic appetunt agere et pati secundum utrumque partem quidam magis desiderant agere seu fricari [Cesena, BCM D.XXIV.2: fricare] quidam vero magis pati [Cesena, BCM D.XXIV.2 adds: et fricari] secundum quod humiditas stimulativa magis est nata confluere in virgam vel in anum."

40. "Erfurt" *Problemata*, CLM 4710, [268]rb: "Hoc autem fit ex consuetudine aliquando quod autem in ano superfluitas colligatur."

41. George of Trebizond, *Collectanea Trapezuntiana*, §157, 4.26, p. 647.

42. PA first version, BN 6541, 75rb: "Notandum quod illud nefandum opus sodomiticum quidam exercent manu fricando virgam. Alii puerorum inter coxas confricando quod et plurimi agunt hodie. Alii autem fricationem faciendo circa anum et virga in ipsum imponendo sicut in vulva [sic] imponitur. Et de talibus videtur sermo Aristotelis esse magis."

43. PA, Venice, 2465, 41vb, reads "opus sodomitarum."

44. PA, BNF lat. 6540, 64va: "In quibus humiditas confluit ad anum ad hoc ut convenientius condescendat ad illos qui ad hoc operantur prevaricatione turpissima."

45. JJ, BAV Ottob. 1764, 60rb: "exercent manu fricando virgam."

46. The manuscript BAV lat. 2175 (fifteenth century) of PA's second version includes "manu," but its scribe used manuscripts of both versions to prepare his copy. See Coucke, "Edition of Peter of Abano, 'Exp.,' IV", xl–xli. As with many such changes, it is thus unlikely (though possible) that it was a scribe early in the transmission chain of the second version who eliminated the phrase, rather than Pietro.

47. PA, first version, BNF lat. 6541, 73ra: "Quidam autem non alterum seorsum exercent ut aparebit circa finem sermonis." The JJ manuscripts consistently read "solum" for "seorsum."

48. "Prague" *Problemata*, NK VIII.A.19, 79rb: "Et in illis fluit materia spermatica versus culum et emittunt sperma in coytu. Et illi volunt fricari per impositionem virge virilis ad oculos [*sic*]."

49. One reader seems to have taken exception to historicizing these acts. In JJ, Paris, Arsenal 723, 80vb, "plurimi" and "hodie" have been scratched out.

50. BM, BNF lat. 15081, 12vb: "propter consuetudinem autem sicut apti nati desiderant pati."

51. PA, BNF lat. 6540, 64vb: "ut delectatio inmensa et emissio seminis ad anum."

52. BM, BNF lat. 15081, 12vb: "Quecumque enim utique facere accidit ipsis gaudere et emittere genituram. Ita desiderant autem facere quibus hec fiunt et magis consuetudo quemadmodum natura fit."

53. BM, BNF lat. 15081, 12vb. At 879b34–6, Louis (ed. and trans., *Problèmes*), keeping the active/passive distinction in the foreground, says that pleasure and ejaculation result from either role: "Quelque soit le rôle qu'ils jouent, la jouissance survient et ils répandent ainsi leur semence. C'est ce que désirent faire ceux qui sont ainsi." Hett (ed. and trans., *Problems*), closely following Forster, reads the Greek text as saying simply that men enjoy (and ejaculate as a result of) what they are accustomed to do: "For men are accustomed to enjoy what they normally do, and to emit semen accordingly. So they desire to do that by which this may occur." Mayhew concurs and, similarly, Gohlke translates it: "Was sie gerade tun, macht ihnen Vergnügen und entfacht den Wunsch, die Samenflüssigkeit auf die Weise loszuwerden," Aristotle, *Probleme*, trans. Paul Gohlke, Aristoteles, Die Lehrschriften, 16 (Paderborn: Ferdinand Schöning, 1961).

54. GT, BL add. 21978, 18vb: "Nam que faciunt his gaudent et sic sperma emittere gliscunt. Quibus igitur hec accidunt hi sic facere cupiunt [et?] magis propter assuetudinem que quasi natura fit."

55. Faith Wallis, "Giulio Guastavini's Commentary on Ps.-Aristotle's Account of Male Same-Sex Coitus, 'Problemata' 4.26," in Borris and Rousseau, *The Sciences of Homosexuality in Early Modern Europe*, 70.

56. TG in *Extra ordinem naturalium libri quibus nonnulli etiam additi sunt Aristoteli ascripti*, 19rb: "Fit enim, ut tam gestiant, quam cum agunt, utique608 genituram nihilominus ita emittere [BL add. 21978 adds: emictere] valeant. Ergo agere cupiunt, quibus haec ipsa usu evenerunt." Gaza's wording could be interpreted to say that by experiencing pleasure and ejaculation from either the active or the passive role, men will desire to take the active role (*cupiunt agere*), but this reading would not accord well with what follows.

57. It is not impossible that the action they desire to take is engagement in a sexual act in which they are the receptive partners, but there is generally a strong correlation between the grammatically passive and the sexually passive, as the rare passive form "coitur" at the opening of Pietro's commentary attests. Cf. Jan Ziolkowski, *Alan of Lille's Grammar of Sex: The Meaning of Grammar to a Twelfth-Century Intellectual*, Speculum Anniversary Monographs, 10 (Cambridge, MA: Medieval Academy of America, 1985), e.g., 69.

58. PA, BNF lat. 6540, 64vb: "Qui moventur ad hunc actum propter delectationem quoniam 'quecumque accidit facere gaudere' et delectare illis [Cesena, BCM D.XXIV.2: illos] qui 'emittunt genituram' recte coitum exercendo, 'ita' [Cesena, BCM D.XXIV.2, for recte . . . 'Ita': Item coitum exercendo ita] accidit illis quibus talia fiunt. Propter hunc enim actum profanum appetunt exercere ad hoc ut gaudeant et delectentur. Vel aliter quia 'quecumque' et cetera, id est tota eorum intentio non est nisi 'emittere genituram' id est humiditatem illam seminalem ut inde eos 'accidat gaudere' et delectari."

59. JJ, Sorbonne 122, 47rb: "Appetit enim virga alios fricare et cum virgis aliorum in ano in eis fricari." Cf. PA, BNF lat. 6540, 64vb: "in ano fricari ab eis."

60. "Aachen" *Problemata*, Erfurt, CA F. 263, 23rb: "iam consuetudo est eis natura quod semper quasi fricant illas partes propter delectationem vel volunt fricari ab aliis et maxime isti sodomite."

61. In classical Latin and/or in a structure containing a contrasting "hic" or "ille," it might conceivably be possible to interpret "iste" as "the former" or "the latter," but in

medieval Latin and in the absence of a parallel structure, no such decision is authorized. See Keith Sidwell, *Reading Medieval Latin* (Cambridge: Cambridge University Press, 1995), § 11.c, p. 365; Alison Goddard Elliott, "A Brief Introduction to Medieval Latin Grammar," in K. P. Harrington, *Medieval Latin*, rev. Joseph Pucci, 2nd ed. (Chicago: University of Chicago Press, 1997), § 5.2, p. 33.

62. "Prague" *Problemata*, NK VIII.A.19, 79rb–va: "Et etiam aliqui per visum et consuetudinem a iuventute volunt tantum fricare, quia recordantur de dilatacione [sic] perverta. Sed alii sunt qui volunt fricari et fricare, qui habent materiam spermaticam versus membrum virile. Et in eis non est displanatio [*sic*; for displasmatio?], sed vie spermatice sunt recte et aperte. Et ipsi coheunt cum mulieribus aliquando. Et forsan materia spermatica in ipsis partitur, ita quod aliqua pars fluit versus culum et alia versus priapum."

63. Avicenna, *Liber canonis medicine*, bk. 3, fen 20, tr. 1, ch. 42, 358ra. It is likewise conceivable that an earlier version attributed to this group the desire "to be rubbed [*fricari*]" rather than "to rub [*fricare*]".

64. BM, BNF lat. 15081, 12vb: "non ante pubertatem sed circa pubertatem."

65. PA, BNF lat. 6541, 75rb: "Consueta convertuntur in naturam et magis quam illis qui non ante pubertatem incipiunt." In the margin: "Sodomitati ante pubertatem etc. Sunt illi qui efficiuntur postea sodomite etc."

66. Table for PA, BNF lat. 6543, 46rb: "Sodomitati ante pubertatem et tempus quo possint emittere sperma sunt illi qui efficiuntur postea sodomite propter memoriam et cetera, quia consuetudo vertitur in naturam." The placement of the marginal notation "*sodomite*" in BNF lat. 6540, the manuscript for which BNF lat. 6543 is the index, is a little more ambiguous. The word occurs not at Pietro's introduction of it in connection with habit but at top of next column, where Pietro is elaborating on Aristotle's putative doubling back to those anatomically suited to both rubbing and being rubbed.

67. Michael Rocke, *Forbidden Friendships: Homosexuality and Male Culture in Renaissance Florence* (New York: Oxford University Press, 1996), 94–111.

68. PA, BNF lat. 6540, 65ra: "Ita hoc modo desiderant illi qui hec operantur nephanda. . . . In quorum extirpationem lex inclamat: 'Insurgant leges armentur vita [Cesena, BCM D.XXIV.2: iura] ultore gladio feriantur.' Unde Avicenna 3° *Canonis* 20ª [tr. 1, ch. 11, 353ra]: 'Coitus cum infantibus fetidus [Cesena, BCM D.XXIV.2: fedus] est apud multitudinem gentium et prohibitus in lege.' "

69. *Codex Justinianus*, 9.9.31, and *Codex Theodosii*, 9.7.3.

70. It appears in Ioannes Saresberiensis, *Policraticus, I–V*, ed. K. S. B. Keats-Rohan, Corpus Christianorum Continuatio Mediaevalis 118 (Turnhout: Brepols, 1993), bk. 3, ch. 13, 219, and thus also in the fourteenth-century French translation of it: Denis Foulechat, *Le Policratique de Jean de Salisbury (1372), Livres I–III*, ed. Charles Brucker, Publications Romanes et Françaises 209 (Geneva: Droz, 1994), 240. See also John Boswell, *Same Sex Marriages in Premodern Europe* (New York: Villard, 1994), 85–86, citing Peter the Chanter, and n. 163.

71. PA, BNF lat. 6540, 65ra: "Et primo ostendit quomodo se habeant ad hunc actum incipientes ipsum exercere tempore pubertatis dicens et ideo 'quicumque non ante pubertatem' et tempore [Cesena, BCM D.XXIV.2: tempus] in quo possunt emittere sperma 'sed

circa' hanc consueverunt suponi et fricari circa anum eo quod in consuetudine talis eis relinquitur mero [Cesena, BCM D.XXIV.2: memoria]."

72. JJ, BNF lat. 15454, 54va: "non ante pubertatem et in tempore in quo non possunt emittere sperma."

73. All PA manuscripts read "tempus" except BNF lat. 6540, whose careful scribe may have preferred what he saw in a JJ manuscript, some of which read "tempore." With the word "time" in a different grammatical form, they seem to say, " 'not before puberty,' and at the time when they are capable of emitting semen."

74. BM, BNF lat. 15081, 12vb: "Si autem luxuria accidat existens et mollis et cito singula horum accidunt."

75. TG in *Extra ordinem naturalium libri quibus nonnulli etiam additi sunt Aristoteli ascripti*, 19rb: "Sed si accidat, ut idem et salax, et mollis sit, longe expeditius haec omnia evenire posse putandum est"; GT, BL add. 21978, 18vb: "Si vero libidinosus etiam et mollis fuerit facilius singula horum accidunt." See also the translations of Louis, Hett, and Mayhew, and Wallis, "Guilio Guastavini's Commentary," 61 and 70, the latter in Wallis's English translation.

76. Neal, "Disorder of Body, Mind, or Soul," 44–46.

77. PA, BNF lat. 6540, 65ra: "Ostendit quomodo se habeant ad hoc illi qui 'ante pubertatem' supponuntur dicens quod si huiusmodi turpis 'luxuria accidat' tempore quo 'mollis' et tenera est natura istorum qui supponuntur ut quando non possunt sperma emittere, omnia predictorum istis 'cito' adveniunt ut delectatio inmensa et emissio seminis ad anum et cetera. Et consueta convertuntur in naturam et magis quam illis qui [BAV lat. 2175 adds: non] ante pubertatem incipiunt." In PA, BNF lat. 6541A "turpis" is replaced by "temporis."

78. PA, BN 6541, 75rb, margin: "Sodomitati ante pubertatem etc."

79. See, for example, Jordan, *Invention of Sodomy*, 143–47, on Thomas Aquinas.

80. Pietro d'Abano, *Conciliator*, diff. 26, 39vb: "sunt iuvenes et non pueri, cum sint superflue molles et lenes."

81. Luke Demaitre, "The Idea of Childhood and Child Care in Medical Writings of the Middle Ages," *Journal of the History of Psychology* 4 (1977), 467.

82. Christine de Pizan, *Le livre des trois vertus: Édition critique*, intro. and notes, Charity Cannon Willard, ed. in collaboration with Eric Hicks, Bibliothèque du XVe siècle, 50 (Paris: Honoré Champion, 1989), bk. I, ch. 15, 61, speaking here especially of girls: "que on lui admenistre livres de devocion ou qui parlent de bonnes meurs; ne nulz de choses vaines, de folies ou de dissolucions ne souffrera que devant elle soyent portéz, pour ce que la doctrine et enseignement que l'enfant retient en sa premiere jonnesce il en a communement recort toute sa vie."

83. *Mollis* occurs at various points in Pietro's *Physiognomia* with no visible sexual associations. Boswell, *Christianity, Social Tolerance, and Homosexuality*, 106–7, interprets the senses of the Greek, ranging from "gentle" to "debauched" in early Christian texts. See also Halperin, "How to Do the History of Male Homosexuality," in *How to Do the History of Homosexuality* (Chicago: University of Chicago Press, 2002), 111–12; and Carol Clover, "Regardless of Sex: Men, Women, and Power in Early Northern Europe," *Speculum* 68 (1993): 363–87.

84. Aristotle, *Ethica nicomachea*, VII.6, 47b23, 48a12, and 50a14. Similarly, Walter Burley, *Expositio Gualteri Burlei super decem libros Ethicorum Aristotelis* (Venice: Andrea Torresano de Asula, 1500), bk. 7, tr. 1, ch. 7, 115va–b. Elsewhere in the *Ethics* softness (*mollities*) appears in connection with cowardice and laziness. At II.2, 03b25 Aristotle mentions the importance of early training, "ex iuvene" in the medieval Latin (fasc. 3, 164). See also Aristotle, *Politica*, VII.17–VIII.4.

85. Aristotle, *Ethica nicomachea*, fasc. 3, VII.10, 50b14. Some of Pietro's medical readers may also have been familiar with the lengthy discussion of the Scythians in the Hippocratic *Airs, Waters, and Places* (17–22), where Scythian men are said to be "soft" (20–21). The sexual consequence is impotence, not excessive or misdirected sexual activity; the demographic consequence is low fertility; and the gender consequence is that some men take on the social roles of women. The work was available but not part of the medical curriculum: Kibre, *Hippocrates Latinus*, 25–28.

86. Pierre J. Payer, *Sex and the Penitentials: The Development of a Sexual Code, 550–1150* (Toronto: University of Toronto Press, 1984), 41–42 and 62, locates the Latin in various contexts, including masturbation and homosexual acts. See Allen J. Frantzen, *Before the Closet: Same-Sex Love from "Beowulf" to "Angels in America"* (Chicago: University of Chicago Press, 1998), 165–66.

87. George of Trebizond, *Collectanea Trapezuntiana*, text 157, 4.26, p. 647: "Ac ideo humor huiusmodi ad podicem divertitur. . . . Unde natura molles efficiuntur. Hi sunt qui pati nimium gestiunt, nec agere ullo modo possunt."

88. "Bavarian" *Problemata*, Ghent UB 178, 129rb–va: "Et hic dicit Aristoteles quod pueris, interim quod sunt tenelli et molles, cito advenit delectatio inmensa ex emissione seminis, et emissio seminis ad anum, quia consueta convertuntur in naturam et magis quam in illis qui non ante pubertatem incipiunt."

89. "Bavarian" *Problemata*, Krakow, BJ 2095, 227vb: "*nondum* pubertatem incipiunt" (emphasis added). The reading at CLM 12021, 135ra, is more ambiguous but appears to agree.

90. The puzzling omission of "not" occurs in all the manuscripts of PA's second version except BAV lat. 2175, and in several of the manuscripts of the JJ redaction (Paris, Arsenal 723, Sorbonne 122). Thus the apparent misreading is not likely to have been due to a single scribal error.

91. "Prague" *Problemata*, NK VIII.A.19, 79rb.

92. George of Trebizond, *Collectanea Trapezuntiana*, text 157, 4.26, p. 647: "Et de hac causa feruntur ad hanc turpitudinem magis quam memoria. Qui vero maiusculi se viris subiecerunt, sola voluptatis huiusmodi memoria pati cupiunt."

93. PA, BNF lat. 6540, 65ra: "Et consueta convertitur in naturam."

94. Christine de Pizan, *Le livre du corps de policie*, ed. Robert H. Lucas, Textes Littéraires Français [145] (Geneva:Droz, 1967), pt. 1, ch. 24, 76–77: "Il disoit que le mouvement du ciel le tenoit si fermement qu'il ne se pouoit partir. . . . Mais en ce qui est subgect a l'ame, c'est assçavoir deliberacion de voulenté les influences du ciel n'ont point de seigneurie, non obstant qu'il peut estre vray que l'action du ciel donne a l'omme pluseurs inclinacions, si comme aux aucuns joliveté, luxure, ou autres mouvemens naturelz. Mais non obstant ce,

l'omme y peut mettre frain par raison et resister quant au fait a toutes teles inclinacions." Like Pietro, she associates the idea with Aristotle's *Ethics*. For Nicole Oresme, see, for example, "Livre de divinacions," ch. 6 in G. W. Coopland, *Nicole Oresme and the Astrologers: A Study of His Livre de divinacions* (Liverpool: Liverpool University Press, 1952), 66.

95. Pietro d'Abano, *Lucidator dubitabilium astronomiae*, in *Tratti de astronomia*, ed. Federici Vescovini, diff. 1, 126–28, and Federici Vescovini commentary, 72–75.

96. See Graziella Federici Vescovini, "L'antropologia naturale di Pietro d'Abano," *Paradigmi* 15 (1997): 525–41. In addition to the *Lucidator*, see his *Compilatio phisonomie*, especially pt. 3, [37]r–[50]v.

97. Nancy Siraisi, "The 'Expositio problematum Aristotelis' of Peter of Abano," *Isis* 61 (1970): 321–39, 325–27.

98. George of Trebizond commentary on Ps.-Ptolemy, *Centiloquium*, in Guido Bonatus *De astronomia tractatus X*, aphorism 80, col. 49: "Totam enim vim huius formativae virtutis ita gubernat [Saturnus]."

99. "Bavarian" *Problemata*, Krakow, BJ 2095, 227vb; George of Trebizond, *Collectanea Trapezuntiana*, text 157, 4.26, p. 647, cites the *Centiloquium* (upon which he himself wrote a commentary) in the context of *Problemata* IV.26. For Despars, see Neal, "Disorder of Body, Mind, or Soul," 51, and Danielle Jacquart, "Theory, Everyday Practice, and Three Fifteenth-Century Physicians" *Osiris*, 2nd ser. 6, *Renaissance Medical Learning: Evolution of a Tradition* (1990): 140–60; for Evrart de Conty, see below, Chapter 5.

100. All the extant manuscripts of PA's first version except one (BAV lat. 2174) include "inclinatio." I do not have a reading for Paris, Arsenal 723, but of the other manuscripts of JJ only BAV lat. 2176 does so. Johannes de Janduno, *Quaestiones super 8 libros Physicorum Aristotelis*, bk. 2, q. 14, 39va, on Aristotle, *Physica*, II.8, 199b3–5, citing *De generatione et corruptione*, II.

101. There are as many as four different versions of which houses of Jupiter are involved. In addition, there is an easily understandable confusion between "80" and "8°" in the reference to the *Centiloquium*.

102. PA, BNF lat. 6540, 65ra: "Ptholomeus vero in 8. verbo *Centiloquii* [*Centiloquium* 80] et in 4° *Quadripartiti* assignat causam ex coniunctione vel aspectu malorum cum Venere in signo malorum et maxime in domo 7ª et 6ª Iovis coniunctione vel aspectu ab hiis remoto et cetera."

103. He cites the Ptolemaic works throughout his *Lucidator*. Both circulated in Arabic and entered Latin in the company of the commentaries. John of Seville translated the Pseudo-Ptolemy *Centiloquium*. There were several medieval translations of Ptolemy's *Quadripartitum*, of which the two that circulated most widely were those of Plato of Tivoli and Egidius de Thebaldis. Charles Homer Haskins, *Studies in the History of Mediaeval Science* (Cambridge, MA: Harvard University Press, 1924, repr. New York: Frederick Ungar, 1960), 110–12; Francis J. Carmody, *Arabic Astronomical and Astrological Sciences in Latin Translation: A Critical Bibliography* (Berkeley: University of California Press, 1956), 16, 18–19. The translation of Egidius included the commentary, of which Pietro made ample use (see list of references in *Lucidator*, 415), so he undoubtedly used that translation. He may also have been familiar with Plato of Tivoli's version. The commentator "Ali Rodoan" or

"Haly Heben Rodan" is identified by modern scholars as 'Alī ibn Riḍwān. See Federici Vescovini, "Nota introduttiva" to differentia 1 in Pietro d'Abano, *Lucidator*, 61, n. 7.

104. Federici Vescovini, "Nota introduttiva" to differentia 1 in Pietro d'Abano, *Lucidator*, 73. The relevant passage in Pietro's text is differentia 1, 126–28.

105. 'Alī ibn Riḍwān, commentary on Ps.-Ptolemy, *Centiloquium*, trans. John of Seville (Venice: Bonetus Locatellus, 1493), § 40, 110va: "Et ego quidem vidi quendam divitem cuius nativitatis domini erant Mars et Saturnus et erant fortes et delectabatur in commestione marinarum beluarum que erant in salse et fetentes et coquebat illas in furno. Iurabat etiam quod quando coibat cum formosis et suaviter redolentibus deficiebat eius virga et cum vili et olente confortabatur."

106. Ptolemy, *Quatripartitum*, trans. Egidius Tebaldi (Venice: Bonetus Locatellus, 1493), bk. 4, ch. 4, "De coniugiis," 94rb: "Et si in signis fuerint femininis, substinebit quod res eadem fieret de eo. Et si ambo matutinales fuerint, diliget pueros solummodo. Et si in signis masculinis, diliget iacere cum viris cuiuscumque etatis fuerint."

107. 'Alī ibn Riḍwān, in Ptolemy, *Quadripartitum*, bk. 4, ch. 4, 94rb: "Et si aliquo horum duorum statuum fuerit Mars et Venus in signis femininis dabunt nato magnam superationem luxurie in voluntate, ita quod patietur fieri de seipso id quod facit de mulieribus. Et si Mars et Venus fuerint matutinales, natus cum pueris amabit iacere solummodo. Et idem si in quartis fuerint masculinis et si in aliquo horum duorum statuum fuerint Venus et Mars in signis masculinis, natus diliget iacere cum viris cuiuscumque etatis fuerint, nec diliget mulieres."

108. Ptolemy, *Quadripartitum*, trans. Plato of Tivoli, bk. 4, ch. 4, 94ra: "Quia si signa feminina fuerint, ipse forsitan ad turpes actus supponetur. Et si uterque matutinalis exciterit cum masculis solummodo scelus illud perpetrabit. Et si signa fuerint masculina, omnibus modis circa masculos agere sollicitabitur." The Greek was "only boys" (παιδικὰ μόνα). Ptolemy, *Le pervisioni astrologiche (Tetrabiblos)*, ed. and trans. Simonetta Feraboli, Scrittori Greci e Latini ([n.p.]: Fondazione Lorenzo Valla, Arnoldo Monadori, 1985), § 17, 318.

109. 'Alī ibn Riḍwān, commentary, in Ps.-Ptolemy, *Centiloquium*, trans. John of Seville, § 80, 114va: "incitabit natum ut preferat stuprum posteriorem anteriori et in excitatione spermatis cum manu. Sed si fuerit Saturnus orientalis, diliget confricationem cum pueris et preponit negros albis et flavis. Quod si fuerit in infortunium ex incorporatione Martis, significabit quod non sit zelotipus et quod coeat cum pluribus pueris qui et coeant cum illo et cum uxore eius. Et perveniet ut fornicetur cum propriis filiis et agat cum eis contra legem et cetera que pertinent ad immunditiam."

110. George of Trebizond, commentary on Ps.-Ptolemy, *Centiloquium*, aphorism 80, cols. 49–50: "Quare fit ut quando ei coniungitur, obstruat claudatque seminales hominum meatus." The commentary was based on his own translation from the Greek.

111. On George's fight with Poggius Bracciolini, whom he compared to a "Florentina femina," see Monfasani, *George of Trebizond*, 109–111. Poggio is the protagonist of Stephen Greenblatt's *The Swerve: How the World Became Modern* (New York: W. W. Norton, 2011).

112. George of Trebizond, commentary on Ps.-Ptolemy, *Centiloquium*, aphorism 80, col. 50: "ut quasi iam decrepitus ageret quod iuvenis facere nunquam potuit." On his versions, see Monfasani, *Collectanea Trapezuntiana*, § 161, pp. 690, 692. There is a seventeenth-century

English translation of the commentary that contains this story: Georgius Trapezuntius, *Commentary of Georgius Trapezuntius upon the Aphorismes of Ptolemeus*, trans. Thomas Blackbourne, London, Wellcome Library, ms. 139 on aphorism 80. Monfasani, *Collectanea Trapezuntiana*, § 161, p. 692. On Poggio Bracciolini (1380–1459), see also *Dizionario biografico degli italiani*, vol. 13 (Rome: Istituto della Enciclopedia Italiana, 1971), 640–46.

113. Alanus de Insulis, "Enchiridion de planctu nature," § 1, met. 1, lines 27–28, fasc. 2, p. 807.

114. Ptolemy, *Quadripartitum*, IV.4, 93vb. According to the translation of Plato of Tivoli, "Erit eunuchus . . . aut erit sterilis vel sine virge foramine"; according to that of Egidius, "Erit castratus, algogerus spadus vel manerus qui filium non habebit aut erit clausus." For reasons that may reside in the Arabic, or in the choices of the Latin translators, the Latin of the commentary by Ibn Riḍwān reads, "Erit castratus vel manerus. Et si fuerit mulier erit clausa."

115. The association appears not only in *Problemata* XXX.1 but also in medieval texts on lovesickness. Cf. Raymond Klibansky, Erwin Panofsky, and Fritz Saxl, *Saturn and Melancholy: Studies in the History of Natural Philosophy, Religion, and Art* (London: Thomas Nelson, 1964), 17, 30, 35, 84, etc.

CHAPTER 3. "JUST LIKE A WOMAN"

1. PA, Cesena, BCM D.XXIV.2, 36va: "Redacte sunt in naturam et mores mulierum." PA, BNF lat. 6541A and JJ, Sorbonne 122 also read "redacte," in the feminine plural. "Redacti," in the masculine plural, occurs in the other manuscripts.

2. "Erfurt" *Problemata*, CLM 4710, [268]rb.

3. EC, BNF fr. 24281, 116v.

4. "Custodia" appears in all mss. of PA's first version and JJ; none of PA's second. Aristotle, *History of Animals, Books VI–X*, ed. and trans. David M. Balme, Loeb Classical Library, 439 (Cambridge, MA: Harvard University Press, 1991), IX [= XI in other modern editions], 581b 9–19.

5. Various signs of attention can be found at BM, BAV Ottob. 2083, 14rb (one of three at IV.26); PA with "custodia," BNF lat. 6541, 74vb (two of twelve marginalia at IV.26 by one reader, two of four by another, and one of two by a third); JJ, BNF lat. 15454, 54ra (one of many at IV.26); Seville, 7.7.9, 52vb (one annotator commented only on this section; another [53ra–b] only on the section concerning habit), Sorbonne 122, 47ra (two of the three nonverbal marks).

6. JJ, Seville 7.7.9, 52vb, margin.

7. Rocke, *Forbidden Friendships*, 106–9.

8. Ruth Mazo Karras and David Lorenzo Boyd, "'Ut cum muliere': A Male Transvestite Prostitute in Fourteenth-Century London," in Fradenburg and Freccero, *Premodern Sexualities*, 101–16. For the text of the court record, see Boyd and Karras, "The Interrogation of a Male Transvestite Prostitute in Fourteenth-Century London," *GLQ: A Journal of Lesbian and Gay Studies* 1 (1995): 459–65.

9. Karma Lochrie, *Covert Operations: The Medieval Uses of Secrecy*, The Middle Ages (Philadelphia: University of Pennsylvania Press, 1999), esp. ch. 5, "Sodomy and Other Female Perversions," 177–227.

10. Gulielmus Peraldus, *Summa de viciis*, 197ra–357vb, in *Summa virtutum et vitiorum* (Venice: Paganinus de Paganinis Brixiensis, 1497), reproduced in Italian Books Before 1601, roll 460, item 3 (Watertown, MA: University Microfilms, n.d.), 204va: "Sic deus scit illum quem ipse fecit homo, cum videt quod ipse se fecerit feminam, non reputat creaturam suam." Jordan has called attention to this dimension of Peraldus's work (*Invention of Sodomy*, 109–113), and to this particular passage (112). Kelly appears to be mistaken on this point: Henry Ansgar Kelly, "The Pardoner's Voice, Disjunctive Narrative, and Modes of Feminization," in *Essays in Honor of V. A. Kolve*, ed. R. F. Yaeger and Charlotte C. Morse (Asheville, N.C.: Pegasus Press, 2001), 442, n. 70.

11. Lochrie, *Covert Operations*, 183–86.

12. Peraldus, *Summa virtutum et vitiorum*, 203vb: "Ut cum quis procurat vel consentit ut semen alibi quam in loco ad hoc a natura deputato effundatur."

13. Arnaud de Verniolle, in Goodich, trans., *Other Middle Ages*, 134.

14. Pierre Recort, in Goodich, trans. *Other Middle Ages*, 127–28.

15. Rocke, *Forbidden Friendships*, esp. ch. 3, 87–111.

16. Puff, "Localizing Sodomy: The 'Priest and Sodomite.'"

17. BM, BNF lat. 15081, 12vb: "Quare necesse est locum hunc secari ipsorum. Secatio aut hec quidem omnino facit corruptionem, hec autem perversionem. Illa quidem igitur non est. Mulier enim itaque fieret."

18. PA, BNF lat. 6540, 63vb.

19. BM, BNF lat. 6327, 21va, gloss: "possibile"; Escorial f. I.11, 7ra, gloss: "faciendum."

20. "Aachen" *Problemata*, Erfurt, CA F.263, 23 rb: "Incurreunt [sic] mortem vel perversionem mentis vel totaliter fierent sicud mulier."

21. Aristotle, *Problemata*, IV.26, 879b28–29: "Διὸ καὶ ἄπληστοι, ὥσπερ αἱ γυναῖκες."

22. PA, BNF lat. 6540, 64rb–va: "Ostendit quomodo tales se habeant ad curam istius male plasmationis pororum quibus sperma in virgam defertur. Et primo ostendit quod non possint curari per cyrurgiam secando. . . . Aliquando inducit 'perversionem' et mutationem in deterius in partibus pororum, sicut accidit non parum secantibus plagas."

23. PA, BNF lat. 6540, 64va: "Item si fieret, qui vir erat primo mutaretur in mulierem et sic mutaretur in peius. Unde bene dictum est ex ea 'perversionem' fieri. . . . Non tamen intelligendum quod [Cesena, BCM D.XXIV.2 adds: vir] essentialiter mutetur in mulierem sed potius accidentaliter acquiriendo actus et mores muliebres."

24. PA, BNF lat. 6540, 63va: "Aut propter id quod hii meatus qui terminantur ad virgam extendantur [Cesena, BCM D.XXIV.2: excecantur], materia opilativa interveniente, sicut oculi excecantur cum humor viscosus cadit in nervos opticos, sicut est videre . . . in illis quorum complexiones infrigidate sunt et humectate in tantum quod redacti [Cesena, BCM D.XXIV.2: redacte] sunt in naturam et mores muliebres."

25. PA, BNF lat. 6540, 64rb; George of Trebizond commentary on Ps.-Ptolemy, *Centiloquium*, aphorism 80, col. 49.

26. EC, BNF fr. 24281, 116v: "Aristote voelt dire que aucun sont qui desirent et voelent

acomplir ceste chose pour ataindre la desus dite delectation en compaignment de fame en la maniere que nature et raisons y enclinent, et non autrement."

27. PA, BNF lat. 6540, 63rb: "Querit Aristoteles in hoc capitulo duas questiones. Et prima est quare est quod gaudent sive delectantur illi 'cum quibus' actus coitus peragitur, ut mulieres aut eis proportionati."

28. These do not exhaust medieval configurations of the womanly that find their way into the literature of the *Problemata*. For example, there are echoes below of the fascination with the hidden womb explored in Katharine Park, *Women's Secrets: Gender, Generation, and the Origins of Human Dissection* (New York: Zone, 2006), and with the polluting potential explored in Dyan Elliott, *Fallen Bodies: Pollution, Sexuality, and Demonology in the Middle Ages*, The Middle Ages (Philadelphia: University of Pennsylvania Press, 1999).

29. James A. Schultz, *Courtly Love, the Love of Courtliness, and the History of Sexuality* (Chicago: University of Chicago Press, 2006), 18.

30. Monica H. Green, *Making Women's Medicine Masculine: The Rise of Male Authority in Pre-Modern Gynaecology* (Oxford: Oxford University Press, 2008); Green, "The Development of the *Trotula*," *Revue d'Histoire des Textes* 26 (1996): 119–203, reprinted in *Women's Healthcare in the Medieval West*, Variorum (Aldershot, U.K.: Ashgate, 2000), § V; and Green, ed. and trans., *The Trotula: A Medieval Compendium of Women's Medicine*, The Middle Ages (Philadelphia: University of Pennsylvania Press, 2001).

31. *Liber de sinthomatibus mulierum*, in Green, ed. and trans., *The Trotula*, §§ 1–44, 70–83.

32. Green, *Making Women's Medicine Masculine*, 163–203. See also Green, "Obstetrical and Gynecological Texts in Middle English," *Studies in the Age of Chaucer* 14 (1992): 53–88, reprinted in *Women's Healthcare in the Medieval West*, § IV.

33. José Pablo Barragán Nieto, ed., *El "De secretis mulierum" atribuido a Alberto Magno: Estudio, edición crítica y traducción*, Textes et Études du Moyen Âge 63 (Porto: Fédération Internationale des Instituts d'Études Médiévales, 2012). I am grateful to Helen Rodnite Lemay for having lent me photoreproductions of the 1508 and 1580 editions of this work.

34. Lochrie, *Covert Operations*. On the "secrets" literature more generally, see William Eamon, *Science and the Secrets of Nature: Books of Secrets in Medieval and Early Modern Culture* (Princeton: Princeton University Press, 1994).

35. By the end of the fifteenth century, versions or adaptations of the work were available in French, German, Dutch, and even Czech. In addition to Margaret Rose Schliessner, "Pseudo-Albertus Magnus: 'Secreta mulierum cum commento,' Deutsch: Critical Text and Commentary," Ph.D. diss., Princeton University (Ann Arbor: University Microfilms International, 1987, microfilm), 5–6, and "A Fifteenth-Century Physician's Attitude Toward Sexuality: Dr. Johann Hartlieb's 'Secreta mulierum' Translation," in *Sex in the Middle Ages: A Book of Essays*, ed. Joyce E. Salisbury, Garland Medieval Casebooks 3; Garland Reference Library in the Humanities 1360 (New York: Garland, 1991), 110–25; cf. Monica H. Green, "'Traittie tout de mençonges': The '*Secrés des dames*,' 'Trotula,' and Attitudes Toward Women's Medicine in Fourteenth- and Early Fifteenth-Century France," in *Christine de Pizan and the Categories of Difference*, ed. Marilyn Desmond, Medieval Cultures, 14 (Minneapolis: University of Minnesota Press, 1998); reprinted in *Women's Healthcare in the Medieval West*, § 6, 171, nn. 11–13, and 173, nn. 28–30.

36. Blair, "The 'Problemata' as a Natural Philosophical Genre," 181–185; Margaret Rose Schleissner, "Sexuality and Reproduction in the Late Medieval 'Problemata Aristotelis,'" in *Licht der Natur: Medizin in Fachliteratur und Dichtung: Festschrift für Gundolf Kiel zum 60. Geburtstag*, ed. Josef Domes et al., Göppinger Arbeiten zur Germanistik 585 (Göppingen: Kümmerle, 1994), 383–98.

37. L. R. Lind, ed., *Problemata varia anatomica: The University of Bologna, MS 1165*, Humanistic Studies 38 (Lawrence: University of Kansas Publications, 1968), 65.

38. Lind, ed., *Problemata varia*, 70.

39. Lind, ed., *Problemata varia*, 53.

40. Park, *Women's Secrets*; Green, *Making Women's Medicine Masculine.*

41. Lind, ed., *Problemata varia*, 53, 61, 63–5; 60; 71.

42. Aristotle, *Generation of Animals*, III.5, 756a6–30. But Aristotle observed that in certain insects the female inserts a member into the male (I.16, 721a2–17).

43. BM, BN Marciana 2488, 204rb: "Propter quid quidam cum quibus [Gloss: scilicet passive ad modum femellarum] coitur gaudent et hii simul quidem agentes [Gloss: ut mares] hii autem non." Italian ownership is attested from the first half of the fifteenth century. The glosses show features in common with other manuscripts clearly situated in fourteenth-century Italy, and contain technical medical terms. See Guido Billanovich and Luigi Olivieri, "Pietro d'Abano e il codice Antoniano XVII 270," *Italia medioevale e umanistica* 28 (1985): 221–94, on 92, n. 41; 101; 103, n. 55, et passim. AL #1603.

44. PA, BNF lat. 6540, 65ra: "illi qui ante pubertatem supponuntur." On coital positions, see, for example, Brundage, *Law, Sex, and Christian Society*, 367, 452–53, 508; Karras, *Sexuality in Medieval Europe*, 83.

45. Michael Camille, "Manuscript Illumination and the Art of Copulation," in *Constructing Medieval Sexuality*, ed. Karma Lochrie, Peggy McCracken, and James A. Schultz, Medieval Cultures, 11 (Minneapolis: University of Minnesota Press, 1997), 58–90.

46. PA, IV, prologue, BNF lat. 6540, 53ra: "Coytus est mutua actio maris et femelle qua tertie digestionis expellit superfluum quod delectationis causa exercetur."

47. PA, BNF lat. 6540, 63va: "cum gaudent [Cesena, BCM D.XXIV.2: congaudent]."

48. Even an unusual exception adds complexity: The scribe of a thirteenth-century Italian copy that Pietro may have seen and annotated omitted the word "with," and someone corrected the text, probably before Pietro would have seen it. That circumstance could have prompted him to notice and gloss it (Padua, Antoniana Scaff. XVII, 370, 12rb). Billanovich and Olivieri, "Pietro d'Abano e il codice Antoniano," 242, n. 40, and 268–69. The Greek text leaves no doubt about the phrase's passive sense. It contains no prepositional phrase equivalent to "with whom"; the idea is conveyed compactly by a past participle, "ἔνιοι ἀφροδισιαζόμενοι." Aristotle, *Problemata*, IV.26, 879a36.

49. PA's first version and JJ mss. read "a quibus," with the exception of BNF lat. 6542, which reflects uncertainty, offering "a cum quibus." PA's second version reads "aliquibus," with the exception of the hybrid ms. BNF lat. 6540. Scribal conventions may come into play here, however. The distinction between *a quibus* and *aliquibus* can be as little as a single pen stroke—between *a* and *a*ⁱ. But if it was an error, it was one that made some sense.

50. PA, BNF lat. 6540, 65ra: "Alii autem fricationem faciendo circa anum et virgam in ipsum imponendo sicut in vulvam imponitur."

51. PA, BNF lat. 6540, 64va–b: "Quidam sunt qui appetunt pati et fricari. . . . Hii appetunt fricare et fricari secundum utramque istarum partium. Appetunt enim virga alios fricare et cum virgis aliorum in ano fricari ab eis."

52. "Prague" *Problemata,* Prague, NK VIII.A.19, 79rb: "Propter quid aliqui sunt viri qui volunt tantum fricari et quidam tantum fricare et quidam volunt fricari et etiam fricare in coytu."

53. "Prague" *Problemata,* Prague, NK VIII.A.19, 79va: "Et ipsi coheunt cum mulieribus aliquando."

54. As Patricia Simons argues for a later period, although penetration was neither central nor simple, it constituted a persistent motif in the representation of male bodies in action: *The Sex of Men in Premodern Europe: A Cultural History,* Cambridge Social and Cultural Histories (Cambridge: Cambridge University Press, 2011), 112–22. She observes (216) that sodomy charges against men do not necessarily involve penetration.

55. Arnaud of Verniolle, in Goodich, trans., *Other Middle Ages,* 134.

56. Aristotle, *Generation of Animals,* IV.4, 770a6–7, trans. Scot, 184: "Et universaliter est melius opinari quod causa est in materia." Cf. trans. Moerbeke, 134: "Totaliter autem magis causam putandum in materia." See also IV.4, 770a31.

57. PA, IV.10, BNF lat. 6540, 56vb, where he uses "masculus orbatus vel occasionatus sicut dicit in sermo *De generatione animalium,*" and IV.26, 64ra, where he uses "orbatus." BM, BNF lat. 6327, 21va, gloss. Aristotle, *Generation of Animals,* I.3, 737a29: "ἄρρεν πεπηρωμένον." Both terms occur among scholastics. William of Moerbeke's Latin translation reads, "orbatus masculus" (*De generatione animalium,* 55).

58. In addition to PA, IV.10, BNF lat. 6540, 56vb, see BN Marciana 2488, 204rb, where, at "ut accidit eunuchis et effeminatis" in BM, the gloss is "scilicet occasionatus." See also Albert Mitterer, "*Mas occasionatus* oder zwei Methoden der Thomasdeutung," *Zeitschrift für katholische Theologie* 72 (1950): 80–103, for examples from Thomas Aquinas and Albertus Magnus.

59. Aristotle, *Generation of Animals,* IV.3, 769b29–30, trans. Scot, 183: "Filius monstruosus est etiam occasionatus." Cf. trans. Moerbeke, 133: "Monstrum orbatio quedam est."

60. Barragán Nieto, ed., *Il "Secretis mulierum,"* ch. 1, 240–44. In this edition, the Latin text occupies the even-numbered pages, the Spanish translation the odd.

61. Barragán Nieto, ed., *Il "Secretis mulierum,"* ch. 6, 390–420.

62. Barragán Nieto, ed., *Il "Secretis mulierum,"* ch. 6, 416–18: "Et quod omnes modi speciales in monstris possunt reduci ad duos generales manifeste patet ex intencione Avicenna in secundo *Sufficiencie* sue, ubi eciam addit quod diminucio et turpitudo est in privacione operis propter inobedienciam materie vel insufficientiam."

63. Barragán Nieto, ed., *Il "Secretis mulierum,"* ch. 8, 448, 450, and 454. Cf. *Liber de sinthomatibus mulierum,* § 47, and *De curis mulierum,* § 141, in Green, ed. *The Trotula,* 84 and 85; 120 and 121.

64. See, for example, Jole Agrimi and Chiara Crisciani, "Savoir médicale et

anthropologie religieuse: Les représentations et les fonctions de la *vetula* (XIIIe–XVe siècle)," *Annales: Économies, Sociétés, Civilisations* 48 (1993): 1281–1308; and Fernando Salmón and Montserrat Cabré, "Fascinating Women: The Evil Eye in Medical Scholasticism," in *Medicine from the Black Death to the French Disease*, ed. Roger French, Jon Arrizabalaga, Andrew Cunningham, and Luis García-Ballester, History of Medicine in Context (Aldershot, U.K.: Ashgate, 1998), 53–84.

65. Green, "'Traittié tout de mençonges'," 156.

66. EC, BNF fr. 24281, 117r: "de perverse nature, effeminé et imparfait."

67. PA, BNF lat. 6540, 63vb–64ra: "secundum naturam eius monstruosam"; 64rb: "'secundum naturam' suam monstruosam"; "hunc 'locum' monstruose"; and "humidi [Cesena, BCM D.XXIV.2: huiusmodi] monstruositatis."

68. PA, BNF lat. 6540, 64vb: "qui magis appent [Cesena, BCM D.XXIV.2: apparent] apud naturam monstruosi"; cf. Avicenna, *Canon*, bk. 3, fen 20, tr. 1, ch. 43, 358ra.

69. Aristotle, *Generation of Animals*, IV.4, 772b26–773a2.

70. Questions surrounding hermaphrodites had not, however, taken on the public visibility and significance they acquired in the early modern period. See Lorraine Daston and Katharine Park, "The Hermaphrodite and the Orders of Nature: Sexual Ambiguity in Early Modern France," in Fradenburg and Freccero, *Premodern Sexualities in Europe*, 419–38.

71. PA, BNF lat. 6540, 63va: "Sanguis superfluus melancholicus multis superfluitatibus aliorum humorum admixtus non autem naturalis. Talis autem est sanguis menstruus et per ymoroydas emissus." PA also advances the comparison of hemorrhoids with menstruation at X.2. See Maaike van der Lugt, "Aristotle's 'Problems' in the West," 80–81

72. PA's first version included "vel nasi," but both JJ and PA in his second version appear to have dropped it. It was restored BAV lat. 2175 (PA's second version) and BNF lat. 6542 (JJ). The term "melcholic" is absent from PA in the two Cesena copies and BNF lat. 6541A, and from JJ, Sorbonne 122.

73. PA, BNF lat. 6540, 64rb; *Generation of Animals*, I.20, 728a9–20; Moerbeke trans., 33. Cf. Scot trans., 46.

74. "Monstrositas" was sometimes written (and, presumably, pronounced) "monstruositas," compounding the invitation to confuse the two.

75. At four of the six places where monster-related terms appear in PA, BN Marciana 2465 registers menstrum-related terms (spelled "mestr-," reflecting the Italian pronunciation) and BAV lat. 2175 gives such readings twice. Of the manuscripts of the Jean de Jandun redaction (all but one of which omit one of the relevant instances), three (CU Peterhouse 79, Paris, Bibliothèque Mazarine 3520, and Sorbonne 122) refer to "naturam eius menstruosam," and one (Ghent, UB 72) refers to an anatomy "menstruose plasmatum." Only occasionally in some manuscripts are the relevant words abbreviated (e.g., BNF lat. 6541; Seville, Biblioteca Capitular y Colombina 7-7-9). This evidence suggests that the slip occurred more than once.

76. Lind, *Problemata varia*, 67, records the confusion elsewhere, in the context of hermaphrodites: "Mulier est uir occasionatus et menstruum [monstrum] in natura, ut patet in libro de animalibus." The transposition occurs in more distant contexts as well. For example, in a twelfth-century manuscript of Aristotle's *Physics* at II.8, 199b3–5: *Physica:*

Translatio vetus, ed. Fernand Bossier and Jozef Brams, 2 fascicules, Aristoteles Latinus, 7.1, Corpus Philosophorum Medii Aevi Academiarum Consociatarum Auspiciis et Concilio Editum, Union Académique Internationale (Leiden: Brill, 1990), fasc. 2, p. 88.

77. Aristotle, *Problemata* IV.13, for example, deals with monstrous births and mentions semen and the womb; Pietro's commentary introduces "menstruum" there as well. Manuscripts of the Bartholomeus text and the commentaries usually do not show signs of confusion at that point. Exceptions are in the two manuscripts that employ ambiguous abbreviations in IV.26: PA, IV.13, BNF lat. 6540, 58vb, and 6541, 67va.

78. For example, Helen Rodnite Lemay, ed. and trans., *Women's Secrets: A Translation of Pseudo-Albertus Magnus' "De Secretis Mulierum" with Commentaries*, SUNY Series in Medieval Studies (Albany: State University of New York Press, 1992), ch. 1 (Commentator B), 71 and 73–74 and Cecco d'Ascoli, "Commentary," in Lynn Thorndike, ed. *The "Sphere" of Sacrobosco and Its Commentators*, Corpus of Medieval Scientific Texts 2 (Chicago: University of Chicago Press, [1949]), 409: "Et post mortem Christi omnes homines Iudei ut mulieres menstrua patiunt," referred to and discussed by Peter Biller, "Views of Jews from Paris Around 1300: Christian or 'Scientific'?" in *Christianity and Judaism: Papers Read at the 1991 Summer Meeting and the 1992 Winter Meeting of the Ecclesiastical History Society*, ed. Diana Wood (Oxford: Blackwell for the Ecclesiastical History Society, 1992), 199 and n. 38. See also Willis Johnson, "The Myth of Jewish Male Menses," *Journal of Medieval History* 24 (1998): 273–95; and Gianna Pomata, "Menstruating Men: Similarity and Difference of the Sexes in Early Modern Medicine," in *Generation and Degeneration: Tropes of Reproduction in Literature and History from Antiquity Through Early Modern Europe*, ed. Valeria Finucci and Kevin Brownlee (Durham, NC: Duke University Press, 2001), 109–52.

79. PA, IV.25, BNF lat. 6540, 62va: "Et hoc confirmat auctoritate poete [Hesiod, *Work and Days,* l.586] qui dixit feminas maxime esse luxuriosas."

80. Aristotle, *Generation of Animals*, IV.5, 774a1–3. See also IV.4, 773a16–9.

81. PA, BNF lat. 6540, 64vb: "Iuvencule 14 annorum appetunt coitum et alii de quibus hic est sermo." The comment occurs some distance beyond the text to which it refers, however, suggesting the possibility that the marginalia were themselves copied from another manuscript.

82. BM in BN lat. 6307, 201vb, margin: "Nota quod mulieres sunt insatiabiles in coitu." The manuscript belonged at some time to Charles, Duke of Orleans (1394–1465). AL #562.

83. "Prague" *Problemata*, Prague, NK VIII.A.19, 79rb.

84. "Vatican" *Problemata*, BAV lat. 2481, at the folio marked "93"rb, also excluded it, following his general practice of reproducing only the opening of each *problema*.

85. BM, BAV Chigi G.V.131, 75rb: "Necesse est igitur perverti et alibi moveri seminalis expurgationis. Propter quod insatiabiles quemadmodum mulieres. Modica enim humiditas non cogitur exire et infrigidatur cito."

86. BM, BNF lat. 15081, 12vb: "Et quibus quidem in anum hii pati desiderant; quibuscumque autem in utraque hii et agere et pati. In utrum autem plus hoc magis desiderant."

87. Pseudo-Albertus, *De secretis mulierum* (1508), 124–25: "Ille enim iuvenes mulieres, quando multum abundant in tali materia, multum appetunt coitum, propter materiae

abundantiam. Ideo peccatum est in natura illas retrahere, et coitum prohibere ad illum quem diligunt, quamvis sit peccatum in moribus, de quo nihil ad propositum." On constricted passages, ch. 1 (Commentary B), 73 (cf. porousness, ch. 5 [Commentary B], 104); on humidity, desire, and intercourse, ch. 11 (Text and Commentary A), 132.

88. PA, BNF lat. 6540, 64va: "Sicut et 'mulieres' sunt 'insatiabiles' et maxime iuvencule quarum meatus adhuc sunt angusti. Eo quod 'modica' est 'humiditas' que 'cogitur' expelli mota et liquata per fricationem que quidem [Cesena, BCM D.XXIV.2: quia] 'modica' est 'cito infrigidatur' non potens expelli. Propter quod iterum appetunt fricari ut expellatur."

89. The "Aachen" commentator (Erfurt, CA F. 263, 23vb) introduces an added dimension when he suggests that girls' breasts as well as genitals are stimulated.

90. JJ, Sorbonne 122, 47ra: "mulieres sunt precipue insatiabiles et maxime iuvencule."

91. PA, BNF lat. 6540, 64va: "Notandum est quod mulieres insatiabiles sunt precipue iuvencule circa annum 14[m] existentes de quibus fere est sermo in 9 de *Ystoriis animalium* et cetera. Iuvencule 'maxime incitantur ad venereorum usum' cum incipiunt et 'venerea passe incontinentes fiunt.' Cuius quidem causa est quia tempore illo propter humoris spermatici et menstrui descesum [Cesena, BCM D.XXIV.2: descensum] incipiunt inspissari genitalia et fissura vulve claudi et labia eius mollificari etiam ingrossari, pilosa fieri. Et incipit tunc ad coitum insanire. Non tamen emittit in desiderio et quanto plus coit aut manu se fricaverit tanto plus appetit eo quod per talem fricationem humor attrahitur sed non emittitur. Et cum humore attrahitur calor et ventositas consurgit sed cum corpus frigidum sit et clausum in poris non cito semen emittit." Aristotle, *Historia animalium,* IX.1 [= VII, part 1, in most modern editions], 581a32–b19. See Balme, ed., *History of Animals.*

92. See Laurence Moulinier, "Le corps des jeunes filles dans les traités médicaux du Moyen Âge," in *Le corps des jeunes filles de l'Antiquité à nos jours,* ed. Louise Bruit Zaidman, Gabrielle Houbre, Christiane Klapisch-Zuber, and Pauline Schmitt Pantel, Collection pour l'Histoire (Paris: Perrin, 2001), 80–109.

93. Albertus Magnus, *De animalibus libri XXVI,* ed. Hermann Stadler, Beiträge zur Geschichte der Philosophie und Theologie des Mittlealters, 15 and 16 (Münster: Aschendorff, 1916 and 1920), bk. 9, tr. 1, ch. 1, §§ 7–11, 676–78. See Danielle Jacquart and Claude Thomasset, "Albert le Grand et les problèmes de la sexualité," *History and Philosophy of the Life Sciences* 3 (1981): 73–93.

94. JJ, BAV Ottob. 1764, 59vb: "custodia circa hoc tempus indigent."

95. JJ, Seville, Colombina 7.7.9, 52vb, margin.

96. "Bavarian" *Problemata,* Ghent, UB 178, 109ra.

97. Table to PA, BN Marciana 2672, 13va.

98. JJ, Seville 7.7.9, 52vb, margin: "Quando debet custodiam mulier ne abutatur cum viro alio."

99. *De curis mulierum,* in Green, *The Trotula,* 120 and 121. For an explanation of suffocation of the uterus and its relation to menstruation, see Green's "Introduction," 22–31.

100. WB, prologue, Bodl. Digby 153, 110a, alluding to Proverbs 30:15–16. At least in his first version, at IV.15, PA attributes to Galen the view that "in vulva est fortis virtus appetitiva" (Nuremburg SB, Cent.III.38, 30ra; also in BAV lat. 2174). He apparently omitted it from his second version. It appears in JJ as well as in two of the three manuscripts of the

first version and is absent only from one manuscript of the second version (the often hybrid BAV lat. 2175). Cf. PA, IV.10.

101. "Prague" *Problemata*, NK VIII.A.19, 79va: "Nota quod mulieres habent maiorem delectationem in coytu extensive. Hoc est quod mulier per longius tempus delectatur quam vir, quia delectatur in recipiendo sperma viri et in emittendo sperma proprium. Sed vir magis delectatur intensive in coytu, id est, per intentiorem delectationem quam mulier licet per minus tempus durat quam mulier."

102. Cadden, *Meanings of Sex Difference*, 150–62. PA alludes to it in his exposition of *Problemata* IV.15, BNF lat. 6540, 60ra.

103. PA, BNF lat. 6540, 64vb: "propter consuetudinem que potius coruptela est."

104. On this basis William Burgwinkel distinguishes the sodomites in Dante's *Purgatory* from those in the *Inferno*: " 'The Form of Our Desire': Arnaut Daniel and the Homoerotic Subject in Dante's 'Commedia,' " *GLQ: A Journal of Lesbian and Gay Studies* 10 (2004): 584–85.

105. "Aachen" *Problemata*, Erfurt, CA F. 263, 23rb: "Iam consuetudo est eis natura quod semper quasi[?] fricant illas partes propter delectationem vel volunt fricari ab aliis."

CHAPTER 4. "BEYOND THE BOUNDARIES OF VICE"

1. "Arnaud of Verniolle," in Goodich, trans., *Other Middle Ages*, 142.

2. In addition to the famous cases of the Knights Templar and Hugh Dispenser, examples of such associations are apparent in the enforcement of antisodomy laws in Florence (see Rocke, *Forbidden Friendships*) and by the Crown of Aragon (see James A. Brundage, "The Politics of Sodomy: Rex v. Pons Hugh de Ampurias [1311]," in *Sex in the Middle Ages: A Book of Essays*, ed. Joyce E. Salisbury, Garland Medieval Casebooks [New York: Garland, 1991], 239–46).

3. WB, prologue to IV, Bodl. Digby 153, 110r.

4. Pietro's repeated reference to the *Ethics* at *Problemata* IV.26 is especially remarkable since he declines to cite that work when dealing with later sections of the *Problemata* that would seem to call for such citations. See Matthew Klemm, "Medicine and Moral Virtue in the *Expositio Problematum Aristotelis* of Peter of Abano," *Early Science and Medicine* 11 (2006), 332–33.

5. The four invocations occur at PA, BNF lat. 6540, 63vb (*Ethics*, X.5, 1175b15); 64vb: "Et hoc est quod dicit quod ita hoc modo desiderant illi qui hec operantur nephanda quos Aristoteles 7° *Ethicorum* [VII.7, 1149a1] ait fore 'extra terminos malicie' "; 64vb (II.1, 1103a20–24); 65ra: "Causam vero istius actus nephandi dicit Aristoteles in 7° *Ethicorum* [VII.7, 1149a1] esse bestialitatem omnia [Cesena, BCM D.XXIV.2: omnimodam] et perversionem anime."

6. The first, via Arabic, rendition was a simplified summary of the work; the second was the more sophisticated paraphrase and commentary by Ibn Rushd (Averroes), the Muslim Aristotelian most influential in Christian Europe. See Aristotle, *Ethica nicomachea*, fasc. 1, *Praefatio*, and Auguste Pelzer, "Les versions latines des ouvrages de morale conservés sous

le nom d'Aristote en usage au XIIIe siècle," *Revue néo-scolastique de philosophie* 23 (1921): 316–41 and 378–412. On the Alexandrine version: Concetto Marchesi, *L'"Etica Nicomachea" nella tradizione latina medievale (documenti ed appunti)* (Messina: Ant. Trimarchi, 1904), 105–28. See also H. Paul F. Mercken, *The Greek Commentaries on the "Nicomachean Ethics" of Aristotle in the Latin Translation of Robert Grosseteste, Bishop of Lincoln († 1253)*, 3 vols., Corpus Latinum Commentariorum in Aristotelem Graecorum, pt. 6, vols. 1–3 (Leuven: Leuven University Press, 1973–91). The Grosseteste translation as represented by *A. Recensio pura*, fasc. 3 in the Gauthier edition, corresponds most closely to PA's quotations and is thus the version referred to in this book.

7. Commentaries on the *Ethics* and the *Sentences* are the core sources for the most extensive treatment of Aristotelian ethics in the Middle Ages. Odin Lottin, *Psychologie et morale aux XIIe et XIIIe siècles*, 2nd ed., 6 vols. in 7 (Gembloux, Belgium: J. Duculot, 1957–60).

8. Martin Grabmann, "Das Studium der aristotelischen Ethik an der Artistenfakultät der Universität Paris in der ersten Hälfte des 13. Jahrhunderts," *Philologisches Jahrbuch* 53 (1940), 339–42.

9. On some of the sore spots, see Piché, ed., *La condamnation parisienne de 1277*, §§ 22 (173), 144 (170), 159 (164), 176 (170), on pp. 86, 122, 126, 132; cf. Roland Hissette, *Enquête sur les 219 articles condamnés à Paris le 7 mars 1277* (Louvain: Publications Universitaires, 1977), 263–70; Martin W. F. Stone, "Moral Psychology After 1277: Did the Parisian Condemnation Make a Difference to Philosophical Discussions of Human Agency?" in *Nach der Verurteilung von 1277: Philosophie und Theologie an der Universität von Paris im letzten Viertel des 13. Jahrhunderts: Studie und Texte/After the Condemnation of 1277: Philosophy and Theology at the University of Paris in the Last Quarter of the Thirteenth Century: Studies and Texts*, ed. Jan A. Aertsen, Kent Emery Jr., and Andreas Speer, Miscellanea Mediaevalia, Veröffentlichungen des Thomas-Instituts der Universität zu Köln, 28 (Berlin: Walter de Gruyter, 2001), 795–826. Luca Bianchi, *Il vescovo e i filosofi: La condanna parigina del 1277 e l'evoluzione dell'aristotelismo scolastico*, Quodlibet 6 (Bergamo: Pierluigi Lubrina, 1990), 153–54, argues that the "true center" of the 1277 conflicts in Paris was the recovery of the *Nicomachean Ethics*.

10. See Nancy G. Siraisi, *Taddeo Alderotti and His Pupils: Two Generations of Medical Learning* (Princeton: Princeton University Press, 1981), 27–42 and 72–95, and "The *Libri morales* in the Faculty of Arts and Medicine at Bologna: Bartolomeo da Varignana and the Pseudo-Aristotelian 'Economics,'" *Manuscripta* 20 (1976): 105–18.

11. Brunetto Latini, *Li livres dou tresor*, ed. Francis J. Carmody (Berkeley: University of California Press, 1948). The sections drawn from Aristotelian ethics occupy 175–223.

12. Marchesi, *L'"Etica nicomachea" nella tradizione latina medievale*; Nicole Oresme, *Le livre de Ethiques d'Aristote Published from the Text of MS. 2902, Bibliothèque Royale de Belgique*, ed. Albert Douglas Menut (New York: G. E. Stechert, 1940). For the Spanish reception, see Anthony Pagden, "The Diffusion of Aristotle's Moral Philosophy in Spain, ca. 1400–ca. 1600," *Traditio* 31 (1975): 286–313.

13. The index of AL lists the manuscripts of the main Latin versions, yielding the following numbers (excluding fragments, uncertain items, and items listed in the Supplement):

Nicomachean Ethics, 229; *Metaphysics*, 195; *Politics*, 103; *Problemata*, 52; *Prior Analytics*, 255; *Posterior Analytics*, 276. These numbers are helpful indicators of cultural currency but no more than suggestive. For example, the *Prior Analytics* was better known by more scholars than the *Posterior Analytics*, which may have been copied and commented on more frequently precisely because of its greater inaccessibility.

14. See, for example, three thirteenth-century texts edited in Lottin, *Psychologie et morale*, vol. 3, pt. 2.1, 621–50.

15. Charles Lohr, "Medieval Latin Aristotle Commentaries," *Traditio* 23 (1967): 313–413; 24 (1968): 149–245; 26 (1970): 135–216; 27 (1971): 251–351; 28 (1972): 281–396; 29 (1973): 91–197; 30 (1974): 119–44.

16. I am grateful to Mark Jordan for this observation. See his "Homosexuality, *Luxuria*, and Textual Abuse," in *Constructing Medieval Sexuality*, ed. Karma Lochrie, Peggy McCracken, and James A. Schultz, Medieval Cultures 11 (Minneapolis: University of Minnesota Press, 1997), 35.

17. Klemm, "Medicine and Moral Virtue."

18. Aristotle, *Ethica nicomachea*, fasc. 3, II.1, 1103a19: "neque una moralium virtutum, natura nobis infit."

19. The medieval struggle to understand the relations between nature and virtue predates the full availability of the Aristotelian text. See Cary J. Nederman, "Nature, Ethics, and the Doctrine of 'Habitus': Aristotelian Moral Philosophy in the Twelfth Century," *Traditio* 45 (1989–90): 87–110, and Marcia Colish, "*Habitus* Revisited: A Reply to Cary Nederman," *Traditio* 48 (1993): 77–92.

20. Thomas Aquinas, *Sententia libri Ethicorum*, 2 parts, vol. 47 of *Opera omnia iussu Leonis XIII P. M. edita*, ed. Order of Preachers (Rome: Sancta Sabina, 1969), bk. 2, lect. 1; pt. 2, p. 77b: "Et ideo neque in his quae sunt secundum naturam neque in his quae sunt contra naturam consuetudo aliquid facit."

21. Averroes, *In Moralia nicomachia expositio*, vol. 3 of Aristotle, *Libri moralem totam philosophiam complectentes* (Venice: Iunctas, 1562; repr. Frankfurt am Main: Minerva, 1962), bk. 2, ch. 1, determination, 18vb: "Tunc virtutes non sunt in nobis secundum naturam neque preter naturam, scilicet involuntarie vel coacte." Cf. Albertus Magnus, perhaps thinking more literally of the stone in Aristotelian physics: "Nihil eorum quae sunt secundum 'naturam' vel preter naturam sicut violenta, sunt ex assuetudine." Albertus Magnus, *Super Ethica commentum et quaestiones*, ed. Wilhelm Kübel, vol. 14 of *Opera omnia* in 2 parts (Münster: Aschendorff, 1968–87), bk. II, ch. 1, lect. 1, § 105, pt. 2, p. 93.

22. Aristotle, *Ethica nicomachea*, fasc. 3, III.8, 1114a23–29: "Propter naturam quidem enim turpes, nullus increpat, set eos qui propter desidiam et negligenciam. Similiter. . . . Nullus enim utique inproperabit ceco natura vel ex infirmitate vel ex plaga, set magis miserebitur. Eum autem qui ex vini potacione vel alia incontinencia, omnis utique increpabit."

23. WB, Bodl. Digby 153, 10r: "Nullus enim est vituperandus propter naturalia nec etiam laudandus. *Ethicorum*."

24. Walter Burley, *Expositio Gualteri Burlei super decem libros Ethicorum Aristotelis* (Venice: Simon de Luere, printer; Andrea Torresano de Asula, publisher, 1500), III, tr. 1,

cap. 5, 48va–49ra: "Nullus iustus increpatur propter illud quod inest ei a natura sed solum propter illud quod inest voluntarie. Verbi gratie cecum vel turpem a nativitate nullus increpat sed magis miseretur."

25. Clemens Stroick, *Heinrich von Friemar: Leben, Werke, philosophische-theologische Stellung in der Scholastik* (Freiburg I. Br.: Herder, 1954), 53–59 and 77–95. On the presence of his work in Italy around 1400, David A. Lines, "The Commentary Literature on Aristotle's 'Nicomachean Ethics' in Early Renaissance Italy: Preliminary Considerations," *Traditio* 54 (1999) 246.

26. Heinrich von Friemar, *Commentum super libros Ethicorum Aristotelis*, ms. Leipzig, UB 1439, 47va: "Si enim virtus moralis precise nobis inesset a natura tunc . . . tractare de virtutibus non spectaret ad moralem sed ad philosophiam naturalem." Similarly, the anonymous gloss on the *Ethics* in ms. Salamanca, Biblioteca Universitaria, 2705, 10rb.

27. Aristotle, *Ethica nicomachea*, fasc. 3, VII.6, 1148b20–24.

28. Aristotle, *Ethica nicomachea*, fasc. 3, VII.1, 1145a19–20: "Ad bestialitatem autem maxime utique congruit dicere super nos virtutem, heroicam quandam et divinam."

29. Aristotle, *Ethica nicomachea*, fasc. 3, VII.1, 1145a26–28: "Set hec quidem honorabilius virtute, hec autem alterum quoddam genus malicie."

30. Aristotle, *Ethica nicomachea*, fasc. 3, VII.7, 1148b15–19: "Quia autem quedam quidem delectabilia natura, et horum hec quidem simpliciter, hec autem secundum genera animalium et hominum, hec autem non sunt, set hec quidem propter passiones, hec autem propter consuetudines fiunt, hec autem propter perniciosas naturas."

31. See Jordan, "Homosexuality, *Luxuria*, and Textual Abuse," 33–35.

32. Jean Buridan, *Questiones super decem libros Ethicorum Aristotelis ad Nicomachum* (Paris: Poset Le Preux, 1513; repr. Frankfurt: Minerva, 1968), bk. 7, q. 1, 140va: "Sed preter naturam delectabilia vocantur que delectant animalia ratione cuiusdam depravationis ipsorum. Concedo igitur quod talia sunt convenientia et sic naturalia talibus depravationibus. Sed non sequitur ergo simpliciter naturalia vel simpliciter convenientia, sicut non sequitur amarum habenti gustum infectum ergo simpliciter amarum."

33. Albertus Magnus, *Super Ethica*, bk. 7, lect. 5, § 637, pt. 2, 544–45: "Et dicit, quod cum, 'quaedam sint natura delectabilia,' vel 'simpliciter' omnibus sicut cibus in communi et huiusmodi omnibus animalibus, vel 'secundum genera animalium,' quia naturaliter alicui est delectabilis hic cibus et alius asino et alius homini, 'et' similiter secundum genera 'hominum,' quia alia naturaliter sunt delectabilia melancholicis et alia cholericis. . . ." Jean Dunbabin, "The Two Commentaries of Albertus Magnus on the 'Nicomachean Ethics,'" *Recherches de Théologie ancienne et médiévale* 30 (1963): 232–50; Henryk Anzulewicz, "Fatum: Das Phänomen des Schicksals und die Freiheit des Menschen nach Albertus Magnus," in *Nach der Verurteilung von 1277: Philosophie und Theologie an der Universität von Paris im letzten Viertel des 13. Jahrhunderts. Studien und Texte*, ed. Jan A. Aertsen, Kent Emery Jr., and Andreas Speer, Miscellanea Medievalia, Veröffentlichungen des Thomas-Instituts der Universität zu Köln 28 (Berlin: Walter de Gruyter, 2001), 507–34.

34. Aristotle, *Ethica nicomachea*, fasc. 3, VII.7, 1148b31–1149a1: "Quantis quidem igitur natura causa, hos utique nullus diceret incontinentes, quemadmodum neque mulieres quoniam non ducunt, set ducuntur. Similiter autem et egrotative habentibus, propter

consuetudinem. Habere quidem igitur singula horum, extra terminos malicie, quemadmodum et bestialitas.

35. Buridan, *Super decem libros Ethicorum*, bk. 7, q. 4, 141va–b. See also Albertus Magnus, *Super Ethica*, bk. 7, lect. 5, § 635, pt. 2, p. 542.

36. According to the text of Roman law accorded most authority in the late Middle Ages, *The Digest of Justinian*, ed. Theodor Mommsen with Paul Krueger and trans. Alan Watson, 4 vols. (Philadelphia: University of Pennsylvania Press, 1985), bk. 3, ch. 1, § 6, p. 80: "Remouet autem a postulando pro aliis et eum, qui corpore suo muliebria passus est. si quis tamen ui praedonum uel hostium stupratus est, non debet notari."

37. Albertus Magnus, *Super Ethica*, bk. 7, lect. 5, § 637, pt. 2, pp. 543–45; Thomas Aquinas, *Sententia libri Ethicorum*, bk. 7, lect. 5, pt. 2, pp. 76–78.

38. Oresme, *Livre de Ethiques*, bk. 7, ch. 9, 380, gloss 3: "erreur de nature."

39. Albertus Magnus, *Super Ethica*, bk. 7, lect. 5, q. 2, § 631, pt. 2, pp. 539–40: "Contingit autem, quod illud quod est contra naturam bene institutam, efficitur sibi naturale aliquo modo. . . . quia quamvis huiusmodi non sint naturalia quasi convenientia ex principiis naturalibus speciei, sunt tamen naturalia quantum ad esse, . . . sicut alicui est naturale, quod habeat sex digitos aut aliquid huiusmodi."

40. Albertus Magnus, *Super Ethica*, bk. 7, lect. 5, q. 5, § 634, pt. 2, p. 542.

41. Oresme, *Livre de Ethiques*, II.9, glosses 2, 4, and 5, 380.

42. Sears, *The Ages of Man*, 25–31.

43. Thomas Aquinas, *Sententia libri Ethicorum*, bk. VII, lect. 5; pt. 2, p. 399a: "Similiter etiam inter homines colericis delectabilia sunt naturaliter frigida, quae temperant eorum complexionem, fleumaticis vero calida."

44. Albertus Magnus, *Super Ethica*, bk. 7, lect. 15, § 686, pt. 2, p. 586 (1154b11–12): "Et dicit, quod quia melancholicus umor 'mordet,' ideo 'corpus melancholicorum' est semper quasi corruptum et ideo 'semper indiget medicine' contra tristitiam, scilicet delectatione."

45. See Bert Hansen, *Nicole Oresme and the Marvels of Nature: A Study of His " De Causis Mirabilium"* (Toronto: Pontifical Institute of Mediaeval Studies, 1985); Daston and Park, *Wonder and the Order of Nature*, 21–66; and Gregory B. Stone, *The Ethics of Nature in the Middle Ages: On Boccaccio's Poetaphysics*, The New Middle Ages (New York: St. Martin's, 1998), 61–83.

46. On these subjects, scholastics generally made reference to Aristotle's *Generation of Animals*. See *De generatione animalium translatio Guillelmi de Moerbeka*, II.1, 731b18–733a25, and IV.4, 772b14–773a32.

47. Aristotle, *Ethica nicomachea*, fasc. 3, VII.7, 1148b31–1149a1.

48. Aristotle, *Ethica nicomachea*, fasc. 3, VII.7, 48b20–30: "Dico autem bestiales, puta eam hominem quam dicunt pregnantes rescindentem pueros devorare, vel qualibus gaudere aiunt quosdam silvestrium circa Pontum, hos quidem crudis, hos autem hominum carnibus, hos autem pueros comodare ad invicem in convivium, vel circa Phalarin dictum. Hii quidem bestiales, hii autem propter egritudines fiunt et maniam quibusdam, quemadmodum matrem sacrificans et comedens, et conservi epar. Hii autem egritudinales vel ex consuetudine, puta pilorum evulsiones et unguium corrosiones [Gloss: seu comesciones]. Adhuc autem carbonum et terre. Cum hiis autem que venereorum masculis."

49. Albertus Magnus, *Super Ethica*, bk. 7, lect. 5, § 637, pt. 2, p. 545.

50. *Greek Commentaries on the Nicomachean Ethics*, ed. Mercken, bk. 7, ch. 7, vol. 3, 41: "masculis autem abuti" and "masculos corrumpere."

51. The widely circulated revision of it makes no change: Aristotle, *Ethica Nicomachea translatio Roberti Lincolnensis sive "Liber Ethicorum": B. Recensio recognita*, fasc. 4 (1973) of *Ethica nicomachea*, ed. René Antoine Gauthier, 4 fascicules in 5, Aristoteles Latinus, 26, 1–3, Corpus Philosophorum Medii Aevi, Union Académique Internationale (Leiden: Brill, 1972–74), VII.5, 1148b29–30.

52. Albertus Magnus, *Super Ethica*, bk. 7, lect. 5, § 637, pt. 2, p. 545; Averroes, *Liber Nichomachye Aristotelis id est Ethicorum*, trans. Hermanus Allemanus, bk. 7, ms. Toledo, Biblioteca Capitular 94.14, 46r. Thomas Aquinas, *Sententia libri Ethicorum*, bk. 7, lect. 5; pt. 2, p. 400a. Thomas also uses "masculorum concubitus," *Tabula libri Ethicorum* in *Opera omnia iussu Leonis XIII*, ed. Order of Preachers, vol. 48 (Rome: Sancta Sabina, 1971), s.v. "comestio carbonum," p. B83.

53. Anonymous, "Il compendio Alessandrino-Arabo 'Liber Ethicorum,'" edited in Concetto Marchesi, *L'Etica Nicomachea nella tradizione latina medievale*, lxx: "delectationes vero propter naturam malam sunt ut concubitus cum masculis et cetera huiusmodi ignominiosa." See also "malitie crudeles ferine" (lxx). On authorship, manuscripts, and influence, see 105–28.

54. Heinrich von Friemar, *Commentum sive lectura super libros Ethicorum Aristotelis*, Berlin, SB lat. fol. 584, bk. 7, 212va.

55. Oresme, *Livre de Ethiques*, bk. 7, ch. 9, 381: "abus de delit charnel avecques enfans masles"; and gloss 11: "C'est ce que nous appellon sodomie."

56. Brunetto, *Livres dou tresor*, bk. 2, ch. 40, 207: "gesir avec les malles, et des autres choses deshonorables." Here he closely follows the anonymous compendium edited by Marchesi, *L'Etica Nicomachea*, lxx: "concubitus cum masculis et cetera huiusmodi ignominiosa."

57. Albertus de Saxonia, *Super decem libros Ethicorum*, Leipzig, UB 1445, bk. 7, lect. 5, 97r: "Accidit aliquando aliquibus propter animam, corpus et complexionem perversam et vilissimam." See Georg Heidingsfelder, *Albert von Sachsen: Seine Lebensgang und sein Kommentar zur "Nikomachischen Ethik" des Aristoteles*, 2nd ed., Beiträge zur Geschichte der Philosophie des Mittelalters: Texte und Untersuchungen 22, fasc. 3–4 (Münster: Aschendorff, 1927), including his dependence on Walter Burley's *Ethics* commentary, 83–90.

58. Albertus Magnus, *Super Ethica*, bk. 7, lect. 5, q. 6, § 635, pt. 2, p. 542: "Praeterea concubitus masculorum dicitur vitium contra naturam; ergo videtur, quod natura nullo modo moveat in ipsum."

59. Jordan, *Invention of Sodomy*, on Albert, 132–33 and 164–65; on Thomas, 149–57. On Thomas, see also Alain de Libera, *Penser au Moyen Âge*, Chemin de Pensée (Paris: Éditions du Seuil, 1991), 205–11.

60. Serge Lusignan, "Nicole Oresme traducteur et la pensée de la langue française savante," in P. Souffrin and A. Ph. Segonds, eds., *Nicole Oresme: Tradition et innovation chez un intellectuel du XIVe siècle*, Science et Humanisme (Paris: Les Belles Lettres, 1988), 93–104, and, more generally, *"Parler vulgairement": Les intellectuels et la langue française aux XIIIe et XIVe siècles* (Paris: J. Vrin 1986).

61. The Latin *Ethics* uses "masculus" only in this phrase; elsewhere "puer," "iuvenis," and "adolescens" occur.

62. See *Novum glossarium mediae latinitate* and instances in *Thesaurus linguae latinae*, s.v. "masculus." Du Cange, s.v. "masculorum concubitores," for which early medieval legal sources are cited, gives the Greek (παιδερασταί) and (ἀρσενοκοταί).

63. Genesis 34:25 and Exodus 1:16.

64. *Greek Commentaries on the "Nicomachean Ethics,"* ed. Mercken, bk. 7, ch, 7, vol. 3, 41: "exercitati 'ex pueris' masculos corrumpere, isti assuescunt hoc a pueritia cum malis hominibus conversantes."

65. Avicenna, *Canon*, bk. 3, fen 20, ch. 11, 353ra.

66. Heinrich von Friemar, *Commentum super libros Ethicorum*, Leipzig, UB 1439, 340ra.

67. Jordan, *Invention of Sodomy*, 154.

68. Aristotle, *Ethica nicomachea*, fasc. 3, VII.7, 1148b6–12.

69. Aristotle, *Ethica nicomachea*, fasc. 3, VII.7, 1148b15–35.

70. Buridan, *Super libros Ethicorum*, bk. 7, quest. 17, 148 [*sic* for 151]ra–va, and q. 4, 141vb. See also Nicolaus Theodorici de Amsterdam (d. 1460), *Questiones super libros Ethicorum*, ms. Leipzig, UB 1451, 147vb–148ra and Albertus Magnus, *Super Ethica*, bk. 5, lect. 4, § 383, pt. 2, pp. 324–25.

71. Aristotle, *L'Éthique à Nicomaque*, trans. and comm. René Antoine Gauthier and Jean Yves Jolif, 2 vols. in 4, Aristote: Traductions et Études, Collection Publiée par l'Institut Superieur de Philosophie de l'Université de Louvain (Louvain: Publications Universitaires, 1970), VII.6, 1148b29–30: "à quoi nous ajouterons: faire l'amour avec les mâles; ces pratiques sont tantôt fruit de la nature et tantôt fruit de l'habitude (ainsi pour ceux dont on a abusé dès la plus tendre enfance)."

72. Jordan, *Invention of Sodomy*, 149–50.

73. Aristotle, *Ethica nicomachea*, fasc. 3, VII.7, 1148b34–1149a1: "Similiter autem et egrotative habentibus, propter consuetudinem. Habere quidem igitur singula horum, extra terminos malicie, quemadmodum et bestialitas." The Greek phrasing sets bestiality more clearly apart.

74. Heinrich von Friemar, *Super libros Ethicorum*, ms. Berlin, SB lat. fol. 584, 212ra: "Nota quod membra istius divisionis in translatione arabica clarius exprimuntur ubi dicitur sic delectationes quedam sunt naturales, quedam bestiales et quedam ferinales, et istarum quedam innascuntur ratione temporis, quedam ratione egritudinis, quedam per consuetudinem, quedam propter malas naturas." Cf. "Compendio Alessandrino-arabo," ed. Marchesi, *L' "Etica Nicomachea" nella tradizione latina medievale*, lxix.

75. Albertus Magnus, *Super Ethica*, bk. 7, lect. 5, § 637, pt. 2, p. 545: "Et isti sunt bestiales in affectu propter malam naturam et quidam 'propter aegritudinem.'" Oresme, *Livre de Ethiques*, bk. 7, ch. 9, 381 with glosses 8 and 10; 382 with gloss 14.

76. Albertus Magnus, *Super Ethica*, bk. 7, lect. 5, q. 2, § 631, pt. 2, pp. 539–40: "Sed hoc potest esse dupliciter: aut quod conversum sit in naturam huiusmodi, et sic sunt delectationes bestiales, aut quod sit adhuc in motu ad huiusmodi dispositionem, et sic sunt delectationes aegritudinales quaedam."

77. Oresme, *Livres de Ethique*, bk. 7, ch. 9, gloss 10, 381; Heinrich von Friemar, *Super libros Ethicorum*, ms. Berlin, SB lat. fol. 584, 212va. Brunetto Latini (*Livres dou tresor*, bk. 2, ch. 40, § 4, 207) and his source ("Compendio alessandrino-arabo," ed. Marchesi, p. lxix) say that pulling hair, chewing nails, and eating coal or earth can arise from either sickness or habit.

78. Averroes, *Liber Nichomachye Aristotelis*, bk. 7, ch. 5, 101ra: "Et de istis est concubitus masculorum. Etenim hoc accidit quibusdam hominum ex parte naturae malae. Quibusdam vero ex parte consuetudinis."

79. Thomas Aquinas, *Summa theologiae*, ed. Thomas Gilby et al., 61 vols. (London: Eyre and Spottiswoode, 1964–80), pt. 1 of pt. 2, q. 31, art. 7, vol. 20 (ed. Eric d'Arcy), 24.

80. Albert of Saxony, *Lectura super X libros Ethicorum*, ms. Leipzig, UB 1445, 97r: "illud ultimum . . . accidit aliquando aliquibus propter animam, corpus, et complexionem perversam et vilissimam."

81. Dante treats Brunetto as a sodomite in the *Inferno*, but the controversy about what he meant is unlikely to be advanced by Brunetto's phrasing here, since it is closely dependent on his source. Dante, *Inferno*, XV. See Joseph Pequiney, "Sodomy in Dante's 'Inferno' and 'Purgatorio,'" *Representations* 36 (1991): 22–42; John Freccero, "The Eternal Image of the Father," in *The Poetry of Allusion: Virgil and Ovid in Dante's Commedia*, ed. Rachel Jacoff and Jeffrey T. Schnapp (Stanford: Stanford University Press, 1991), 62–76 and 264–65; Michael Camille, "The Pose of the Queer: Dante's Gaze and Brunetto Latini's Body," in *Queering the Middle Ages*, ed. Glenn Burger and Steven F. Kruger, Medieval Cultures 27 (Minneapolis: University of Minnesota Press, 2001), 57–86.

82. Albertus Magnus, *Super Ethica*, bk. 7, lect. 5, § 637, pt. 2, p. 545: "et quidam etiam huiusmodi sunt propter 'aegritudines' aliquas alias 'vel' propter 'consuetudinem,' sicut quidam comedunt 'ungues' et evellunt 'pilos' et comedunt eos et quaedam alia quae sunt minus innaturalia quam praedicta, sicut comestio 'carbonum et terrae' et concubitus 'masculorum.' Et haec omnia 'accidunt' propter causas praedictas."

83. Aristotle, *Ethica nicomachea*, fasc.3, III.5, 1111b6–9, on will and choice; III.15, 1152b5–21, on children and animals; VI.9, 1142a11–13, on geometry and prudence. The ages of those being referred to are indeterminate. In addition to the words for "child" and "childhood," the Latin version of the *Ethics* speaks of "youth" (in the comparison of geometry and prudence). That term often referred to the stage of life that followed adolescence and, in a scheme compatible with Aristotelian thought, extended from the age of twenty-one to thirty-five. Sears, *Ages of Man*, 38–53; see also plate 65.

84. Aristotle, *Politica (libri I–II.11): Translatio prior imperfecta interprete Guillelmo de Moerbeke (?)*, Aristoteles Latinus 29.1, Corpus Philosophorum Medii Aevi, Union Académique Internationale (Bruges: Desclée de Brouwer, 1961), I.13, 1259b30–60a33. In this work, e.g., VII.17 and VIII.4, puberty divides the two phases of education. The *Politics* was less widely read than the *Ethics*.

85. See, for example, Aristotle, *Ethica nicomachea*, fasc. 3, II. 2,1104a20–25, and VII.13, 1151b8–14.

86. Perhaps because they had other contexts in mind, Latins do not seem to have taken note of Averroes's example of boys raised in slavery. Averroes, *Liber Nichomachye*

Aristotelis, bk. 7, ms. Toledo, Biblioteca Capitular 94.14, 46r: "Ut illis qui consueverunt in carceracionem statim a puericia."

87. Aristotle, *Ethica nicomachea*, fasc. 3, VII.14, 1152a30–32: "Facilius enim consuetudinem transmutare natura. Propter hoc enim consuetudo difficilis, quoniam nature assimulatur." The syntax of the first sentence is curious but medieval commentators emended it and agreed on its sense. See for example, Albertus Magnus, *Super Ethica*, bk. 7, lect. 12, § 666, pt. 2, p. 570, and Thomas Aquinas, *Sententia libri Ethicorum*, bk. 7, lect. 10; pt. 2, p. 422b.

88. Albertus Magnus, *Super Ethica*, bk. 7, lect. 11, § 666, pt. 2, p. 570.

89. Thomas Aquinas, *Sententia libri Ethicorum*, bk. 2, lect. 1; pt. 1, p. 77a: "Quae quidem inclinatio vel est a natura quae inclinat in id quod est sibi conveniens, vel est ex consuetudine quae vertitur in naturam."

90. Thomas Aquinas, *Sententia libri Ethicorum*, bk. 7. lect. 5; pt. 2, p. 400a–b: "Quibusdam enim accidunt ex natura corporalis complexionis quam acceperunt a principio, quibusdam vero accidunt ex consuetudine, puta quia assuefiunt ad huiusmodi a pueritia. Et simile est de his qui in hoc incidunt ex aegritudine corporali, nam prava consuetudo est quasi quaedam aegritudo animalis."

91. Oresme, *Livre de Ethiques*, bk. 7, ch. 9, gloss 15, 382: "Les uns bestials sont telz par nature, les autres par maladie, les autres par acoustumance, qui est une maladie de l'ame."

92. "Bavarian" *Problemata*, Krakow, BJ 2095, 227vb: "Causam illius nephandi dicit Aristotiles in septimo *Ethicorum* esse bestialitatem omnimodam et perversionem anime. Propter quod non immerito vult eos ibidem extra terminos omnis malicie existere." See also the margins of PA at this point in his commentary: BNF lat. 6541, 75rb; and BNF lat. 6540, 65ra.

93. Thomas Aquinas, *Sententia libri Ethicorum*, bk. 7, lect. 5; pt. 2, p. 401a: "Et eadem ratio videtur esse de his qui aegrotative se habent, id est qui habent corruptam dispositionem propter malam consuetudinem, quae etiam opprimit iudicium rationis ad modum perversae naturae."

94. Albertus Magnus, *Super Ethica*, bk. 10, lect. 17, § 932, pt. 2, p. 779.

95. Thomas Aquinas, *Summa theologiae*, pt. 1 of pt. 2, q. 9, art. 2, vol. 17, 68. Elsewhere he treats other aspects of these questions, for example, *Summa theologiae*, q. 10, vol. 17, 82–96, and pt. 1 of pt. 2, q. 21, vol. 18, 104–18. See also anonymous texts edited by Lottin, "La liberté chez trois maîtres ès arts de Paris au dernier quart du XIIIe siècle," *Psychologie et Morale*, vol. 3, pt. 2.1, appendix 2, 621–50.

96. See, for example, the work cited by Martin Grabmann, "Das Studium der aristotelischen Ethik der Artistenfacultät der Universität Paris," 353. From the theological side, commentaries on Peter Lombard's *Sentences* were the most important site.

97. Heinrich von Friemar, *Super libros Ethicorum Aristotelis*, ms. Berlin, SB lat. fol. 584, bk. 2, 45rb.

98. Albertus Magnus, *Super Ethica*, bk. 6, lect. 4, § 486, pt.2, p. 417.

99. Albertus Magnus, *Super Ethica*, bk. 7, lect. 5, § 635, pt. 2, p. 542: "aut oportebit ponere errorem geneaticorum, qui dicunt, quod homines secundum signa nativitatis sunt tales vel tales naturaliter, quod non est solum contra fidem, sed contra omnem civilitatem et bonos mores, ut probat Tullius." The reference is to Cicero's *De divinatione*, 1.2.

100. For example, Albertus Magnus, *Super Ethica*, bk. 10, lect. 17, §§ 931–32, vol. 2, pp. 778–80; and bk. 7, lect. 11, § 666, pt. 2, p. 570.

101. Thomas Aquinas, *Summa theologiae*, pt. 1 of pt. 2, q. 21, art. 2, vol. 18, 110: "Actus naturales non sunt in potestate naturalis agentis, cum natura sit determinata ad unum; et ideo, licet in actibus naturalibus sit peccatum, non tamen est ibi culpa." The reference to the *Ethics* is on 108.

102. Thomas Aquinas, *Summa theologiae*, pt. 1 of pt. 2, q. 10, art. 3, vol. 17, 90–92.

103. Johannes Buridanus summarizes this position, *Quaestiones super libros ethicorum*, bk. 7, q. 3, 141ra. See also q. 4, 141vb.

104. Thomas Aquinas, *Sententia libri Ethicorum*, bk. 7, lect. 5; pt. 2, p. 400b: "Illud autem quod est modicum quasi nihil esse videtur nec contingit de facili quod modica vis rationis concupiscentias fortes repellat." Oresme seems to be following Aquinas here: *Livre de Ethiques*, bk. 7, ch. 9, gloss 12, 381.

105. This absurdity was one argument against the possibility of a vacuum existing in nature. Edward Grant, *Much Ado About Nothing: Theories of Space and Vacuum from the Middle Ages to the Scientific Revolution* (Cambridge: Cambridge University Press, 1981), 24–66.

106. Buridan, *Super libros Ethicorum*, bk. 7, q. 4, 141va–b: "Dicendum est breviter quod ad continentiam et incontinentiam proprie dictas prout de his intendit Aristoteles in isto septimo requiritur primo rectitudo rationis et secundo impetus passionis obvians ipsi rationi. . . . Requiritur etiam ulterius ad continentiam et incontinentiam proprie acceptas efficacia seu valitudo quedam tam rationis quam passionis."

107. Medieval reasoning about the types and effects of resistance to local motion was very sophisticated and cannot be fully treated here. See Marshall Clagett, *The Science of Mechanics in the Middle Ages*, Publications in Medieval Science 4 (Madison: University of Wisconsin Press, 1961), 421–44.

108. Buridan, *Super libros Ethicorum*, bk. 7, q. 4, 141vb: "Et hoc videbatur Aristoteles intendere primo *Politice*, dicens 'femina quidem habet consilium sed invalidum; puer autem habet quidem sed imperfectum.' Et quod 'similiter existimandum habere circa virtutes morales. Existimandum enim oportere habere omnes sed non eodem modo sed quantum unicuique ad sui ipsius opus.' Et addit quod non est 'eadem temperantia viri et mulieris neque fortitudo neque iustitia.'" The translation of "opus" as "role" is loose. I quote the context of the sentence to suggest why "work" in the sense of "deed" is not applicable here. The quotation is from Aristotle, *Politica*, I.13, 1260a13–23.

109. Buridan, *Super libros Ethicorum*, bk. 7, q. 1, 140ra–va.

110. Buridan, *Super libros Ethicorum*, bk. 7, q. 1, 140rb: "Alii autem sunt peius nati carentes demotica et habent pravos mores naturales inclinantes ad malitias. Modo non est possibile prave natos ad tantum pervenire gradum excellentie ad quantum possunt pervenire bene nati. Vel si hoc est possibile propter libertatem voluntatis, sicut dicebatur in quindecima questione quarti libri, tamen hoc est difficile et invisum. . . . Ex his ergo satis apparet quod bestialitas et heroica sunt possibiles hominibus, non omnibus, sed heroica bene natis et bestialitas male natis." In keeping with the sense and context of the passage, I read "invisum" as the opposite of "visum" rather than as the participle of "invideo." On this

sense of "demotic," see Oresme, *Le livre de Ethiques d'Aristote*, VII.16, 397–98, and *Lexikon latinitatis nederlandicae medii aevi*.

111. Buridan, *Super libros Ethicorum*, bk. 7, q. 1, 140rb: "Non enim pertingere possunt ad eam nisi bene nati et optime erudite."

112. Walter Burley, *Expositio super decem libros Ethicorum*, bk. VII, tr. 1, ch. 5, 112ra: "et quidam delectantur in coitu masculorum."

113. Walter Burley, *Expositio super decem libros Ethicorum*, bk. VII, tr. 1, ch. 5, 112rb: "Circa delectabilia contra naturam non est continentia simpliciter. . . . Prima ratio talis est: Bestie non dicuntur continentes nec incontinentes, qui non habent universalem op.[?] sed fantasiam et memoriam singularium. Tunc arguit sic: Illi qui habent tam modicam rationem universalem que non sufficit ad repellendum fortes concupiscentias non dicuntur continentes aut incontinentes, quia tam continentes quam incontinentes simpliciter habent rationem fortem universalem. Sed homines qui assimilantur bestiis in malitia circa delectabilia contra naturam non habent rationem universalem fortem que possit repellere concupiscentias fortes sive illa malitia fit ex natura perversa sive ex egritudine aut ex consuetudine. Ergo huiusmodi homines non sunt simpliciter incontinentes aut incontinentes et per consequens circa delectabilia contra naturam non est incontinentia simpliciter.

114. Walter Burley, *Expositio super decem libros Ethicorum*, bk. VII, tr. 1, ch. 5, 112rb: "Et dicit quod habere singula horum, scilicet delectationes circa delectabilia contra naturam ex egritudine aut consuetudine, est extra terminos malitie supple humane sicut et bestialitas."

115. EC, BNF fr. 24281, 117r: "Il n'est mie doubte qu'ils se porroient bien retraire de ceste tres fole et tres abhominable oppinion . . . s'il voloient bien fort et bien vertueusement resister a lors inclinacions des le premier commencement. Et pource dit Ptholomees que li saiges seignourist sur les estoiles. [Margin: In prologo *Almagesti* {I.i}] Et ausi dit Aristote ailleurs que nous sommes seigneur de tous nos fais des le commencement de nostre vie dusques en la fin. [Margin: 7° *Ethicorum*]."

116. Aristotle, *Ethica nicomachea*, fasc. 3, III.8, 1114b30–31. It is in another context, the distinction between singular acts and habits, that Aristotle makes this comment about control.

117. Walter Burley, *Expositio super decem libros Ethicorum*, bk. VII, tr. 1, ch. 5, 112va: "In omnibus enim his potest esse malitia extra terminos malitie humane."

118. See Klemm, "Medicine and Moral Virtue."

119. Aristotle, *Ethica nicomachea*, fasc. 3, VII.9, 1149b35–1150a6: "Non enim habet eleccionem neque raciocinacionem, set discessit a natura quemadmodum insanientes hominum. Minus autem bestialitas malicia, terribilius autem. . . . Innocencior enim pravitas semper que non habentis principium. Intellectus autem principium." Gauthier, *L'Ethique à Nicomaque*, vol. 1, pt. 2, 202, takes this as a mistranslation of the Greek.

120. Thomas Aquinas, *Summa theologiae*, pt. 1 of pt. 2, q. 31, art. 7; vol. 20, 24: "Contingit enim in aliquo individuo corrumpi aliquod principiorum naturalium speciei; et sic id quod est contra naturam speciei, fieri per accidens naturale huic individuo; sicut huic aquae calefactae est naturale quod calefaciat. Ita igitur contingit quod id quod est contra naturam hominis . . . fiat huic homini connaturale, propter aliquam corruptionem naturae in eo existentem. Quae quidem corruptio potest esse vel ex parte corporis . . . ; vel etiam

ex parte animae, sicut propter consuetudinem aliqui delectantur in comedendo homines, vel in coitu bestiarum aut masculorum, aut aliorum huiusmodi, quae non sunt secundum naturam humanam."

121. Piché, ed., *La condamnation parisienne de 1277*, § 49 (66), 96.

122. Piché, ed., *La condemnation parisienne de 1277*, § 166 (206), 128, and § 183 (205), 134; see also 172 (207), 130. See Libera, *Penser au Moyen Âge*, 181–211.

123. Palémon Glorieux, *La littérature quodlibétique de 1260 à 1320*, vol. 1, Bibliothèque Thomiste 5 (Le Saulchoir, Kain: Revue des Sciences Philosophiques et Théologiques, 1925), 138.

124. Piché, ed., *La condamnation parisienne de 1277*, prologue, 72; and § 166 (206), 128: "Nonnulli Parisius studentes . . . quosdam manifestos et execrabiles errores . . . tractare et disputare presumunt. . . . Quod peccatum contra naturam, utpote abusus in coitu, licet sit contra naturam speciei, non tamen est contra naturam individui."

125. If there was an increased emphasis on the will, it does not seem to have redirected the discussion of the specific issues addressed here. Cf. Martin W. F. Stone, "Moral Psychology After 1277."

126. See Edward Grant, *Science and Religion, 400 B.C. to A.D. 1550: From Aristotle to Copernicus*, Greenwood Guides to Science and Religion (Westport, Conn.: Greenwood Press, 2004), 181–202.

127. BM, Escorial f.I.11, 7ra, margin: "Sodomite sunt in extremitate malitie. Aristoteles in *Eticis*." It is possible the author of the glosses was familiar with Pietro's work. At another point (IV.1, 6ra) he cites the same source as Pietro, but he might have arrived at it independently.

CHAPTER 5. WHAT'S WRONG?

1. The issue is not "bienséance" in the modern sense, and Evrart uses problems of translation to avoid complex details that do not involve moral issues (e.g., in geometry), but his discomfort is clear from the context. See Jacquart, *La Médecine médiévale dans le cadre parisien*, 281–82; and Goyens, "Comprendre Aristote au Moyen Âge," 144–46.

2. EC, BNF fr. 24281, 116v: "un probleme qui n'est pas bien seans au dire voir en la langue francoise. . . . ceste matere comme il est ia toucie n'est pas bien gracieuse a traictier en francois nen langaige quelconques pource qu'elle est abhominable meismes a nature."

3. EC, BNF fr. 24281, 117r: "ceste tres fole et detestable volonte."

4. Jacques Chiffoleau, "Dire l'indicible: Remarques sur la catégorie du *nephandum* de XIIe au XVe siècle," *Annales ESC* 45 (1990): 289–324; Puff, *Sodomy*, 50–74; Hergemöller, *Sodom und Gomorrha*, 21–22; Jordan, *Invention of Sodomy*, 147–54; Lochrie, *Covert Operations*, 177–205; Beatrice Michaelis, "Recht verschwiegen: Das 'Tabu' der Sodomie in der Sprache des spätmittelalterlichen Rechts," in *Geschlecht als Tabu: Orte, Dynamiken und Funktion der De/Thematisierung von Geschlecht*, ed. Ute Frietsch, Konstanze Hanitzsch, Jennifer John and Beatrice Michaelis, Gender Codes 5 (Bielefeld: transcript, 2008), 141–54.

5. J. M. M. H. Thijssen, *Censure and Heresy at the University of Paris, 1200–1400*, The

Middle Ages (Philadelphia: University of Pennsylvania Press, 1998), for example, 34–35. See also Mary Martin McLaughlin, *Intellectual Freedom and Its Limitations in the University of Paris in the Thirteenth and Fourteenth Centuries* (New York: Arno Press, 1977).

6. Cadden, "Preliminary Observations on the Place of the 'Problemata,'" 2–8.

7. WB, prologue, Bodl. Digby 153, 102r: "Causa finalis est scire causas problematum collectorum et mirabilium diversorum in eis contentorum. Nec propter quorundam vilitatem est ab eis desistendum, quia non est spurnendum vilia contemplari. In quolibet enim quantocumque vili aliquid est divinum et mirabile aliquod reperitur. Aristoteles 13 *De animalibus* [*De partibus animalium*, I.5, 645a15–19]."

8. WB, Bodl. Digby 153, 110r, "Non vilia contemplando sed operando vel volendo viles [Oj113, BMr: viciosi] efficimur. *Ethicorum*." Aristotle, *Ethica nicomachea*, AL fasc. 3, II.2, 1103b26–30.

9. Cadden, *Meanings of Sex Difference*, 228–58.

10. Probably the most widely disseminated was an anonymous collection of natural questions often labeled "Aristotle's *Problemata*" in the Middle Ages and Renaissance (inc. "Omnes homines naturaliter"). See Blair, "The 'Problemata' as a Natural Philosophical Genre," 181–85.

11. Green, "'Traitté tout de mençonges,'" showing that Christine was reacting to a genre of vernacular writing about women's sexual and reproductive bodies; and Joan Cadden, "Medieval Scientific and Medical Views of Sexuality: Questions of Propriety," *Medievalia et Humanistica*, n.s. 14 (1986): 157–71.

12. PA, IV.1, BNF lat. 6540, 53ra: "Quod delectationis causa exercetur ut plurimum fit et aliquando causa generationis filiorum. Interdum etiam causa sanitatis et ponderis depositionis." Coucke interprets the last phrase plausibly as "release from a burden." "Edition of Peter of Abano, 'Exp.' IV," 266.

13. A maniculum at JJ, IV.1, BNF lat. 6542, 60vb.

14. BM, gloss at IV.2 or IV.4, Erfurt, CA Q. 237, 12r: "Notandum experimentum utrum mulieres sunt fecunde."

15. PA, IV.10, BN 6540, 57ra: "Solum inter homines autem plures moventur ad coytum propter delectationem quam propter liberos procreandos."

16. See, for example, Bernardus de Gordonio, *Practica dicta Lilium* (Venice: Bonetus Locatellus, 1498), pt. VII, chap. [1], 87vb); cf. Constantinus Africanus, *De coitu: El tratado de andrología de Constantino el Africano*, ed. and trans. Enrique Montero Cartelle, Monografías de la Universidad de Santiago de Compostela 77 (Santiago de Compostela: Universidad de Santiago, 1983), 299.

17. EC, BNF fr. 24281, 116vb: "ordenée pour la conservation de toute le espece."

18. PA, IV.1, BNF lat. 6540, 53ra: "Notandum est quod coytus est mutua actio maris et femelle qua tertie digestionis expellitur superfluum, quod delectationis causa exercetur ut plurimum fit et aliquando causa generationis filiorum, interdum etiam causa sanitatis." The definition may be borrowed. Danielle Jacquart and Claude Thomasset, *Sexuality and Medicine in the Middle Ages*, trans. Matthew Adamson (Princeton: Princeton University Press, 1988), 212, n. 93, report its occurrence in BNF lat. 16195.

19. An occasional author or copyist resists the change from Bartholomeus's rubric. The

tables of contents in two manuscripts of the Bavarian *Problemata* (which follows Pietro's very closely) offer "De venereis" and (as if to avoid choosing between the authority of the Latin Aristotle and the authority of its chief expositor) "De venereis et coitu," Ghent, UB 178, 157v; and CLM 12021, 118ra. Krakow, BJ 2095, 231vb, does not have a title.

20. EC, BNF fr. 24281, 102v et passim.

21. Joëlle Ducos, "Traduction et lexique scientifique: Le cas des 'Problèmes' d'Aristote traduits par Evrart de Conty," in *Traduction et adaptation en France à la fin du Moyen Âge et à la Renaissance: Actes du colloque organisé par l'Université de Nancy II, 23–25 mars 1995*, ed. Charles Brucker (Paris: Honoré Champion, 1997), 237–47.

22. PA, IV, prologue, BNF lat. 6540, 53ra: "Scripsit enim Aristoteles ad Alexandrum ipsum fore destructionem virtutum et legis transgressorem [BAV lat. 2175: transgressionem] morumque femineorum generativum. Quod etiam verum esse videtur cum superfluit fueritque indecenter peractus," quoting [Pseudo]-Aristotle, *Liber de secretis secretorum* (Reutlingen: Michel Greyff, c. 1483), microfilm, German Books Before 1601, item 6 [*sic* for 4], ch. 15, [9]r: "Coitus est distractio et destructio corporis et abbreviatio vite et corruptio virtutum legis transgressio femineos mores generat." The passage in Aristotle also mentions negative effects on health.

23. PA, prologue, BNF lat. 15454, margin, 45rb: "Nota quomodo coitus est transgressio legis."

24. WB, IV.2, London, Gray's Inn 2, margin, 148va: "Propter destructionem morum anime."

25. WB, Bodl. Digby 206, 104va: "Particula 5 [= IV] fuit totaliter demissa propter materie turpitudinem." I previously asserted that the manuscript had belonged to a French prior (Cadden, "'Nothing Natural Is Shameful,'" 86). Maaike Van der Lugt has corrected this error: "Aristotle's *Problems* in the West," 85.

26. See Cadden, "'Nothing Natural Is Shameful,'" 80–84.

27. "Excusatio lectori propter inhonestatem materie." The term "inhonestas" lacks the force of "turpitudo." Some form of this label occurs in Oxford, Oriel 28, [69]v; CU Peterhouse 220, 54v; CU EE.I.22, 17r; Bodl. Digby 153, 110r; and BL, Royal 12.E.16, 92v; it does not occur in Oxford, St. John's 113.

28. WB, Bodl. Digby 153, 110r: "Nullum naturale turpe sed omnia munda mundo existunt. Non tamen est spurnendum vilia contemplari quia in quolibet quantocumque vili aliquid divinum et mirabile continetur secundum Philosophum .13. *De animalibus* [*De partibus animalium*, I.5, 645a15–19]. Non vilia contemplando sed operando vel volendo viles efficimur. *Ethicorum*." The latter reference is perhaps to *Ethica nicomachea*, II.2, 1103b26–34.

29. WB, prologue, Bodl. Digby 153, 102r: "Felix qui poterit causas cognoscere rerum [Virgil, *Georgics*, 2.490]."

30. WB, IV, prologue, Bodl. Digby 153, 110r: "Bonum autem est scire malum sed malum est malum operari, quia sciendo bonum et malum contingit a malo melius preservari." CU Peterhouse 220 gives a reference to Boethius, *Topics* [III].

31. See Cadden, *Meanings of Sex Difference*, 259–77.

32. WB, IV, prologue, Bodl. Digby 153, 110r: "Nec propter eos qui turpiloquium publicum plus abhorrent quam operari turpiter in privatis."

33. WB, IV, prologue, Bodl. Digby 153, 110r: "In quorum libris viliora quam in medicinalibus vel phisicis [Oxford, St. John's 113: philosophicis] sunt reperta."

34. Exodus 13:12 (cf. Exodus 34:19) and Proverbs 30:15–16.

35. Those with documented connection to Oxford: mss. London, Gray's Inn 2, Bodl. Digby 153, Bodl. Digby 206, Oxford, Oriel 28, Oxford, St. John's 113, Oxford, Magdalen 65. Those containing works of natural philosophy and/or medicine: mss. CU Peterhouse 220; BL Royal 12.E.16 and add. 62127; Oxford, Magdalen 65 and St. John's 113; Bodl. Digby 77, 153, 161, and 206. Those with clerical connections: Bodl. Digby 153 was owned by a Franciscan, John Bruyl, and a Dominican, Thomas Lucas. Also linked to the manuscripts, Richard Hopton, licensed to preach in Canterbury (Oxford, Oriel College 28); Frater Thomas (Oxford, St. John's 113). For further details on ownership of the Burley manuscripts, see Cadden, "'Nothing Natural Is Shameful,'" 86–88, and Van der Lugt, "Aristotle's 'Problems' in the West," 84–87.

36. WB, IV, prologue, BL Royal 12.E.16, 92v–93r: "sed magis siquid dignum reprehensione viderint, corrigant et emendent."

37. WB, BL Royal 12 E. 16, 98r.

38. BM, V.15, BAV lat. 2481, 93ra: "Tanta delectatio est quod mirum est quod theologi non posuerunt ipsam peccatum." See Van der Lugt, "Aristotle's 'Problems' in the West," 95–97 and Appendix 1, 105–7. Readings at IV.26 reflect the Jean de Jandun redaction of Pietro's work.

39. See Madeline H. Caviness, "Obscenity and Alterity: Images That Shock and Offend Us/Them, Now/Then," in *Obscenity: Social Control and Artistic Creation in the European Middle Ages*, ed. Jan M. Ziolkowski, Cultures, Beliefs and Traditions 4 (Leiden: Brill, 1998), 155–75 and figs. 14–36. On "turpiloquium" embedded in "scurrilitas," Carla Casagrande and Silvana Vecchio, *I peccati della lingua: Disciplina ed etica della parola nella cultura medievale*, Bibliotheca Biographica, Sezione Storico-Antropologica (Rome: Istituto della Enciclopedia Italiana, 1987), 369–406, esp. 395–99.

40. See, for example, Michael Camille, "Bestiary or Biology? Aristotle's Animals in Oxford, Merton College, MS 271," in *Aristotle's Animals in the Middle Ages and Renaissance*, ed. Carlos Steel, Guy Guldentops, and Pieter Beullens, Medievalia Lovaniensia, ser. 1, studia 27 (Leuven: Katholieke Universiteit Leuven, Instituut voor Middeleeuwse Studies, 1999), 355–96.

41. Michael Camille, "Manuscript Illumination and the Art of Copulation," in *Constructing Medieval Sexuality*, ed. Karma Lochrie, Peggy McCracken, and James A. Schultz, Medieval Cultures 11 (Minneapolis: University of Minnesota Press, 1997), 58–90.

42. Oxford, Pembroke 193, as discussed in Michael Camille, "Obscenity Under Erasure: Censorship in Medieval Illuminated Manuscripts," in *Obscenity in the Middle Ages: Social Control and Artistic Creation in the European Middle Ages*, ed. Jan Ziolkowski, Cultures, Beliefs and Traditions 4 (Leiden: Brill, 1998), 139–154 and figs. 1–13, on 150–51; late thirteenth-century according to Montague Rhodes James, *A Descriptive Catalogue of the Manuscripts in the Library of Pembroke College, Cambridge* (Cambridge: Cambridge University Press, 1905), 180. Cf. Jan Ziolkowski, *Alan of Lille's Grammar of Sex: The Meaning of Grammar to a Twelfth-Century Intellectual*, Speculum Anniversary Monographs 10 (Cambridge, MA: Medieval Academy of America, 1985).

43. Aristotle, *De animalibus*, IX, Oxford, Merton 271, 65v. See also 69v. On the significance of the clothing, see Michael Camille, "Bestiary or Biology?" 372.

44. BM, Erfurt, CA F.16, 48vb: "Laus tibi sit Christe quoniam liber explicit iste. Detur pro penna scriptori pulchra puella."

45. Padua, Biblioteca Antoniana Scaff. XVII.370, 11ra–b. Guido Billanovich and Luigi Olivieri, "Pietro d'Abano e il Codice Antoniano XVII 370," *Italia medioevale e umanistica* 28 (1985): 221–94, have analyzed the hands found in the manuscript. Olivieri ("Intorno ai 'Problemata' di Aristotele: Le glosse del Codice Antoniano XVII 370 [ff. 1r–2v] e il commento de Pietro d'Abano," 221–75) argues on the basis of a comparison with the text of his *Expositio* that annotations at the beginning of Book I are Pietro's. Billanovich ("Copisti, possessori, e postillatori trecenteschi del Codici Antoniano XVII 370," 276–94) analyzes the hands, and identifies one of the owners, Francesco di Pietro da Sacile (the son of a Paduan professor of grammar who had owned it previously), as the person responsible for almost all the other marginal notations—drawings of heads, hands, and so on, some of which he reproduces. He makes no mention of the figure on 11r, though he notes (on 290) the nude figure at 147rb.

46. PA, IV.23, BNF lat. 15454, 52va: "Habentes testiculos pendentes valde non sunt potentes in coitu." Because of the condition of the manuscript the doodle was not available for reproduction. The manuscript belonged to Petrus Rosselli, who sold it in 1496, and, in addition to corrections and marginal marks, it contains numerous references to Avicenna. On Petrus, see Jacquart's *Supplément* to her revised edition of Wickersheimer's *Dictionnaire biographique des médecins*, s.v. Pierre Rousseau, 242: perhaps the "Rosselli medicus" who bought a manuscript (BNF lat. 16133) in Nov. 1477. The book was later in the possession of a doctor of theology, see Denise Gid, "La bibliothèque de François Guillebon prieur de Sorbonne (†1534)," in *Bibliophiles et reliures: Mélanges offerts à Michel Wittock*, ed. Annie De Coster and Claude Sorgeloos with Marcus de Schepper (Brussels: Fl. Tulkens, 2006), # 1, p. 258. Pietro asserted that he knew the significance of hanging testicles from experience. See his *Compilatio phisonomie,* pt. 2, ch. 9, [31]r.

47. "Bavarian" *Problemata*, Krakow, BJ 2095, 232vb and 233rb. That the names are in red ink indicates that Johannes Stolle de Glogovia (about whom nothing else is known) inserted them when or after he went over the completed manuscript to add titles and rubrics. He did not take similar advantage of the spaces remaining in other parts of the manuscript.

48. Camille, "Bestiary or Biology?" 374–75.

49. See images from literary texts in Camille, "Obscenity Under Erasure," 148; Camille, *Image on the Edge,* illus. 50 (101) and illus. 55 (106). Also Caviness, "Obscenity and Alterity," fig. 14.

50. Rocke, *Forbidden Friendships*, 135–36 and n. 107.

51. Puff, "Localizing Sodomy," 180.

52. PA, BAV lat. 2175, margin, 52rb. JJ, Sorbonne 122 also has a nonverbal mark at this point (47ra), conceivably a sexual doodle.

53. BM, BNF lat. 6307, margin, 201vb: "Nota quod mulieres sunt insatiabiles in coitu."

54. PA, table, BN Marciana 2672, 13va.

55. JJ, Seville, Biblioteca Capitular y Colombina 7-7-9, margin, 52vb: "Quando debet custodium mulier ne abutatur cum viro alio"; " 'Super filiam luxuriosam pone custodiam.' Solamon Inperabolis."

56. PA, BNF lat. 6540, margin, 63vb. The sentence is in the gutter of the tightly bound manuscript, and I have not been able to decipher it completely: "Et [hiis?] etiam aparet causa quare / ille de sabaudia [or: sabandia] 30 / vicibus coibat in die de quo conquesta est [uxor?] paren / tibus et pepigerunt cum / ag[ere?] sol[is?] 10 vicibus in die." I am grateful to Danielle Jacquart for assistance here, though she is not responsible for the paleographical or substantive inferences I have drawn.

57. Avicenna, *Liber canonis medicine*, bk. 3, fen 20, tr. 1, ch. 11, 353ra: "Coitus infantibus fedus est apud multitudinem gentium et prohibitus in lege."

58. EC, BNF fr. 24281, 117ra: "Et de ce sensient il finablement que cil sont bien pervers et bien desnaturé bien bestial et bien effeminé et bien mal elbreus [Hague, Koninklijke Bibliotheek 133A3: mal eureux] et bien infortuné qui sont de ceste secte. . . . Et pource ausi faingnent aucun poete que nature les fait escommenier par genius."

59. Alain de Lille, "De planctu naturae," § 18, prose 9, pp. 878–79.

60. Guillaume de Lorris and Jean de Meun, *Le roman de la rose*, ed. Félix Lecroy, 2 vols., Les Classiques Français du Moyen Âge 92 and 95 (Paris: Honoré Champion, 1983 and 1966), vol. 1, lines 4313–15. The doctrinal and moral propriety of *The Romance of the Rose*, which was widely read in courtly and academic circles, was a matter of serious debate. See Christine de Pizan et al., *Le débat sur "Le roman de la rose,"* ed. Eric Hicks, Bibliothèque du XVe siècle 43 (Paris: Honoré Champion, 1977). Although Evrart seems to have taken the poem's standpoint as unambiguous, it was probably not. See *The Romance of the Rose*, trans. Charles Dahlberg, 3rd ed. (Princeton: Princeton University Press, 1995), notes, p. 374.

61. This is the case for the "Vatican" abbreviation, BAV lat. 2481, and the "St. Victor" version, BNF lat. 14728.

62. "Aachen" *Problemata*, Erfurt, CA F. 263, 23rb.

63. "Bavarian" *Problemata*, Krakow, BJ 2095, 227vb.

64. "Bavarian" *Problemata*, Krakow, BJ 2095, 227va: "carent testiculis aut pravos aut debiles omnino habent."

65. PA, BAV lat. 2175, table, fol. iirb: "Quare coientes ambo gaudent actu illo."

66. *Problema* IV.26 is absent in all extant manuscripts of the "Flemish" *Problemata* (The Hague, KB 70, Brugge, Openbare Bibliotheek 481, Naples, Biblioteca Nazionale VII. AA.39). It is also lacking in one manuscript of WB (BL add. 62127) and in the text that survives only in BAV lat. 901.

67. BM, Erfurt, CA Q 237.

68. Pietro marks IV.16 (on the benefits of intercourse for phlegmatic diseases) and IV.23 (on the causes of erections) as having been treated in connection with earlier *problemata*; the Flemish *Problemata* omits both of these as well. The copyist who abridged WB (BL add. 62127) kept IV.16 and omitted IV.23 along with IV.26 and twelve others. The "Vatican 901" text omits twenty-two *problemata* in Book IV, among them 26.

69. "Prague" *Problemata*, Prague, NK I.C.25, 74vb: "Nota multa nolui scribere propter turpitudinem."

70. "Prague" *Problemata*, Prague, NK VIII.A.19, 79rb–va. In this version, IV.26 is the only *problema* in Book IV in which expansive comments marked by "Nota" have been incorporated into the text.

71. Cadden, *Meanings of Sex Difference*, 153–54.

72. In a manuscript of the Aristotelian text, BM, BNF lat. 6327, 21rb, "quidam cum quibus" is glossed "femine."

73. "Vatican" *Problemata* in BAV lat. 2481. On the text and its environment, see Van der Lugt, "Aristotle's *Problems* in the West," 75–77 and Appendix 1, 105–7.

74. "Vatican" *Problemata*, BAV lat. 2481, 93rb: "in omni emissione superfluitatis accidit gaudium. . . . Igitur alias demonstrat inductione Aristotelis longum verbum et bene clara."

75. Currently BNF lat. 15081, a late fourteenth-century manuscript from St. Victor. See Ouy, *Les manuscrits de l'abbaye de Saint-Victor*, vol. 2, *Texte*, 579, # LLL8; and BN 14725, an early fourteenth-century manuscript containing the fifteenth-century St. Victor coat of arms, on which see Charles Samaran et al., *Catalogue de manuscrits en écriture latine portant des indications de date, de lieu ou de copiste* (Paris: Centre Nationale de la Recherche Scientifique, 1962), vol. 3, 379, and Ouy, *Les manuscrits de l'abbaye de Saint-Victor*, vol. 2, *Texte*, 578; # LLL7.

76. "St. Victor" *Problemata*, BNF lat. 14728, 283vb: "Quare vir et mulier proportionaliter gaudent in peragendo coitum. Quia iii. i° actu evacuatur superfluitas ut patet ex dictis; iii° in aliarum superfluitatum emissione delectantur ut patet de egestione et urina. Quare sequitur propositum." A note (313v) records that the work was acquired by Jean Lamasse, who was prior 1422–48. The numbering of the responses (1 and 3) suggests that this text was an abbreviation of another in which three points were specifically enumerated.

77. "Erfurt" *Problemata*, CLM 4710, [268]rb: "Quare quidam tam active quam passive in coitu delectantur."

78. See Cadden, "'Nothing Natural Is Shameful,'" 81–82. On brevity, WB, prologue, Bodl. Digby 153, 102r.

79. WB, Bodl. Digby 153, 112r: "Quare coeuntes maxime gaudent."

80. WB, Bodl. Digby 153, 112r: "Mulier enim delectatur concipiendo et emittendo [Oxford, St. John's 113 adds: sed vir solum delectatur emittendo]."

81. WB, Bodl. Digby 153, 112r: "Nota appetitus coitus provenit quatuor scilicet ex natura, cibariis ventosis, imaginatione et consuetudine coeundi. Viri effeminati sunt in coeundo insatiabiles et insani sicud mulieres. Incurantur [Oxford, St. John's 113: curantur] per diuretica, laxativa et vomitiva sicud fame, siti, tristitia, vigilia, verbere et labore."

82. PA, IV.1, BNF lat. 6540, 53ra.

83. BM, Erfurt, CA Q. 237.

84. JJ, Paris, Arsenal 723, 80vb: "Notandum quod istud nephandum opus sodomiticum quidam exercent manu fricando virgam, alii puerorum inter coxas confricatione, quod et [erased: plurimi] agunt [erased: hodie].

85. BM, Escorial, f.I.11, 7ra. The lemmas correspond to Pietro's division of the text, and, explicating the same text, Pietro applies the same word not to the condition (*passio*) but to the habit (*passio eis propter pravam consuetudinem contingit*), so it is likely the

glossator had access to Pietro's commentary. There is no evidence that this individual lectured on the *Problemata*, but he had also marked textual divisions within the first dozen questions of Book I. On glossator's hand, AL, # 1216. The marks at IV.26 (but not elsewhere in IV) in Padua, Antoniana Scaff. XVII, 370, 12rb–va, are somewhat similar.

86. JJ, BNF lat. 15454, margin, 54va: "Nota turpia"; "Turpissimum et fedissimum dictu."

87. BM, Escorial f.I.11, margin, 7ra: "Sodomite sunt in extremitate malitie. Aristoteles in *Eticis.*"

88. BM, BNF lat. 6327, 21va–b.

89. PA, BAV lat. 2175, 52va and PA, BNF lat. 6540, 64vb.

90. PA, table, BNF lat. 6543, 46ra–b, which corresponds to the marginal notations at IV.26 in PA, BNF lat. 6541. The same table is contained in BN Marciana 2672.

91. Göttweig, Benediktanerstiftsbibliothek 138, 238v.

92. Taddeo Alderotti, *In Ioannitii Isagogarum libellum:* "Item sunt alie species coitus que fiunt penes modum coeundi et penes diversitatem corporum coeuntium quas demitto, tum quia sunt inumerabiles quasi et tum quia spectat magis ad theologos vel morales gratia vitiorum et tum quia turpe est de talibus sermonem efficere." In *Expositiones. In arduum Aphorismorum Ypocratis volumen. In divinum pronosticorum Ypocratis volumen. In preclarum regiminis acutorum Ypocratis opus. In subtilissimum Ioannitii Isagogarum libellum* (Venice: Iannes Baptista Nicolinus, 1527), quoted in Coucke, "Edition of Peter of Abano, 'Exp. IV,' " 219, n. 124. In fact, Taddeo himself wrote about moral philosophy and even translated a version of Aristotle's *Ethics* into Italian. Siraisi, *Taddeo Alderotti and His Pupils,* 72–95.

93. EC BNF fr. 24281, 117r: "malvaise et perverse volontes." Evrart uses "volonte" once elsewhere in IV.26 (116v); The Hague, KB 133A3, 1 137va, substitutes "ymagination." Although some of his references are borrowed from Pietro, those to Aristotle's *Metaphysics* II and *Ethics*, X; to the prologue of Ptolemy's *Almagest*; and to Ovid and Jean de Meun are his own.

94. EC BNF fr. 24281, 116v: "en compaignment de fame en la maniere que nature et raisons y enclinent." He refers at this point to IV.15, the *problema* upon which Burley also drew in explicating IV.26.

95. EC, BNF fr. 24281, 116v: "Car la delectation conferme le oeuvre et le fait meilleur et plus durable si comme dit Aristote. [Margin: 10° *Ethicorum* (X.4, 1174b24–75a3)]"; "abusant de l'oeuvre desus dite." The latter phrase might contain an echo of Étienne Tempier's condemnation of "abusus in coitu," Piché, ed., *La condamnation parisienne de 1277*, #166 (206), 128.

96. EC, BNF fr. 24281, 117r: "tres fole et destestable"; "perverse nature effeminé et imparfait"; "voie deplaisant a nature et a toute raison." The Hague, KB 133A3, 1, 137va, enhances the language: "ville et perverse nature."

97. EC, BNF fr. 24281, 117r: "Appart il que Aristote voelt ramener ceste tres fole et detestable volonte a raison naturele."

98. EC, BNF fr. 24281, 117r: "raison bien acertes laquele selont la verite doit l'omme gouverner et du tout estre dame des vertus sensitives et meisme de coustume et de nature quant a choses qui sont soubmises a nostre poissance."

99. JJ, BNF lat. 15454, margin, 54va: "Turpissimum et fedissimum dictu." Cf. Burley's reference to *turpiloquium* in his prologue to IV.

100. EC, BNF fr. 24281, 116v: "Et pource n'y voeil ge pas arrester longuement ne poursievir le texte. Et si ne le voeil mie ausi du tout trespasser oultre afin con ne cuide mie que il y ayt negligence."

101. EC, BNF fr. 24281, 116v: "i'en toucherai un petit ci apres en general et assez legierement."

102. EC, BNF fr. 24281, 116v: "Et dispose es membres de generation si que la matere ne sen poet passer et courre prestement au lieu deu comme nature le desire et demande"; 117r: "malvaise coustume."

103. Evrart is referring to Aristotle, *Metaphysics*, I.1, but also cites the same line from Virgil with which Burley had opened his work: "Felix qui poterit rerum cognoscere causas" (*Georgics* 2.490).

104. EC, BNF fr. 24281, 116v: "Ains est grans eurs et grans perfections a l'entendement de bien ainsi comprendre les causes desus dites des choses de nature si comme il fu dit en commencement de cest livre."

105. EC, BNF fr. 24281, 116v: "Et pource que ceste matere regarde bien la general matere de ceste quarte partie pource ne le vault mie li prince des philosophes oublier ne trespasser qu'il n'en fast aucune mention."

106. EC, BNF fr. 24281, 116v: "Pour laquelle chose phylosophie est appellée science de verite, si comme Aristote dit ailleurs. Et au voir dire riens n'est lait quant a ceste consideration, combien que la chose considerée ne soit en soy pas belle." In the margin, he cites "2° *Methaphysice* [II.1, 993b20]. Vocari vero philosophiam veritatis scientiam recte habet."

107. EC, BNF fr. 24281, 116v: "Et si ne le voeil mie ausi du tout trespasser oultre afin con ne cuide mie que il y ayt negligence. Il samble dont briefment qu'il voeille demander ainsi pour quoy c'est que on se delite generalment en le desus dite oeuvre de generation en quelconques maniere que ce soit fait."

EPILOGUE

1. *Perry v. Schwarzenegger*, 2010 U.S. Dist. (N.D. Cal., Aug. 4, 2010), 704 F. Supp. 2d 921; 2010 U.S. Dist., pp. 931, 999, 966; LEXIS 78817, https://vpn.lib.ucdavis.edu/hottopics/lnacademic/,DanaInfo=www.lexisnexis.com+? (consulted Sept. 1, 2012).

Manuscripts Consulted

Earlier modern shelf marks appear in parentheses; texts consulted in each manuscript appear in square brackets. See Abbreviations, pp. 219–20.

Angers
 Bibliothèque Municipale 450 (435) [BM]
Assisi
 Biblioteca Comunale (formerly Biblioteca del Convento di S. Francesco), Fondo Antico 663 [2 copies of BM]
Berlin
 Staatsbibliothek
  lat. fol. 482 [WB, *Commentum super libros Ethicorum Aristotelis*]
  lat. fol. 584 [Heinrich von Friemar, *Commentum super libros Ethicorum Aristotelis*]
Brugge
 Openbare Bibliotheek
  476 [BM]
  481 ["Flemish" *Problemata*]
Cambridge
 Peterhouse (at the University Library)
  79 [JJ]
  121 [BM]
  220 [WB]
 Sidney Sussex College
  68 (Delta.4.6.) [BM]
 University Library
  Ee.I.22 [WB]
  Kk.II.5 [BM]
Cesena
 Biblioteca Comunale Maletestiana

S.VI.2 [BM; PA]
S.VI.3 [BM; PA]
D.XXII.2 [BM]
D.XXIV.2 [BM; PA]

Erfurt
Universitäts- und Forschungsbibliothek Erfurt/Gotha
Codices Amploniani F. 15 [JJ]
Codices Amploniani F. 16 [BM]
Codices Amploniani F. 26 [BM]
Codices Amploniani F. 236 [Anon. *Problemata,* I (attrib. PA)]
Codices Amploniani F.263 [BM; anon. *Problemata* fragments]
Codices Amploniani Q. 16 ["Erfurt" *Problemata*]
Codices Amploniani Q. 237 [BM]

Florence
Biblioteca Medicea Laurentiana
Plut. 89 sup. 59 [TG]
Plut. 89 sup. 84 [GT]

Ghent
Bibliotheek van de Rijksuniversiteit te Gent
72 [JJ]
178 ["Bavarian" *Problemata*]
357 [BM]

Göttingen
Niedersächsische Staats- und Universitätsbibliothek theol. 124 ["Erfurt" *Problemata*]

Göttweig
Benediktanerstiftsbibliothek 138 (131) [table to PA or JJ]

The Hague
Koninklijke Bibliotheek
70E9 ["Flemish" *Problemata*]
133A3, 1 and 2 [EC]

Jena
Thüringer Universitäts- und Landesbibliothek El.f.81 (Gall., f.81) [EC]

Krakow
Biblioteka Jagiellońska
654 ["Erfurt" *Problemata*]
2095 ["Bavarian," *Problemata*, etc.]

Leipzig
  Universitätsbibliothek
    1435 ["Erfurt" *Problemata*]
    1439 [Heinrich von Friemar, *Commentum super libros Ethicorum Aristotelis*]
    1445 [Albert of Saxony, *Lectura super X libros Ethicorum*]
London
  British Library
    Royal 12 E. 16 [WB]
    Additional 21978 [BM; GT; TG]
    Additional 62127 [WB]
  Gray's Inn Library 2 [WB]
  Wellcome Library 139 [GT, *Commentary of Georgius Trapezuntius upon the Aphorismes of Ptolemeus*, trans. Thomas Blackburn]
Munich
  Bayerische Staatsbibliothek
    Codex Latinus Monacensis 4710 ["Erfurt" *Problemata*]
    Codex Latinus Monacensis 12021 ["Bavarian" *Problemata*]
Naples
  Biblioteca Nazionale Vittorio Emanuele III, VII.AA.39 ["Flemish" *Problemata*]
Nuremberg
  Stadtbibliothek
    Cent. III.19 [Avicenna, *Liber canonis medicine*]
    Cent. III.38 [PA]
Oxford
  Balliol College Library 250 [BM]
  Bodleian Library
    Bodley 463 [BM]
    Digby 77 [WB]
    Digby 153 [WB]
    Digby 161 [WB]
    Digby 206 [WB]
  Magdalen College Library 65 [WB]
  Merton College Library 270 (O.3.2) [BM]
  Oriel College Library 28 [WB]
  St. John's College Library 113 [WB]

Padua
- Biblioteca Antoniana Scaffale XVII, 370 [BM]
- Biblioteca Universitaria 1418 [BM]

Paris
- Site Arsenal, Bibliothèque Nationale de France 723 [JJ]
- Bibliothèque Nationale de France
  - fonds latin 3121A [BM]
  - fonds latin 6307 [BM]
  - fonds latin 6327 [BM]
  - fonds latin 6540 [BM; PA]
  - fonds latin 6541 [PA]
  - fonds latin 6541A [BM; JJ]
  - fonds latin 6542 [JJ]
  - fonds latin 6543 [table to PA]
  - fonds latin 14725 [BM]
  - fonds latin 14728 ["St. Victor" *Problemata*]
  - fonds latin 15081 [BM]
  - fonds latin 15454 [JJ]
  - fonds latin 16633 [BM]
  - fonds latin 16089 [Pietro d'Abano, *Physionomia*]
  - fonds français 210 [EC]
  - fonds français 211 [EC]
  - fonds français 563 [EC]
  - fonds français 564 [EC]
  - fonds français 24281 [EC]
  - fonds français, nouvelles acquisitions 3371 [EC?]
- Bibliothèque Mazarine
  - 774 (495) [BM]
  - 3520 [JJ]
- Bibliothèque Universitaire de la Sorbonne 122 [JJ]

Prague
- Národní Knihovna České Republiky
  - lat. I.C.25 ["Prague" *Problemata*]
  - lat. VIII.A.19 ["Prague" *Problemata*]

Salamanca
- Biblioteca Universitaria, 2055 (Madrid, Biblioteca del Palacio Nacional, 259) [BM]

San Lorenzo de El Escorial
    Real Biblioteca del Monasterio de San Lorenzo de El Escorial
        lat. f.I.11 [BM; Alexander of Aphrodisias, *Problemata*]
        lat. V.III.21 [Aristotle, *Ethica nicomachea*]
Seville
    Biblioteca Capitular y Colombina
        7-7-9 [JJ]
Toledo
    Archivo y Biblioteca Capitulares, 94.14 [Averroes, *Liber Nichomachye Aristotelis id est Ethicorum*]
The Vatican
    Biblioteca Apostolica Vaticana
        Borghesiani 37 [BM]
        Chisiani E.VIII.254 [BM]
        Chisiani G.V.131 [BM]
        Ottoboniani latini 1764 [JJ]
        Ottoboniani latini 2083 [BM]
        Reginensi latini 747 [Alexander of Aphrodisias, *Problemata*]
        Vaticani latini 901 ["Vatican 901" *Problemata*]
        Vaticani latini 2112 [BM]
        Vaticani latini 2174 [BM; PA]
        Vaticani latini 2175 [PA]
        Vaticani latini 2176 [BM; JJ]
        Vaticani latini 2481 ["Vatican" *Problemata*]
        Vaticani latini 2993 [BM]
        Vaticani latini 3008 [BM]
        Vaticani latini 10452 [BM]
Venice
    Biblioteca Nazionale di San Marco
        lat. 2465 (VI.126) [PA]
        lat. 2488 (VI.43) [BM]
        lat. 2672 (VI.156) [table to PA or JJ]
        lat. 2846 (VI.44) [BM]

Works Cited

PRIMARY SOURCES

Alain de Lille. "De planctu naturae," ed. Nikolaus M. Häring. *Studi medievali*, ser. 3, 19/2 (1978): 797–879.

Albertus Magnus. *De animalibus libri XXVI*. Ed. Hermann Stadler. 2 vols. Beiträge zur Geschichte der Philosophie des Mittelalters 15 and 16. Münster: Aschendorff, 1916 and 1920.

———. *Quaestiones super De animalibus*. Ed. Ephrem Filthaut. Vol. 12 of *Opera omnia*, ed. Bernhard Geyer. Münster: Aschendorff, 1955.

———. *Super Ethica commentum et quaestiones*. Ed. Wilhelm Kübel. Vol. 14 of *Opera omnia* in 2 parts. Münster: Aschendorff, 1968–87.

[Pseudo-]Albertus Magnus. *De secretis mulierum*. Lyon: Johannes Quadratus, 1580.

———. *El "De secretis mulierum" atribuido a Alberto Magno: Estudio, edición crítica y traducción*. Ed. José Pablo Barragán Nieto. Textes et Études du Moyen Âge 63. Porto: Fédération Internationale des Instituts d'Études Médiévales, 2012.

Aristotle. *De animalibus: Michael Scot's Arabic-Latin Translation*, part 2, *Books XI–XIV: Parts of Animals*. Ed. Aafke M. I. Van Oppenraaij. Aristoteles Semetico-Latinus 5. Leiden: Brill, 1998.

———. *De generatione animalium translatio Guillelmi de Moerbeka*. Ed. H. J. Drossaart Lulofs. Aristoteles Latinus 17, 2.v. Union Académique Internationale, Corpus Philosophorum Medii Aevi. Bruges : Desclée de Brouwer, 1966.

———. *Ethica nicomachea: Translatio Roberti Grosseteste Lincolniensis sive "Liber ethicorum": A. Recensio pura*. Fasc. 3A (1972) of *Ethica nicomachea*. Ed. René Antoine Gauthier. 4 fascicules in 5. Aristoteles Latinus 26, 1–3. Corpus Philosophorum Medii Aevi, Union Académique Internationale. Leiden: Brill, 1972–74.

———. *Ethica nicomachea: Translatio Roberti Grosseteste Lincolniensis sive "Liber Ethicorum": B. Recensio: recognita*. Fasc. 4 (1973) of *Ethica nicomachea*. Ed. René Antoine Gauthier. 4 fascicules in 5. Aristoteles Latinus 26, 1–3. Corpus Philosophorum Medii Aevi, Union Académique Internationale. Leiden: Brill, 1972–74.

———. *L'Éthique à Nicomàque*, trans. and comm. René Antoine Gauthier and Jean Yves Jolif. 2 vols. in 4. Aristote: Traductions et Études, Collection Publiée par l'Institut Supérieur de Philosophie de l'Université de Louvain. Louvain: Publications Universitaires, 1970.

———. *History of Animals, Books VII–X.* Ed. and trans. David M. Balme. Loeb Classical Library 439. Cambridge, MA: Harvard University Press, 1991.

———. *Physica: Translatio vetus.* Ed. Fernand Bossier and Jozef Brams. 2 fascicules, fasc. 2. Aristoteles Latinus 7.1. Corpus Philosophorum Medii Aevi Academiarum Consociatarum Auspiciis et Concilio Editum, Union Académique Internationale. Leiden: Brill, 1990.

———. *Politica (libri I–II.11): Translatio prior imperfecta interprete Guillelmo de Moerbeka (?).* Aristoteles Latinus 29.1. Corpus Philosophorum Medii Aevi Academiarum Consociatarum Auspiciis et Concilio Editum, Union Académique Internationale. Bruges: Desclée de Brouwer, 1961.

———. *Politique.* Ed. and trans. Jean Aubonnet. 3 vols. Collection des Universités de France. Paris: Les Belles Lettres, 1960–89.

———. *Problemata.* Trans. Theodorus Gaza. In *Extra ordinem naturalium libri quibus nonnulli etiam additi sunt Aristoteli ascripti. Alexandri Problematum libri duo,* 1ra–98rb. Vol. 7 (1562) of *Aristotelis opera cum Averrois commentariis.* Venice: Junctas, 1562–74. Reprinted Frankfurt am Main: Minerva, 1962.

———. *Problemata Aristotelis cum duplici translatione antiqua videlicet et nova scilicet Theodori Gaza cum expositione Petri Aponi.* Venice: Bonetus Locatellus, 1501. Reproduced in Italian Books before 1601, roll 169. Cambridge, MA: General Microfilm, n.d.

———. *Problemata physica.* Trans. Hellmut Flashar. Aristoteles Werke in deutscher Übersetzung 19. Berlin: Akademie, 1975.

———. *Problemata physica.* In *Aristoteles pseudepigraphus.* Ed. Valentine Rose, 666–76. Leipzig: B. G. Teubner, 1863. Reprinted Hildesheim: Georg Olms, 1971.

———. *The "Problemata physica" Attributed to Aristotle: The Arabic Version of Hunain ibn Ishaq and the Hebrew Version of Moses ibn Tibbon.* Ed. Lou S. Filius. Aristoteles Semitico-Latinus 11. Leiden: Brill, 1999.

———. *Probleme.* Trans. Paul Gohlke. Aristoteles, Die Lehrschriften 16. Paderborn: Ferdinand Schöning, 1961.

———. *Problèmes.* Ed. and trans. Pierre Louis. 3 vols. Collection des Universités de France. Paris: Les Belles Lettres, 1991.

———. *Problems.* Ed. and trans. W. S. Hett. Rev. ed., 2 vols. Loeb Classical Library. Cambridge, MA: Harvard University Press, 1970.

———. *Problems.* Ed. and trans. Robert Mayhew. 2 vols. Loeb Classical Library. Cambridge, MA: Harvard University Press, 2011.

Averroes [Ibn Rushd]. *In Moralia nicomachia expositio.* Vol. 3 of Aristotle, *Libri moralem totam philosophiam complectentes.* Venice: Iunctas, 1562. Reprinted Frankfurt am Main: Minerva, 1962.

Avicenna [Ibn Sīnā]. *De animalibus.* Trans. Michael Scot. In *Logica, Sufficientia, De celo et mundo, De anima, De animalibus, De intelligentii, Alfarabius De intelligentiis, Philosophia prima,* 29ra–64rb. Venice: [Putzbach,] 1508. Reprinted Frankfurt am Main: Minerva, 1961.

———. *Liber canonis medicine.* Trans. Gerardus Cremonensis. Venice: Pagininis, 1507. Reprinted Hildesheim: Georg Olms, 1964.

———. *Sextus naturalium*. In *Liber de anima seu Sextus de Naturalibus: Édition critique de la traduction latine médiévale*. Ed. S. van Riet. 2 vols. Avicenna Latinus. Louvain: E. Peeters, 1972.

Bernardus de Gordonio. *Practica dicta Lilium*. Venice: Bonetus Locatellus, 1498.

Borris, Kenneth, ed. *Same-Sex Desire in the English Renaissance: A Sourcebook of Texts, 1470–1650*. New York: Routledge, 2004.

Brunetto Latini. *Li livres dou tresor*. Ed. Francis J. Carmody. Berkeley: University of California Press, 1948.

Buridan, Jean. *Questiones super decem libros Ethicorum Aristotelis ad Nicomachum*. Paris: Poset Le Preux, 1513. Reprinted Frankfurt am Main: Minerva, 1968.

Burley, Walter. *Expositio Gaulteri Burlei super decem libros Ethicorum Aristotelis*. Venice: Simon de Luere, 1500.

———. *In Physicam Aristotelis expositio et quaestiones*. Venice: Simone de Luere, Andrea Torresano de Asula, 1501. Reprinted Hildesheim: Georg Olms, 1972.

Christine de Pizan. *Le livre du corps de policie*. Ed. Robert H. Lucas. Textes Littéraires Français [145]. Geneva: Droz, 1967.

———. *Le livre des fais et bonnes meurs du sage roy Charles V*. Ed. S. Solente. Société de l'Histoire de France, Série antérieur à 1789. 2 vols. Paris: Librairie Ancienne Honoré Champion, 1936–40.

———. *Le livre des trois vertus: Édition critique*. Ed. Charity Cannon Willard in collaboration with Eric Hicks. Bibliothèque du XVe Siècle 50. Paris: Honoré Champion, 1989.

Christine de Pizan et al. *Le débat sur "Le roman de la rose."* Ed. Eric Hicks. Bibliothèque du XVe Siècle 43. Paris: Honoré Champion, 1977.

Constantinus Africanus. *Liber de coitu: El tratado de andrología de Constantino el Africano*. Ed. and trans. Enrique Montero Cartelle. Monografias de la Universidad de Santiago de Compostela. Santiago de Compostela: Universidad de Santiago, 1983.

———. *Pantegni*. In Isaac [Israeli], *Opera omnia*. Lyon: Andreas Turinus, 1515. 2nd foliation.

Coucke, Gijs, ed. "Edition of Peter of Abano, 'Exp.,' IV, with Synthesis, Annotations and Essays" Vol. 2 of *Philosophy Between Text and Tradition: The Reception of Aristotle's "Problemata" in the Middle Ages: Peter of Abano's "Expositio Problematum."* Ph.D. dissertation. Katholieke Universiteit Leuven, 2008.

Evrart de Conty. *Le Livre des Eschez amoureux moralisés*. Ed. Françoise Guichard-Tesson and Bruno Roy. Bibliothèque du Moyen Français 2. Montreal: Ceres, 1993.

Foerster, Richard, ed. *Scriptores physionomonici graeci et latini*. 2 vols. Bibliotheca Scriptorum Graecorum et Latinorum Teubneriana. Leipzig: B. G. Teubner, 1893.

Foulechat, Denis. *Le Policratique de Jean de Salisbury (1372), Livres I–III*. Ed. Charles Brucker. Publications Romanes et Françaises 209. Geneva: Droz, 1994.

Galen. *De semine*. In *Opera omnia*. Ed. Carl Gottlob Kühn, 4, 512–651. 20 vols. in 22. Leipzig: C. Cnobloch, 1821–33. Reprinted Hildesheim: Georg Olms, 1964–65.

Gentilis Fulginatis. *Expositiones cum textu Avicenne*. Venice: Baptista de Tortis, 1492. Reproduced in Italian Books Before 1601, roll 307. Cambridge, MA: General Microfilm, n.d.

George of Trebizond. *Claudii Ptolemaei Centiloquium sive Aphorismi a Georgio Trapezuntio ex graeco in latinum versi et commentariis illustrati.* In Guido Bonatus, *De astronomia tractatus X, Universum quod iudiciariam rationem nativitatum, aëris, tempestatum, attinet, comprehendentes. Adiectus est Cl. Ptolemaei Liber fructus cum commentariis utilissimus Georgii Trapezuntii.* Basel: Nicolaus Pruknerus, 1550.

———. *Collectanea Trapezuntiana: Texts, Documents, and Bibliographies of George of Trebizond.* Ed. John Monfasani. Medieval and Renaissance Texts and Studies 25; Renaissance Society of America, Renaissance Texts and Studies 8. Binghamton, NY: Medieval and Renaissance Texts and Studies with the Renaissance Society of America, 1984.

Goodich, Michael, ed. *Other Middle Ages: Witnesses at the Margins of Medieval Society.* The Middle Ages. Philadelphia: University of Pennsylvania Press, 1998.

Green, Monica H., ed. and trans. *The Trotula: A Medieval Compendium of Women's Medicine.* The Middle Ages. Philadelphia: University of Pennsylvania Press, 2001.

Guillaume de Lorris and Jean de Meun. *Le roman de la rose.* Ed. Félix Lecroy. 2 vols. Les Classiques Français du Moyen Âge 92 and 95. Paris: Honoré Champion, 1983 and 1966.

Heidingsfelder, Georg. *Albert von Sachsen: Seine Lebensgang und sein Kommentar zur "Nikomachischen Ethik" des Aristoteles.* 2nd ed. Beiträge zur Geschichte der Philosophie des Mittelalters: Texte und Untersuchungen 22, fasc. 3–4. Münster: Aschendorff, 1927.

Hissette, Roland. *Enquête sur les 219 articles condamnés à Paris le 7 mars 1277.* Louvain: Publications Universitaires, 1977.

Ibn Riḍwān, ʿAlī ["Haly Heben Rodan"]. [*Commentary on the "Centiloquium"*]. In Pseudo-Ptolemy, *Centiloquium*, trans. John of Seville.

Ibn Rushd. See Averroes.

Ibn Sīnā. See Avicenna.

Jacopo da Forlì and Dino del Garbo. *De generatione embrionis.* Comp. Bassanius Politus. Venice: Bonetus Locatellus, 1502.

Joannes de Janduno. *Quaestiones super 8 libros Physicorum Aristotelis.* Venice: Iuntas, 1551. Reprinted Frankfurt am Main: Minerva, 1969.

———. *Super libros Aristotelis de anima.* Venice: Hieronymus Scotus, 1587. Reprinted Frankfurt am Main: Minerva, 1966.

Joannes Saresberiensis. *Policraticus, I–V.* Ed. K. S. B. Keats-Rohan. Corpus Christianorum Continuatio Mediaevalis 118. Turnhout: Brepols, 1993.

———. *Le Policratique de Jean de Salisbury (1372), Livres I–III.* Trans. Denis Foulechat, ed. Charles Brucker. Publications Romanes et Françaises 209. Geneva: Droz, 1994.

Justinian. *The Digest of Justinian.* Ed. Theodor Mommsen with Paul Krueger and tr. Alan Watson. 4 vols. Philadelphia: University of Pennsylvania Press, 1985.

Lawn, Brian, ed. *The Prose Salernitan Questions Edited from a Bodleian Manuscript (Auct. F.3.10.).* Auctores Britannici Medii Aevi 5. London: British Academy at Oxford University Press, 1979.

Lind, L. R., ed. *Problemata varia anatomica: The University of Bologna, MS. 1165.* Humanistic Studies 38. Lawrence: University of Kansas Publications, 1968.

Lottin, Odin. *Psychologie et morale aux XIIe et XIIIe siècles.* 2nd ed. 6 vols. in 7. Gembloux, Belgium: J. Duculot, 1957–60.

Mercken, H. Paul F. *The Greek Commentaries on the "Nicomachean Ethics" of Aristotle in the Latin Translation of Robert Grosseteste, Bishop of Lincoln (†1253)*. 3 vols. Corpus Latinum Commentariorum in Aristotelem Graecorum, pt. 6, vols. 1–3. Leuven: Leuven University Press, 1973–91.

Oresme, Nicole. "Livre de divinacions." In G. W. Coopland, *Nicole Oresme and the Astrologers: A Study of "His Livre de divinacions."* Liverpool: Liverpool University Press, 1952.

———. *"Le livre de Ethiques" d'Aristote Published from the Text of MS. 2902, Bibliothèque Royale de Belgique*. Ed. Albert Douglas Menut. New York: G. E. Stechert, 1940.

———. *Nicole Oresme and the Wonders of Nature: A Study of his "De causis mirabilium."* Ed. Bert Hansen. Studies and Texts 68. Toronto: Pontifical Institute of Mediaeval Studies, 1985.

Peraldus, Gulielmus. *Summa virtutum et vitiorum*. Venice: Paganinus de Paganinis Brixiensis, 1497. Reproduced in Italian Books Before 1601, roll 460, item 3. Watertown, MA: University Microfilms, n.d.

Perry v. Schwarzenegger, 704 F. Supp. 2d 921; 2010 U.S. Dist., [N.D. Cal., Aug. 4, 2010]; LEXIS 78817, https://vpn.lib.ucdavis.edu/hottopics/lnacademic/,DanaInfo=www.lexisnexis.com+? (consulted Sept. 1, 2012).

Pietro d'Abano. *Compilatio phisonomie*. Padua: Petrus Maufer, 1474. Reproduced in Manuscripta: Microfilms of Rare and Out-of-Print Books, list 14. Saint Louis: Pius XII Memorial Library, Saint Louis University, n.d.

———. *Conciliator controversiarum quae inter philosophos et medicos versantur Petro Abano Patavino*. Venice: Iuntas, 1565. Facsimile in *Opere de Pietro d'Abano*, ed. G. I. Ludwig. Padua: Il Glifo, 1982.

———. "Edition of Peter of Abano, 'Exp.,' IV, with Synthesis, Annotations and Essays." Ed. Gijs Coucke. Vol. 2 of *Philosophy Between Text and Tradition: The Reception of Aristotle's "Problemata" in the Middle Ages: Peter of Abano's "Expositio Problematum."* Ph.D. dissertation. Katholieke Universiteit Leuven, 2008.

———. *Expositio Problematum Aristotelis*. Ed. Stephanus Illarius. Mantua: Paulus Johannis Puzpach, 1475.

———. *Problemata Aristotelis cum duplici translatione antiqua videlicet et nova scilicet Theodori Gaza cum expositione Petri Aponi. Problemata Alexandri Aphrodisei. . . .* Venice: Bonetus Locatellus, 1501. Reproduced in Italian Books before 1601, roll 169. Cambridge, MA: General Microfilm, n.d.

———. *Trattati di astronomia: "Lucidator dubitabilium astronomiae," "De motu octavae sphaerae" e altre opere*. Ed. Graziella Federici Vescovini. 2nd ed. Il Mito e la Storia, 3. Padua: Programma, 1992.

Ptolemy. *Le pervisioni astrologiche (Tetrabiblos)*. Ed. and trans. Simonetta Feraboli. Scrittori Greci e Latini. [N.p.:] Fondazione Lorenzo Valla, Arnoldo Monadori, 1985.

———. *Quatripartitum*. Trans. Egidius Tebaldi and trans. John of Seville. In *Liber Ptholemei quattuor tractatuum cum Centiloquio eiusdem et commento Haly*. Venice: Bonetus Locatellus, 1493.

Pseudo-Ptolemy. *Centiloquium*. Trans. John of Seville. In *Liber Ptholemei quattuor tractatuum cum Centiloquio eiusdem et commento Haly*. Venice: Bonetus Locatellus, 1493.

Rose, Valentine, ed. *Aristoteles pseudepigraphus.* Leipzig: B. G. Teubner, 1863. Reprinted Hildesheim: Georg Olms, 1971.

Seligsohn, Rudolf. *Die Übersetzung der pseudo-aristotelischen "Problemata" durch Bartholomaeus von Messian: Text und textkritische Untersuchungen zum ersten Buch.* Berlin: E. Ebering, 1934.

Thomas Aquinas. *Summa theologiae.* Ed. Thomas Gilby et al. 61 vols. London: Eyre and Spottiswoode, 1964–80.

———. *Sententia libri Ethicorum.* 2 parts. Vol. 47 of *Opera omnia iussu Leonis XIII P. M. edita.* Ed. Order of Preachers. Rome: Sancta Sabina, 1969.

———. *Tabula libri Ethicorum.* Vol. 48 of *Opera omnia iussu Leonis XIII P. M. edita.* Ed. Order of Preachers. Rome: Sancta Sabina, 1971.

Thorndike, Lynn, ed. *The "Sphere" of Sacrobosco and Its Commentators.* Corpus of Medieval Scientific Texts 2. Chicago: University of Chicago Press, [1949].

SECONDARY SOURCES

Agrimi, Jole, and Chiara Crisciani. "Savoir médicale et anthropologie religieuse: Les représentations et les fonctions de la *vetula* (XIIIe–XVe siècle)." *Annales: Economies, Sociétés, Civilisations* 48 (1993): 1281–1308.

d'Alverny, Marie-Thérèse. "Pietro d'Abano et les 'naturalistes' à l'époque de Dante." In *Dante e la cultura veneta: Atti del convegno de studi organizzato della Fondazione "Giorgio Cini" in collaborazione con l'Istituto Universitario di Venezia, l'Università di Padova, il Centro di Studi Danteschi, e i comuni di Venezia, Padova, Verona: Venezia, Padova, Verona, 30 marzo–5 aprile 1966,* ed. Vittorio Branca and Giorgio Padoan, 107–19. Florence: Leo S. Olschki, 1966.

———. "Pietro d'Abano traducteur de Galien." *Medioevo* 11 (1985): 19–64.

Anzulewicz, Henryk. "Fatum: Das Phänomen des Schicksals und die Freiheit des Menschen nach Albertus Magnus." In *Nach der Verurteilung von 1277: Philosophie und Theologie an der Universität von Paris im letzten Viertel des 13. Jahrhunderts. Studien und Texte,* ed. Jan A. Aertsen, Kent Emery Jr., and Andreas Speer, 507–34. Miscellanea Medievalia, Veröffentlichungen des Thomas-Instituts der Universität zu Köln 28. Berlin: Walter de Gruyter, 2001.

Baswell, Christopher. "Talking Back to the Text: Marginal Voices in Medieval Secular Literature." In *The Uses of Manuscripts in Literary Studies: Essays in Memory of Judson Boyce Allen,* ed. Charlotte Cook Morse, Penelope Reed Doob, and Marjorie Curry Woods, 121–60. Studies in Medieval Culture 31. Kalamazoo, MI: Medieval Institute, 1992.

Benedek, Thomas G. "Beliefs About Sexual Function in the Middle Ages and Renaissance." In *Human Sexuality in the Middle Ages and Renaissance,* ed. Douglas Radcliff-Umstead, 97–116. University of Pittsburgh Publications on the Middle Ages and the Renaissance 4. Pittsburgh: Center for Medieval and Renaissance Studies, University of Pittsburgh, 1978.

Bennett, Judith M. "'Lesbian-Like' and the Social History of Lesbianisms." *Journal of the History of Sexuality* 9 (2000): 1–24.

Bianchi, Luca. *Il vescovo e i filosofi: La condanna parigina del 1277 e l'evoluzione dell'aristotelismo scolastico*. Quodlibet 6. Bergamo: Pierluigi Lubrina, 1990.

Biesterfeldt, Hans Hinrich, and Dimitri Gutas. "The Malady of Love." In *Studies in Islam and the Ancient Near East Dedicated to Franz Rosenthal*, ed. Jeanette Wakin, *Journal of the American Oriental Society* 104 (1984): 21–55.

Billanovich, Guido, and Luigi Olivieri. "Pietro d'Abano e il codice Antoniano XVII 370." *Italia medioevale e umanistica* 28 (1985): 221–94.

Biller, Peter. "Views of Jews from Paris Around 1300: Christian or 'Scientific'?" In *Christianity and Judaism: Papers Read at the 1991 Summer Meeting and the 1992 Winter Meeting of the Ecclesiastical History Society*, ed. Diana Wood, 187–207. Oxford: Blackwell for the Ecclesiastical History Society, 1992.

Blair, Ann. "Authorship in the Popular 'Problemata Aristotelis,'" *Early Science and Medicine* 4 (1999): 189–227.

———. "The 'Problemata' as a Natural Philosophical Genre." In *Natural Particulars: Nature and the Disciplines in Renaissance Europe*, ed. Anthony Grafton and Nancy Siraisi, 171–204. Dibner Institute Studies in the History of Science and Technology. Cambridge, MA: MIT Press, 1999.

Borris, Kenneth, and George Rousseau, eds. *The Sciences of Homosexuality in Early Modern Europe*. London: Routledge, 2008.

Boswell, John. *Christianity, Social Tolerance, and Homosexuality: Gay People in Western Europe from the Beginning of the Christian Era to the Beginning of the Fourteenth Century*. Chicago: University of Chicago Press, 1980.

———. *Same-Sex Unions in Premodern Europe*. New York: Villard, 1994.

Boucher, Caroline. "Des problèmes pour exercer l'entendement des lecteurs: Évrart de Conty, Nicole Oresme et la recherche de la nouveauté." In De Leemans and Goyens, *Aristotle's "Problemata" in Different Times and Tongues*, 174–97.

Brown, Judith. "Lesbian Sexuality in Medieval and Early Modern Europe." In *Hidden from History: Reclaiming the Gay and Lesbian Past*, ed. Martin Duberman, Martha Vicinus, and George Chauncey Jr., 67–75. New York: New American Library, 1989.

Brundage, James A. *Law, Sex, and Christian Society in Medieval Europe*. Chicago: University of Chicago Press, 1987.

———. "The Politics of Sodomy: Rex v. Pons Hugh de Ampurias (1311)." In *Sex in the Middle Ages: A Book of Essays*, ed. Joyce E. Salisbury, 239–46. Garland Medieval Casebooks. New York: Garland, 1991.

Bullough, Vern. "Medieval Medical and Scientific Views of Women." *Viator* 4 (1973): 485–501.

Bullough, Vern L., and James A. Brundage, eds. *Handbook of Medieval Sexuality*. Reference Library of the Humanities 1696. New York: Garland, 1996.

Burger, Glenn, and Steven F. Kruger, eds. *Queering the Middle Ages*. Medieval Cultures 27. Minneapolis: University of Minnesota Press, 2001.

Burgwinkel, William. "'The Form of Our Desire': Arnaut Daniel and the Homoerotic

Subject in Dante's 'Commedia.'" *GLQ: A Journal of Lesbian and Gay Studies* 10 (2004): 565–97.
———. *Sodomy, Masculinity, and Law in Medieval Literature: France and England, 1050–1230*. Cambridge Studies in Medieval Literature. Cambridge: Cambridge University Press, 2004.
Cadden, Joan. *Meanings of Sex Difference in the Middle Ages: Medicine, Science, and Culture*. Cambridge History of Medicine. Cambridge: Cambridge University Press, 1993.
———. "Medieval Scientific and Medical Views of Sexuality: Questions of Propriety." *Medievalia et Humanistica*, n.s., 14 (1986): 157–71.
———. "'Nothing Natural Is Shameful': Vestiges of a Debate About Sex and Science in a Group of Late Medieval Manuscripts." *Speculum* 76 (2001): 66–89.
———. "On a Daughter's Death, Venice, 1477." *Medieval Feminist Newsletter* 24 (1997): 15–17.
———. "Preliminary Observations on the Place of the 'Problemata' in Medieval Learning." In De Leemans and Goyens, *Aristotle's "Problemata" in Different Times and Tongues*, 1–19.
———. "Silences/Sciences: The Natures and Languages of 'Sodomy' in Peter of Abano's 'Problemata' Commentary." In *Constructing Medieval Sexuality*, ed. Karma Lochrie, Peggy McCracken, and James Schultz, 40–57. Medieval Cultures 11. Minneapolis: University of Minnesota Press, 1997.
———. "Trouble in the Earthly Paradise: The Regime of Nature in Late Medieval Christian Culture." In *The Moral Authority of Nature*, ed. Daston and Vidal, 207–31.
Camille, Michael. "Bestiary or Biology? Aristotle's Animals in Oxford, Merton College, MS 271." In *Aristotle's Animals in the Middle Ages and Renaissance*, ed. Carlos Steel, Guy Guldentops, and Pieter Beullens, 355–96. Medievalia Lovaniensia, ser. 1, studia 27. Leuven: Katholieke Universiteit Leuven, Instituut voor Middeleeuwse Studies, 1999.
———. "Manuscript Illumination and the Art of Copulation." In *Constructing Medieval Sexuality*, ed. Karma Lochrie, Peggy McCracken, and James A. Schultz, 58–90. Medieval Cultures 11. Minneapolis: University of Minnesota Press, 1997.
———. "Obscenity Under Erasure: Censorship in Medieval Illuminated Manuscripts." In *Obscenity in the Middle Ages: Social Control and Artistic Creation in the European Middle Ages*, ed. Jan M. Ziolkowski, 139–54 and figs. 1–13. Cultures, Beliefs and Tranditions 4. Leiden: Brill, 1998.
———. "The Pose of the Queer: Dante's Gaze and Brunetto Latini's Body." In *Queering the Middle Ages*, ed. Glenn Burger and Steven F. Kruger, 57–86. Medieval Cultures 27. Minneapolis: University of Minnesota Press, 2001.
Carmody, Francis J. *Arabic Astronomical and Astrological Sciences in Latin Translation: A Critical Bibliography*. Berkeley: University of California Press, 1956.
Casagrande, Carla, and Silvana Vecchio. *I peccati della lingua: Disciplina ed etica della parola nella cultura medievale*. Bibliotheca Biographica, Sezione Storico-Antropologica. Rome: Istituto della Enciclopedia Italiana, 1987.
Caviness, Madeline H. "Obscenity and Alterity: Images That Shock and Offend Us/Them,

Now/Then." In *Obscenity: Social Control and Artistic Creation in the European Middle Ages*, ed. Jan M. Ziolkowski, 155–75 and figs. 14–36. Cultures, Beliefs and Traditions 4. Leiden: Brill, 1998.

Cestaro, Gary P. "Queering Nature, Queering Gender: Dante and Sodomy." In *Dante for the New Millennium*, ed. Teodolinda Barolini and H. Wayne Storey, 90–103. Fordham Series in Medieval Studies 2. New York: Fordham University Press, 2003.

Cherchi, Paolo. "Il quotidiano, i 'Problemata' e la meraviglia: Ministoria di un microgenere." *Intersezioni* 21 (2001): 243–75.

Chiffoleau, Jacques. "*Contra naturam*: Pour une approche casuistique et procédurale de la nature médiévale." *Micrologus* 4 (1996): 265–312.

———. "Dire l'indicible: Remarques sur la catégorie du *nephandum* de XIIe au XVe siècle." *Annales ESC* 45 (1990): 289–324.

Clover, Carol. "Regardless of Sex: Men, Women, and Power in Early Northern Europe." *Speculum* 68 (1993): 363–87.

Colish, Marcia. "*Habitus* Revisited: A Reply to Cary Nederman." *Traditio* 48 (1993): 77–92.

Compton, Louis. "The Myth of Lesbian Impunity." In *The Gay Past: A Collection of Historical Essays*, ed. Salvatore J. Licata and Robert P. Petersen , 11–25. New York: Harrington Park Press, 1985.

Coucke, Gijs. " 'Non adeo est honesta ut delectabilis': Sexual Pleasure in Medieval Medicine: The Case of Peter of Abano." In *Piacere e dolore: Materiali per una storia delle passioni nel Medioevo*, ed. Carla Casagrande and Silvana Vecchio, 117–48. Micrologus 29. Florence: SISMEL Edizioni del Galluzzo, 2009.

Courtenay, William J. "University Masters and Political Power: The Parisian Years of Marsilius of Padua." In *Politische Reflexion in der Welt des späten Mittelalters / Political Thought in the Age of Scholasticism: Essays in Honour of Jürgen Miethke*, ed. Martin Kaufhold, 208–23. Leiden: Brill, 2004.

Daston, Lorraine, and Katharine Park. "The Hermaphrodite and the Orders of Nature: Sexual Ambiguity in Early Modern France." *GLQ: A Journal of Lesbian and Gay Studies* 1 (1995): 419–37. Reprinted in Fradenburg and Freccero, *Premodern Sexualities in Europe*, 117–136.

———. *Wonders and the Order of Nature, 1150–1750*. New York: Zone, 1998.

Daston, Lorraine, and Fernando Vidal, eds. *The Moral Authority of Nature*. Chicago: University of Chicago Press, 2004.

Davidson, Adam. "The Horror of Monsters." In *The Boundaries of Humanity: Humans, Animals, Machines*, ed., James J. Sheehan and Morton Sosna, 36–67. Berkeley: University of California Press, 1991.

De Leemans, Pieter, ed. *Philosophy Between Text and Tradition: Essays on Peter of Abano's "Expositio Problematum Aristotelis."* Forthcoming.

———. "Was Peter of Abano the Translator of Pseudo-Aristotle's 'Problemata Physica'?" *Bulletin de Philosophie Médiévale* 49 (2007): 103–18.

De Leemans, Pieter, and Michèle Goyens, eds. *Aristotle's "Problemata" in Different Times and Tongues*. Mediaevalia Lovaniensia, ser. 1, studia 39. Leuven: Leuven University Press, 2006.

———. "'Et samble qu'il woeille dire . . .': Evrart de Conty traducteur de Pierre d'Abano." In *The Medieval Translator* ed. Jacqueline Jenkins and Olivier Bertrand, 285–302. Traduire au Moyen Âge 10. Turnhout: Brepols, 2007.

Delisle, Léopold. *Le cabinet des manuscrits de la Bibliothèque Nationale: Étude sur la formation de ce dépot comprenant les éléments d'une histoire de la calligraphie, de la miniature, de la reliure, et du commerce des livres à Paris avant l'invention de l'imprimerie.* 4 vols. Histoire Générale de Paris. Paris: Imprimerie Impériale and Imprimerie Nationale, 1868–81. Reprinted Amsterdam: Gérard Th. van Heusden, 1969.

———. *Recherches sur la Librairie de Charles V.* 2 vols. and atlas of plates. Paris: H. Champion, 1907.

Demaitre, Luke. "The Idea of Childhood and Child Care in Medical Writings of the Middle Ages." *Journal of the History of Psychology* 4 (1977): 461–90.

De Wulf, Maurice. *Histoire de la philosophie médiévale.* Vol. 3, *Après le treizième siècle.* 6th ed. Louvain: Institut Supérieur de Philosophie, 1947.

Dexel, Walther. *Untersuchungen über die französischen illuminierten Handschriften der Jenaer Universitätsbibliothek vom Ende des 14. bis zur Mitte des 15. Jahrhunderts.* Zur Kunstgeschichte des Ausland 115. Strasbourg: J. H. Ed. Heitz [Heitz und Mündel], 1917.

Dinshaw, Carolyn. *Getting Medieval: Sexualities and Communities, Pre- and Postmodern.* Durham, NC: Duke University Press, 1999.

Ducos, Joëlle. "Traduction et lexique scientifique: Le cas des 'Problèmes' d'Aristote traduits par Evrart de Conty." In *Traduction et adaptation en France à la fin du Moyen Âge et à la Renaissance: Actes du colloque organisé par l'Université de Nancy II, 23–25 mars 1995,* ed. Charles Brucker, 237–47. Paris: Honoré Champion, 1997.

Dunbabin, Jean. "The Two Commentaries of Albertus Magnus on the 'Nicomachean Ethics.'" *Recherches de Théologie ancienne et médiévale* 30 (1963): 232–50.

Eamon, William. *Science and the Secrets of Nature: Books of Secrets in Medieval and Early Modern Culture.* Princeton: Princeton University Press, 1994.

Economou, George. *The Goddess Natura in Medieval Literature.* Cambridge, MA: Harvard University Press, 1972.

Elliott, Dyan. *Fallen Bodies: Pollution, Sexuality, and Demonology in the Middle Ages.* The Middle Ages. Philadelphia: University of Pennsylvania Press, 1999.

El-Rouayheb, Khaled. *Before Homosexuality in the Arab-Islamic World, 1500–1800.* Chicago: University of Chicago Press, 2005.

Ermatinger, Charles Joseph. "John of Jandun in His Relations with Arts Masters and Theologians." *Arts libéraux et philosophie au moyen âge: Actes du quatrième Congrès International de Philosophie Médiévale, Université de Montréal, . . . 27 août–2 septembre, 1967,* 1173–84. Montreal: Institut d'Études Médiévales, 1969.

Federici Vescovini, Graziella. "L'antropologia naturale de Pietro d'Abano." *Paradigmi* 15 (1997): 525–41.

———. "L''Expositio succincta Problemata Aristotelis' de Pierre d'Abano." In De Leemans and Goyens, *Aristotle's "Problemata" in Different Times and Tongues,* 55–69.

———. "Peter of Abano and Astrology." In *Astrology, Science, and Society: Historical Essays,* ed. Patrick Curry, 19–39. Woodbridge, U.K.: Boydell, 1987.

———. "Pietro d'Abano e Taddeo da Parma sull'imaginazione." In *Intellect et imagination dans la philosophie médiévale / Intellect and Imagination in Medieval Philosophy / Intelecto e imaginação na filosofia medieval, Actes du XIe Congrès International de la Société Internationale pour l'Étude de la Philosophie Médiévale (SIEPM)*, ed. Maria Cândida Pacheco and José F. Meirinhos, vol. 3, 1783–95. 3 vols. Société Internationale pour l'Étude de la Philosophie Médiévale, Rencontres de Philosophie Médiévale 11, 1–3. [Turnhout:] Brepols, 2006.

———. "Pietro d'Abano tra biografia e fortuna: Due 'ritratti' quadrocenteschi." *Medioevo* 16 (1990): 293–321.

Ferrari, Sante. "Per la biografia e per gli scritti di Pietro d'Abano." *Atti della R. Academia dei Lincei: Memorie della classe di scienze morali, storiche e filologiche*, ser. 5, 15 (1915): 629–725.

Foucault, Michel. *The History of Sexuality.* Trans. Robert Hurley. 3 vols. New York: Vintage Books, 1988–90.

Fradenburg, Louise, and Carla Freccero, with Kathy Lavezzo, eds. *Premodern Sexualities in Europe.* New York: Routledge, 1996.

Frantzen, Allen J. *Before the Closet: Same-Sex Love from "Beowulf" to "Angels in America."* Chicago: University of Chicago Press, 1998.

Freccero, John. "The Eternal Image of the Father." In *The Poetry of Allusion: Virgil and Ovid in Dante's Commedia*, ed. Rachel Jacoff and Jeffrey T. Schnapp, 62–76 and 264–65. Stanford: Stanford University Press, 1991.

French, Roger K. *Canonical Medicine: Gentile da Foligno and Scholasticism.* Leiden: Brill 2001.

Geanakoplos, Deno J. "Theodore Gaza, a Byzantine Scholar of the Palaeologan 'Renaissance' in the Italian Renaissance." *Medievalia et Humanistica* 12 (1984): 61–81.

Gid, Denise. "La bibliothèque de François Guillebon prieur de Sorbonne (†1534)." In *Bibliophiles et reliures: Mélanges offerts à Michel Wittock*, ed. Annie De Coster and Claude Sorgeloos with Marcus de Schepper, 256–63. Brussels: Fl. Tulkens, 2006.

Glorieux, Palémon. *La littérature quodlibétique de 1260 à 1320.* 2 vols. Bibliothèque Thomiste 5 and 21, Section historique 18. Le Saulchoir, Kain: Revue des Sciences Philosophiques et Théologiques, 1925 and Paris: J. Vrin, 1935.

Goyens, Michèle. "Comprendre Aristote au Moyen Âge: Le procédé de reformulation dans la traduction des 'Problèmes' par Evrart de Conty." In *Textes et cultures: Réception, modèles, interférences.* Vol. 1, *Réception de l'ántiquité*, ed. Pierre Nobel, 145–63. Collection Littéraire. Besançon: Presses Universitaires de Franche-Comté, 2004.

———. "Le développement du lexique scientifique français et la traduction des 'Problèmes' d'Aristote par Evrart de Conty (c. 1380)." *Thélème: Revista Complutense de Estudios Franceses*, special issue (2003): 189–207.

———. "Évrart de Conty: Traducteur, adapteur et commentateur des 'Problèmes' d'Aristote." In *"Pour acquerir honneur et pris": Mélanges de moyen français offerts à Guiseppi Di Stefano*, ed. Maria Colombo Timelli and Claudio Galderisi, 123–35. Montreal: CERES, 2004.

Goyens, Michèle, and Pieter De Leemans. "Traduire du grec au latin et du latin au français: Un défi à la fidélité." In *Pratiques de traduction au Moyen Âge: Actes du colloque de l'Université de Copenhague 25 et 26 octobre 2002*, ed. Peter Anderson , 204–24. Copenhagen: Museum Tusculanum Press, University of Copenhagen, 2004.

Grabmann, Martin. "Das Studium der aristotelischen Ethik an der Artistenfakultät der Universität Paris in der ersten Hälfte des 13. Jahrhunderts." *Philologisches Jahrbuch* 53 (1940): 339–54.

Grant, Edward. "Nicole Oresme on Certitude in Science and Pseudo-Science." In *Nicole Oresme: Tradition et innovation chez un intellectuel du XIVe siècle*, ed. P. Souffrin and A. Ph. Segonds, 31–43. Paris: Les Belles Lettres, 1988.

———. *Science and Religion, 400 B.C. to A.D. 1550: From Aristotle to Copernicus*. Greenwood Guides to Science and Religion. Westport, CT: Greenwood Press, 2004.

Green, Monica H. "The Development of the 'Trotula.'" *Revue d'Histoire des Textes* 26 (1996): 119–203. Reprinted in Green, *Women's Healthcare in the Medieval West*, V.

———. *Making Women's Medicine Masculine: The Rise of Male Authority in Pre-Modern Gynaecology*. Oxford: Oxford University Press, 2008.

———. "Obstetrical and Gynecological Texts in Middle English." *Studies in the Age of Chaucer* 14 (1992): 53–88. Reprinted in Green, *Women's Healthcare in the Medieval West*, IV.

———. "'Traittié tout de mençonges': The 'Secrés des dames,' 'Trotula,' and Attitudes Towards Women's Medicine in Fourteenth- and Early Fifteenth-Century France." In *Christine de Pizan and the Categories of Difference*, ed. Marilyn Desmond, 146–78. Minneapolis: University of Minnesota Press, 1998. Reprinted in Green, *Women's Healthcare in the Medieval West*, VI.

———. *Women's Healthcare in the Medieval West*. Variorum Collected Studies. Aldershot, U.K.: Ashgate, 2000.

Greenblatt, Stephen. *The Swerve: How the World Became Modern*. New York: W. W. Norton, 2011.

Guichard-Tesson, Françoise. "Evrart de Conty, poète, traducteur et commentateur." In De Leemans and Goyens, eds., *Aristotle's "Problemata" in Different Times and Tongues*, 145–73.

———. "Le métier de traducteur et de commentateur au XIVe siècle d'après Evrart de Conty." *Le Moyen Français* 24–25 (1990): 131–67.

———"La souci de la langue et du style au XIVe siècle: L'autographe des 'Problèmes' d'Evrart de Conty." *Le moyen français*, 33 (1993): 57–84.

Guichard-Tesson, Françoise, and Michèle Goyens. "Comment éditer l'autographe d'une traduction de traduction." *Scriptorium* 63 (2009): 173–205.

Halperin, David M. "Forgetting Foucault: Acts, Identities, and the History of Sexuality." *Representations* 63 (1998): 93–120. Reprinted in Halperin, *How to Do the History of Homosexuality*, 24–47 and 159–70.

———. *How to Do the History of Homosexuality*. Chicago: Chicago University Press, 2002.

——— "How to Do the History of Male Homosexuality." *GLQ: A Journal of Lesbian and Gay Studies*, 6 (2000): 87–123. Reprinted in Halperin, *How to Do the History of Homosexuality*, 104–37 and 185–95.

Halperin, David M., and Valerie Traub, eds. *Gay Shame*. Chicago: University of Chicago Press, 2009.

Hansen, Bert. *Nicole Oresme and the Marvels of Nature: A Study of His "De causis mirabilium" with Critical Edition, Translation, and Commentary.* Studies and Texts 68. Toronto: Pontifical Institute of Mediaeval Studies, 1985.

Haskins, Charles Homer. *Studies in the History of Mediaeval Science.* Cambridge, MA: Harvard University Press, 1924. Reprinted New York: Frederick Ungar, 1960.

Hergemöller, Bernd-Ulrich. *Sodom und Gomorrha: Zur Altagswirklichkeit und Verfolgung Homosexueller im Mittelalter.* Hamburg: MännerschwarmSkript, 1998.

Hissette, Roland. *Enquête sur les 219 articles condamnés à Paris le 7 mars 1277.* Philosophes Médiévaux 22. Louvain: Publications Universitaires, 1977.

Huot, Sylvia. *"The Romance of the Rose" and Its Medieval Readers: Interpretation, Reception, Manuscript Transmission.* Cambridge: Cambridge University Press, 1993.

Hutchenson, Gregory. "The Sodomitic Moor: Queerness in the Narrative of the *Reconquista.*" In *Queering the Middle Ages*, ed. Glenn Burger and Steven F. Kruger, 99–122. Medieval Cultures 27. Minneapolis: University of Minnesota Press, 2001.

Jacquart, Danielle. "L'amour 'héroïque' à travers le traité d'Arnaud de Villeneuve." In *La folie et le corps*, ed. Jean Céard with Pierre Naudin and Michel Simonin, 143–58. Paris: Presses de l'École Normal Supérieur, 1985.

———. *Dictionaire biographique des médecins en France au Moyen Âge: Supplément.* Geneva: Droz, 1979.

———. *La Médecine médiévale dans le cadre parisien, XIVe–XVe siècle.* Penser la Médecine. [N.p.:] Fayard, 1998.

———. "Medical Explanations of Sexual Behavior in the Middle Ages." In *Homo Carnalis: The Carnal Aspect of Medieval Human Life*, ed. Helen Rodnite Lemay, 1–21. Center for Medieval and Early Renaissance Studies, Acta 14 for 1989. Binghamton: State University of New York, 1990.

———. "La réception du 'Canon' d'Avicenne: Comparaison entre Montpellier et Paris aux XIIIe et XIVe siècles." In *Histoire de l'École médicale de Montpellier: Actes du 110e Congrès national des sociétés savantes (Montpellier, 1985): Section d'histoire des sciences et des techniques*, vol. 2, 69–77. Paris: Comité des Travaux Historiques et Scientifiques, 1985.

———. "Theory, Everyday Practice, and Three Fifteenth-Century Physicians." *Osiris*, 2nd ser., 6, *Renaissance Medical Learning: Evolution of a Tradition* (1990): 140–60.

Jacquart, Danielle, and Charles Burnett, eds. *Scientia in margine: Études sur les marginalia dans les manuscrits scientifiques du Moyen Âge à la Renaissance.* École Pratique des Hautes Études, Sciences historiques et philologiques 5, Hautes Études médiévales et modernes 88. Geneva: Droz, 2005.

Jacquart, Danielle, and Claude Thomasset. "Albert le Grand et les problèmes de la sexualité." *History and Philosophy of the Life Sciences* 3 (1981): 73–93.

———. *Sexuality and Medicine in the Middle Ages.* Trans. Matthew Adamson. Princeton: Princeton University Press, 1988.

Jacquart, Danielle, and Gérard Troupeau. "Traduction de l'arabe et vocabulaire médical latin: Quelques exemples." In *La lexicographie du latin médiéval et ses rapports avec les recherches actuelles sur la civilisation du moyen âge, Paris 18–21 octobre 1978*, 367–76.

Colloques Internationaux du Centre National de la Recherche Scientifique 589. Paris: Éditions du Centre National de la Recherche Scientifique, 1981.

James, Montague Rhodes. *A Descriptive Catalogue of the Manuscripts in the Library of Pembroke College, Cambridge*. Cambridge: Cambridge University Press, 1905.

Johnson, Willis. "The Myth of Jewish Male Menses." *Journal of Medieval History* 24 (1998): 273–95.

Jordan, Mark. "Homosexuality, *Luxuria*, and Textual Abuse." In *Constructing Medieval Sexuality*, ed. Lochrie, McCracken, and Schultz, 24–39.

———. *The Invention of Sodomy in Christian Theology*. Chicago: University of Chicago Press, 1997.

Karras, Ruth Mazo. *Sexuality in Medieval Europe: Doing unto Others*. New York: Routledge, 2005.

Karras, Ruth Mazo, and David Lorenzo Boyd. "The Interrogation of a Male Transvestite Prostitute in Fourteenth-Century London." *GLQ: A Journal of Lesbian and Gay Studies* 1 (1995): 459–65.

———. "'Ut cum muliere': A Male Transvestite Prostitute in Fourteenth-Century London." In Fradenburg and Freccero, eds., *Premodern Sexualities in Europe*, 101–16.

Kelly, Henry Ansgar. "The Pardoner's Voice, Disjunctive Narrative, and Modes of Feminization." In *Essays in Honor of V. A. Kolve*, 411–44. Asheville, NC: Pegasus Press, 2001.

Kibre, Pearl. *Hippocrates Latinus: Repertorium of Hippocratic Writings in the Middle Ages*. Rev. ed. New York: Fordham University Press, 1985.

Klemm, Matthew. "Medicine and Moral Virtue in the 'Expositio Problematum Aristotelis' of Peter of Abano." *Early Science and Medicine* 11 (2006): 302–33.

Klibansky, Raymond, Erwin Panofsky, and Fritz Saxl. *Saturn and Melancholy: Studies in the History of Natural Philosophy, Religion and Art*. London: Thomas Nelson, 1964.

Korteweg, Anne S. *Boeken van Oranje-Nassau: De bibliotheek van de graven van Nassau en prinsen van Oranje in de vijftiende en zestiende eeuw*. Exhibition, Museum Meermanno Westreenianum. The Hague: Koninklijke Bibliotheek, Den Haag, 1998.

Kuefler, Mathew. *The Manly Eunuch: Masculinity, Gender Ambiguity, and Christian Ideology in Late Antiquity*. Chicago: University of Chicago Press, 2001.

Kuksewicz, Zdzislaw. "Jean de Jandun et sa conception de la philosophie." In *Was ist Philosophie im Mittelalter? . . . Akten des X. Internationalen Kongresses für mittelalterliche Philosophie der Société Internationale pour l'Étude de la Philosophie Médiévale, 25. bis 30. August 1997 in Erfurt*, ed. Jan A. Aertsen and Andreas Speer, 428–34. Miscellanea Medievalia, Veröffentlichungen des Thomas-Instituts der Universität zu Köln 26. Berlin: Walter de Gruyter, 1998.

———. "Les 'Problemata' de Pietro d'Abano et leur 'rédaction' par Jean de Jandun," *Medioevo* 11 (1985): 113–37.

Kusche, Brigitte. "Zur 'Secreta mulierum'-Forschung." *Janus* 62 (1975): 102–23.

Lacombe, Georges, Aleksander Birkenmajer, and Lorenzo Minio-Paluello. *Aristoteles Latinus: Codices*. 3 vols.: vol. 1, revised edition, Union Académique Internationale, Corpus Philosophorum Medii Aevi. Bruges: Desclée de Brouwer, 1957; vol. 2, Cambridge: Academia, 1955; [vol. 3,] *Supplementa altera*. Bruges: Desclée de Brouwer, 1961.

Lansing, Carol. "Donna con Donna? A 1295 Inquest into Female Sodomy." In *Sexuality and Culture in Medieval and Renaissance Europe*, ed. Philip M. Soergel, 109–22. Studies in Medieval and Renaissance History, 3rd ser., 2. New York: AMS Press, 2005.

Lawn, Brian. *I Quesiti salernitani: Introduzione alla storia della letteratura problematica medica e scientifica nel Medio Evo e nel Rinascimento.* Trans. Alessandro Spagnuolo. [Salerno:] Di Mauro, 1969. [A revised edition of *Salernitan Questions: An Introduction to the History of Medieval and Renaissance Problem Literature.* Oxford: Clarendon Press, 1963.]

Lemay, Helen Rodnite. "The Stars and Human Sexuality: Some Medieval Scientific Views." *Isis* 71 (1980):127–37.

———. "William of Saliceto on Human Sexuality." *Viator* 12 (1981):165–81.

———. "Human Sexuality in Twelfth- Through Fifteenth-Century Scientific Writings." In *Sexual Practices and the Medieval Church*, ed. Vern Bullough and James Brundage, 187–205 and 278–82. Buffalo: Prometheus, 1982.

Liber, Alain de. *Penser au Moyen Âge*. Chemin de Pensée. Paris: Éditions du Seuil, 1991.

Lines, David A. "The Commentary Literature on Aristotle's 'Nicomachean Ethics' in Early Renaissance Italy: Preliminary Considerations." *Traditio* 54 (1999): 245–82.

Lochrie, Karma. *Covert Operations: The Medieval Uses of Secrecy*. The Middle Ages. Philadelphia: University of Pennsylvania Press, 1999.

———. *Heterosyncrasies: Female Sexuality When Normal Wasn't.* Minneapolis: University of Minnesota Press, 2005.

———. "Presumptive Sodomy and Its Silences." *Textual Practice* 13 (1999): 295–310.

Lochrie, Karma, Peggy McCracken, and James A. Schultz, eds. *Constructing Medieval Sexuality*. Medieval Cultures 11. Minneapolis: University of Minnesota Press, 1997.

Lohr, Charles. "Medieval Latin Aristotle Commentaries." *Traditio* 23 (1967): 313–413; 24 (1968): 149–245; 26 (1970): 135–216; 27 (1971): 251–351; 28 (1972): 281–396; 29 (1973): 91–197; 30 (1974): 119–44.

Lusignan, Serge. "Nicole Oresme traducteur et la pensée de la langue française savante." In *Nicole Oresme: Tradition et innovation chez un intellectuel du XIVe siècle*, ed. P. Souffrin and A. Ph. Segonds, 93–104. Science et Humanisme. Paris: Les Belles Lettres, 1988.

———. *"Parler vulgairement": Les intellectuels et la langue française aux XIIIe et XIVe siècles.* Paris: J. Vrin, 1986.

MacClintock, Stuart. *Perversity and Error: Studies on the "Averroist" John of Jandun*. Indiana University Publications, Humanities Series 37. Bloomington: Indiana University Press, 1956.

Malti-Douglas, Fedwa. "Tribadism/Lesbianism and the Sexualized Body in Medieval Arabo-Islamic Narratives." In *Same Sex Love and Desire Among Women in the Middle Ages*, ed. Francesca Canadé Sautman and Pamela Sheingorn , 123–41. Basingstoke, U.K.: Palgrave, 2001.

Manfron, Anna. *La biblioteca di un medico del quattrocento: I codici di Giovanni di Marco da Rimini nella Biblioteca Malatestiana*. Archivi di Bibliofilia. Turin: Umberto Allemandi for the Istituzione Biblioteca Malatestiana, Cesena, 1998.

Marangon, Paolo. *Il pensiero ereticale nella Marca Tevigiana e a Venezia dal 1200 al 1350.* Abano: Francisci, 1984.

Marchesi, Concetto. *L' "Etica Nicomachea" nella tradizione latina medievale (documenti ed appunti)*. Messina: Ant. Trimarchi, 1904.

Marenghi, Gerardo. "Un capitolo dell'Aristotele medievale: Bartolomeo da Messina, traduttore dei 'Problemata physica.'" *Aevum* 36 (1962): 268–83.

Markowski, Miecislaus. *Repertorium commentariorum medii aevi in Aristotelem Latinorum quae in Bibliotheca Amploniana Erffordiae asservantur.* Wrocław: Zakład Narodowy im. Ossolińskich for the Polska Akademia Nauk Instytut Filozofi i Sociologii, 1987.

Martin, Conor. "Walter Burley." In *Oxford Studies Presented to Daniel Callus*, 194–230. Oxford Historical Society, n.s., 16. Oxford: Clarendon Press for the Oxford Historical Society, 1964.

McLaughlin, Mary Martin. *Intellectual Freedom and Its Limitations in the University of Paris in the Thirteenth and Fourteenth Centuries*. New York: Arno Press, 1977.

Michaelis, Beatrice. "Recht verschwiegen: Das 'Tabu' der Sodomie in der Sprache des spätmittelalterlichen Rechts." In *Geschlecht als Tabu: Orte, Dynamiken and Funktion der De/Thematisierung von Geschlecht,* ed. Ute Frietsch, Konstanze Hanitzsch, Jennifer John, and Beatrice Michaelis 141–54. Gender Codes 5. Bielefeld: transcript, 2008.

Mills, Robert. *Pain, Pleasure and Punishment in Medieval Culture*. London: Reaktion, 2005.

Minnis, A. J. *Medieval Theory of Authorship: Scholastic Literary Attitudes in the Later Middle Ages*. 2nd ed. The Middle Ages. Philadelphia: University of Pennsylvania Press, 1988.

Mitterer, Albert. "*Mas occasionatus* oder zwei Methoden der Thomasdeutung." *Zeitschrift für katholische Theologie* 72 (1950): 80–103.

Monfasani, John. *George of Trebizond: A Biography and a Study of His Rhetoric and Logic*. Columbia Studies in the Classical Tradition 1. Leiden: Brill, 1976.

———. "George of Trebizond's Critique of Theodore Gaza's Translation of the Aristotelian 'Problemata.'" In De Leemans and Goyens, *Aristotle's "Problemata" in Different Times and Tongues*, 275–94.

———. "The Pseudo-Aristotelian 'Problemata' and Aristotle's 'De animalibus' in the Renaissance." In Grafton and Siraisi, eds. *Natural Particulars*, 205–47.

———. "Theodore Gaza as a Philosopher: A Preliminary Survey." In *Manuele Crisolora e il ritorno del greco in occidente: Atti del Convegno Internazionale (Napoli, 26–29 giugno 1997)*, ed. Riccardo Maisano and Antonio Rollo, 269–81. Naples: Istituto Universitario Orientale, Dipartimento di Studi dell'Europa Orientale and Dipartimento di Studi del Mondo Classico e del Mediterraneo Antico, 2002.

Moulinier, Laurence. "Le corps des jeunes filles dans les traités médicaux du Moyen Âge." In *Le corps des jeunes filles de l'Antiquité à nos jours,* ed. Louise Bruit Zaidman, Gabrielle Houbre, Christiane Klapisch-Zuber, and Pauline Schmitt Pantel, 80–109. Collection pour l'Histoire. Paris: Perrin, 2001.

Murdoch, John E. "1277 and Late Medieval Natural Philosophy." In *Was ist Philosophie im Mittelalter?* ed. Jan A. Aertsen and Andreas Speer, 111–24. Berlin: de Gruyter, 1998.

Murray, Jacqueline. "Twice Marginal and Twice Invisible: Lesbians in the Middle Ages." In *Handbook of Medieval Sexuality*, ed. Vern L. Bullough and James A. Brundage, 191–222. Reference Library of the Humanities 1696. New York: Garland, 1996.

Musallam, Basim F. *Sex and Society in Islam: Birth Control Before the Nineteenth Century.* Cambridge Studies in Islamic Civilization. Cambridge: Cambridge University Press, 1983.

Nardi, Bruno. "Intorno alle dottrine di Pietro d'Abano." *Nuova revista storica* 4 (1920): 81–97 and 464–81; 5 (1921): 300–313. Reprinted in *Saggi sull'aristotelismo padovano dal secolo XIV al XVI*, 19–74. Studi sulla Tradizione Aristotelica nel Veneto 1. Florence: G. C. Sansoni, 1958.

Nathan, Bassem. "Medieval Arabic Medical Views on Male Homosexuality." *Journal of Homosexuality* 26 (1994): 37–39.

Neal, Derek. "Disorder of Body, Mind, or Soul: Male Sexual Deviance in Jacques Despars's Commentary on Avicenna." In Borris and Rousseau, eds., *The Sciences of Homosexuality in Early Modern Europe*, 42–56.

Nederman, Cary J. "Nature, Ethics, and the Doctrine of 'Habitus': Aristotelian Moral Philosophy in the Twelfth Century." *Traditio* 45 (1989–90): 87–110.

Normore, Calvin G. "Who Was Condemned in 1277?" *Modern Schoolman* 72 (1995): 273–81.

Norpoth, Leo. "Zur Bio- und Wissenschaftslehre des Pietro d'Abano, Mediziners, Philosophen und Astronomen in Padua." *Kyklos: Jahrbuch für Geschichte und Philosophie der Medizin*, 3 (1930): 292–353.

Olivieri, Luigi *Pietro d'Abano e il pensiero neolatino: Filosofia, scienza e ricerca dell'Aristotele greco tra i secoli XIII e XIV.* Centro per la Storia della Tradizione Aristotelica nel Veneto, Saggi e Testi 23. Padua: Antenore, 1988.

Ouy, Gilbert. *Les manuscrits de l'abbaye de Saint-Victor: Catalogue établi sur la base du répertoire de Claude de Grandrue (1514).* 2 vols. Bibliotheca Victorina 10. [Turnhout:] Brepols, 1999.

Pagden, Anthony. "The Diffusion of Aristotle's Moral Philosophy in Spain, ca. 1400–ca. 1600." *Traditio* 31 (1975): 286–313.

Park, Katharine. "Albert's Influence on Late Medieval Psychology." In *Albertus Magnus and the Sciences: Commemorative Essays, 1980*, ed. James A. Weisheipl, 501–35. Studies and Texts 49. Toronto: Pontifical Institute of Mediaeval Studies, 1980.

———. "Nature in Person: Medieval and Renaissance Allegories and Emblems." In *The Moral Authority of Nature*, ed. Daston and Vidal, 50–73.

———. "The Rediscovery of the Clitoris: French Medicine and the Tribade, 1570–1620." In *The Body in Parts: Fantasies of Corporeality in Early Modern Europe*, ed. David Hillman and Carla Mazzio, 170–93. New York: Routledge, 1997.

———. *Women's Secrets: Gender, Generation, and the Origins of Human Dissection.* New York: Zone, 2006.

Paschetto, Eugenia. "La fisiognomica nell'enciclopedia delle scienze di Pietro d'Abano." *Medioevo* 11 (1985): 97–111.

———. *Pietro d'Abano, medico e filosofo*. Florence: Nuovedizioni Enrico Vallechi, 1984.

Payer, Pierre J. *The Bridling of Desire: Views of Sex in the Later Middle Ages.* Toronto: University of Toronto Press, 1993.

———. *Sex and the Penitentials: The Development of a Sexual Code, 550–1150*. Toronto: University of Toronto Press, 1984.

Pelzer, Auguste. "Les versions latines des ouvrages de morale conservés sous le nom d'Aristote en usage au XIIIe siècle." *Revue néo-scolastique de philosophie* 23 (1921): 316–41 and 378–412.

Pequiney, Joseph. "Sodomy in Dante's 'Inferno' and 'Purgatorio.'" *Representations* 36 (1991): 22–42.

Pesenti, Tiziana. "Per la tradizione del testimento di Pietro d'Abano." *Medioevo* 6 (1980): 533–42.

Pomata, Gianna. "Menstruating Men: Similarity and Difference of the Sexes in Early Modern Medicine." In *Generation and Degeneration: Tropes of Reproduction in Literature and History from Antiquity Through Early Modern Europe*, ed. Valeria Finucci and Kevin Brownlee, 109–52. Durham, NC: Duke University Press, 2001.

Puff, Helmut. "Female Sexuality: The Trial of Katherina Hetzeldorfer (1477)." *Journal of Medieval and Early Modern Studies* 30 (2000): 41–61.

———. "Localizing Sodomy: The 'Priest and Sodomite' in Pre-Reformation Germany and Switzerland." *Journal of the History of Sexuality* 8 (1997): 165–95.

———. "Nature on Trial: Acts 'Against Nature' in the Law Courts of Early Modern Germany and Switzerland." In *The Moral Authority of Nature*, ed. Daston and Vidal, 232–53.

———. *Sodomy in Reformation Germany and Switzerland, 1400–1600*. Chicago Series on History, Sexuality, and Society. Chicago: University of Chicago Press, 2003.

Rich, Adrienne. "Compulsory Heterosexuality and Lesbian Existence." *Signs* 5 (1980): 631–60.

Rocke, Michael. *Forbidden Friendships: Homosexuality and Male Culture in Renaissance Florence*. New York: Oxford University Press, 1996.

Rosenthal, Franz. "ar-Rāzī on the Hidden Illness" *Bulletin of the History of Medicine* 52 (1978): 45–60. Reprinted in *Science and Medicine in Islam: A Collection of Essays*, item 9, Collected Studies 330. Aldershot, U.K.: Variorum, 1990.

Roth, Norman. "A Research Note on Sexuality and Muslim Society." In Bullough and Brundage, eds., *Handbook of Medieval Sexuality*, 319–27.

Rowland, Beryl. *Birds with Human Souls: A Guide to Bird Symbolism*. Knoxville: University of Tennessee Press, 1978.

Rowson, Everett K. "Gender Irregularity as Entertainment: Institutionalized Transvestism at the Caliphal Court in Medieval Baghdad." In *Gender and Difference in the Middle Ages*, ed. Sharon Farmer and Carol Braun Pasternack, 41–72. Medieval Cultures 32. Minneapolis: University of Minnesota Press, 2003.

Salisbury, Joyce E. *The Beast Within: Animals and Bestiality in the Middle Ages*. New York: Routledge, 1994.

———. *Medieval Sexuality: A Research Guide*. Medieval Bibliographies 5. New York: Garland, 1990.

———. ed. *Sex in the Middle Ages: A Book of Essays*. Garland Medieval Casebooks 3; Garland Reference Library in the Humanities 1360. New York: Garland, 1991.

Salmón, Fernando, and Montserrat Cabré. "Fascinating Women: The Evil Eye in Medical Scholasticism." In *Medicine from the Black Death to the French Disease*, ed. Roger French, Jon Arrizabalaga, Andrew Cunningham, and Luis García-Ballester, 53–84. History of Medicine in Context. Aldershot, U.K.: Ashgate, 1998.

Samaran, Charles, et al. *Catalogue de manuscrits en écriture latine portant des indications de date, de lieu ou de copiste*. Vol. 3. Paris: Centre Nationale de la Recherche Scientifique, 1962.

Schleissner, Margaret Rose. "A Fifteenth-Century Physician's Attitude Toward Sexuality: Dr. Johann Hartlieb's 'Secreta mulierum' Translation." In Salisbury, ed., *Sex in the Middle Ages*, 110–25.

———. "Pseudo-Albertus Magnus: 'Secreta mulierum cum commento,' Deutsch: Critical Text and Commentary." Ph.D. diss., Princeton University. Ann Arbor: University Microfilms International, 1987.

———. "Sexuality and Reproduction in the Late Medieval 'Problemata Aristotelis.'" In *Licht der Natur: Medizin in Fachliteratur und Dichtung: Festschrift für Gundolf Kiel zum 60. Geburtstag*, ed. Josef Domes et al., 383–98. Göppinger Arbeiten zur Germanistik 585. Göppingen: Kümmerle, 1994.

Schmidtke, Sabine. "Homoeroticism and Homosexuality in Islam: A Review Article." *Bulletin of the School of Oriental and African Studies* 62 (1999): 260–66.

Schmugge, Ludwig. *Johannes von Jandun, 1285/89–1328*. Pariser Historische Studien 5. Stuttgart: Anton Hiersemann, 1966.

Schultz, James A. *Courtly Love, the Love of Courtliness, and the History of Sexuality*. Chicago: University of Chicago Press, 2006.

Sears, Elizabeth. *The Ages of Man: Medieval Interpretations of the Life Cycle*. Princeton: Princeton University Press, 1986.

Seńko, Władysław. *Repertorium commentariorum medii aevi in Aristotelem latinorum quae in bibliothecis publicis parisiis asservantur (Bibliothèque Nationale, Arsenal, Mazarine, Sorbonne, Ste Geneviève)*. Opera Philosophorum Medii Aevi, Textus et Studia 5. 2 fascicules. Warsaw: Akademia Teologii Katolickiej, 1982.

Sherman, Claire Richter. *Imaging Aristotle: Verbal and Visual Representation in Fourteenth-Century France*. Berkeley: University of California Press, 1995.

Simon, Patricia. *The Sex of Men in Premodern Europe: A Cultural History*. Cambridge Social and Cultural Histories. Cambridge: Cambridge University Press, 2011.

Siraisi, Nancy G. *Avicenna in Renaissance Italy: The "Canon" and Medical Teaching in Italian Universities After 1500*. Princeton: Princeton University Press, 1987.

———. "The 'Expositio problematum Aristotelis' of Peter of Abano." *Isis* 61 (1970): 321–39.

———. "The *Libri morales* in the Faculty of Arts and Medicine at Bologna: Bartolomeo da Varignana and the Pseudo-Aristotelian 'Economics.'" *Manuscripta* 20 (1976): 105–18.

———. "Pietro d'Abano and Taddeo Alderotti: Two Models of Medical Culture." *Medioevo* 11 (1985): 139–62.

———. *Taddeo Alderotti and His Pupils: Two Generations of Medical Learning*. Princeton: Princeton University Press, 1981.

Stone, Gregory B. *The Ethics of Nature in the Middle Ages: On Boccaccio's Poetaphysics*. The New Middle Ages. New York: St. Martin's, 1998.

Stone, Martin W. F. "Moral Psychology After 1277: Did the Parisian Condemnation Make a Difference to Philosophical Discussions of Human Agency." In *Nach der Verurteilung von 1277: Philosophie und Theologie an der Universität von Paris im letzten Viertel des 13. Jahrhunderts: Studie und Texte/ After the Condemnation of 1277: Philosophy and Theology at the University of Paris in the Last Quarter of the Thirteenth Century: Studies and Texts*, ed. Jan A. Aertsen, Kent Emery Jr., and Andreas Speer, 795–826. Miscellanea Mediaevalia, Veröffenlichungen des Thomas-Instituts der Universität zu Köln 28. Berlin: Walter de Gruyter, 2001.

Stroick, Clemens. *Heinrich von Friemar: Leben, Werke, philosophische-theologische Stellung in der Scholastik*. Freiburg i. Br.: Herder, 1954.

Swaenepoel, Tine. "Les deux versions des 'Problemata' de Pierre d'Abano." *Filologia Mediolatina* 16 (2009):215–45.

Thijssen, J. M. M. H. *Censure and Heresy at the University of Paris, 1200–1400*. The Middle Ages. Philadelphia: University of Pennsylvania Press, 1998.

———. "What Really Happened on 7 March 1277? Bishop Tempier's Condemnation and Its Institutional Context." In *Texts and Contexts in Ancient and Medieval Science: Studies on the Occasion of John E. Murdoch's Seventieth Birthday*, ed. Edith Sylla and Michael McVaugh, 84–114. Leiden: Brill, 1997.

Thorndike, Lynn. *A History of Magic and Experimental Science during the First Thirteen Centuries of our Era*. Vol. 2. Rev. ed. New York: Macmillan, 1929.

———. "Peter of Abano and Another Commentary on the 'Problems' of Aristotle." *Bulletin of the History of Medicine* 29 (1955): 517–23.

———. "Peter of Abano and the Inquisition." *Speculum* 11 (1936):132–33.

———. "Relations of the Inquisition to Peter of Abano and Cecco d'Ascoli." *Speculum* 1 (1926): 338–43.

Traub, Valerie. "The Present Future of Lesbian Historiography." In *A Companion to Lesbian, Gay, Bisexual, Transgender, and Queer Studies*, ed. George Haggerty and Molly McGarry, 124–45. Companions to Cultural Studies. Oxford: Blackwell, 2007.

———. "The Psychomorphology of the Clitoris, or, The Reemergence of the Tribade in English Culture." In *Generation and Degeneration: Tropes of Reproduction in Literature and History from Antiquity through Early Modern Europe*, ed. Valeria Finucci and Kevin Brownlee, 153–86. Durham, NC: Duke University Press, 2001.

Van der Lugt, Maaike. "Aristotle's 'Problems' in the West: A Contribution to the Study of the Medieval Latin Tradition." In De Leemans and Goyens, eds., *Aristotle's "Problemata" in Different Times and Tongues*, 71–111.

———. "Genèse et postérité du commentaire de Pietro d'Abano sur les 'Problèmes' d'Aristote: Le succès d'un hapax." In *Médecine, astrologie et magie au Moyen Âge: Autour de Pietro d'Abano*, ed. J.-P. Boudet, F. Collard, and N. Weill-Parot, 155–82. Micrologus. Tavarnuzze: SISMEL, Edizioni del Galluzzo. Forthcoming.

———. "L'humanité des monstres et leur accès aux sacrements dans la pensée médiévale." HAL : halshs-00175497, version 1 at http://halshs.archives-ouvertes.fr./ (consulted Dec. 15, 2011). Published without references in Latin in *Monstre et imaginaire social:*

Approches historiques, ed. A. Caiozzo and A.-E. Demartini, 135–61. Paris: Créaphis, 2008.

———. "Pourquoi Dieu a-t-il créé la femme? Différence sexuelle et la théologie médiévale" In *Ève et Pandora: La création de la première femme*, ed. Jean-Claude Schmitt, 89–113. Le Temps des Images. [Paris]: Gallimard, 2001.

———. *Le ver, le démon et la vièrge: Les théories médiévales de la génération extraordinaire: Une étude sur les rapports entre théologie, philosophie naturelle et médecine*. L'Âne d'Or 16. Paris: Les Belles Lettres, 2004.

Ventura, Iolanda. "Aristotelis fuit causa efficiens huius libri: On the Reception of Pseudo-Aristotle's 'Problemata' in Late Medieval Encyclopedic Culture." In De Leemans and Goyens, eds., *Aristotle's "Problemata" in Different Times and Tongues*, 113–44.

———. "*Quaestiones* and Encyclopedias: Some Aspects of the Late Medieval Reception of the Pseudo-Aristotelian 'Problemata' in Encyclopedic and Scientific Culture." In *Schooling and Society: The Ordering and Reordering of Knowledge in the Western Middle Ages*, ed. Alasdair A. MacDonald and Michael W. Twomey. Groningen Studies in Cultural Change 23–42. Leuven: Peeters, 2004.

Wallis, Faith. "Guilio Guastavini's Commentary on Pseudo-Aristotle's Account of Male Same-Sex Coitus, 'Problemata' 4.26." In Borris and Rousseau, eds., *The Sciences of Homosexuality in Early Modern Europe*, 57–73.

Weisheipl, James A. "Classification of the Sciences in Medieval Thought" *Medieval Studies* 27 (1965): 54–90.

———. "Repertorium Mertoniense." *Mediaeval Studies* 31 (1969): 174–224.

Wickersheimer, Ernest. *Dictionnaire biographique des médecins en France au Moyen Âge*. 2 vols. Paris: Droz, 1936. Reprinted Hautes Études Médiévales et Modernes 34, 1–2, Geneva: Droz, 1979.

———. "Henri de Saxe et le 'De secretis mulierum.'" In *3ᵉ Congrès de l'histoire de l'art de guérir, Londres, 17–22 juillet, 1922*, 253–58. Antwerp: De Vlijt, 1923.

Wiesner, Merry E. *Christianity and the Regulation of Sexuality in the Early Modern World: Regulating Desire, Reforming Practice*. Christianity and Society in the Modern World. London: Routledge, 2000.

Ziegler, Joseph. "Sexuality and the Sexual Organs in Latin Physiognomy 1200–1500." In *Sexuality and Culture in Medieval and Renaissance Europe*, ed. Philip M. Soergel, 83–107. Studies in Medieval and Renaissance History, 3rd ser., 2. New York: AMS Press, 2005.

Zimmermann, Albert, and Andreas Speer, eds. *Mensch und Natur im Mittelalter*. 2 vols. Miscellanea Medievalia, Veröffentichungen des Thomas-Instituts der Universität zu Köln 21/1–2. Berlin: Walter de Gruyter, 1991.

Ziolkowski, Jan. *Alan of Lille's Grammar of Sex: The Meaning of Grammar to a Twelfth-Century Intellectual*. Speculum Anniversary Monographs 10. Cambridge, MA: Medieval Academy of America, 1985.

Index

The abbreviation *Probl.* is used below for the titles of various *Problemata*-related works by various authors. It is qualified where necessary to avoid ambiguity.

Acknowledgments

I do not know the names of many of the most important teachers who influenced this book. With their movements that asserted rights and declared liberation they transformed the social and political world that I inhabited—not only the marches I participated in but also the courses I taught. They provided what Aristotle would have called the motive and final causes of this project: the push, the impetus for it, and the end toward which it aspires. As medieval exponents of Aristotelian science were quick to point out, not everything fully realizes its goal. But what was an idea is now a book, and, however imperfect, it seeks both to embody and to honor the participants, past and present, in the changes that made it possible to write a history of sodomy and science in the Middle Ages.

This book also had what Aristotle would have called proximate motive causes. More than twenty years ago, Nancy Siraisi called my attention to Pietro d'Abano's commentary on the little studied Aristotelian work, the *Problems*. At the Morgan Library in New York City I sought out a copy of an early printed edition and found that a sixteenth-century reader had marked extensively the short chapter on the reasons some men enjoy the passive role in intercourse. Since then, the medieval history of that chapter has been a private obsession. But it has also been a collective project, dependent on the expertise, energies, insights, and generosity of many.

The list of manuscripts at the end of the book attests to its reliance on librarians at more than three dozen institutions who have sustained collections and enabled me to have access to them. In the course of my research, I was particularly fortunate to have the assistance of Urs Schoepflin of the Max-Planck Institut für Wissenschaftsgeschichte in Berlin, Ludo Vandamme of the Openbare Bibliotheek Brugge, Geoffrey Waller and Jayne Ringrose of the University Library, Cambridge, Scott Mandelbrote of the Perne Library, Peterhouse, Cambridge, Paola Errani of the Biblioteca Malatestiana, Cesena, Susan Halpert of Harvard's Houghton Library, Kathrin Paasch of the Universitäts- und

Forschungsbibliothek Erfurt/Gotha, Martine De Reu of the Bibliotheek van de Rijksuniversiteit te Gent, Helmut Rohlfinger of the Niedersächsische Staats- und Universitätsbibliothek Göttingen, Anne Korteweg of the Koninklijke Bibliotheek in The Hague, Anna Sobańska of the Biblioteka Jagiellońska, Krakow, Theresa Thom of the Gray's Inn Library, Consuelo Dutschke of Columbia University Library, Martin Kaufman of the Bodleian Library, Sally K. Speirs of Magdalen College Library, Oxford, Angela Williams and Ruth Ogden of St. John's College, Oxford, Milada Svobodová of the Národní Knihovna České Republiky, Prague, and Ramón Gonzálvez Ruiz of the Archivo y Biblioteca Capitulares, Toledo. Special thanks are due to Gary Clark and Jason Newborn of the Interlibrary Loan department at the University of California, Davis, for years of patience, persistence, and resourcefulness.

Sorting through the material and the ideas generated by the research was likewise a collective undertaking. Research groups and seminars that contributed ideas and offered critiques include Abteilung II of the Max-Planck Institut für Wissenschaftsgeschichte, headed by Lorraine Daston, the University of California Medieval Seminar at the Huntington Library, the Society for Medieval Feminist Scholarship, the Society for the Study of Homosexuality in the Middle Ages, members of the project on Gender and Science, Technology, and Medicine at the University of Michigan, the *Problemata* working group centered at the Katholieke Universiteit Leuven under the leadership of Michèle Goyens and Pieter De Leemans, and, at the University of California, Davis, the Cross Cultural Women's and Gender History Program and the Science and Technology Studies Program. Research Assistants at UCD, including Amy Lindgren, Michael Barbezat, and Patricia Turning lent their questions as well as their labor (from manuscript catalogue searches to paleographical databases). Among them Shennan Hutton made special and sustained contributions. Critical assistance came at every stage and with every aspect of the project from the seminar/support group known as the Ladies Who Lunch. The U.S. National Science Foundation under Grant No. 0115556, the Max-Planck Institut für Wissenschaftsgeschichte, and the University of California, Davis, provided the material conditions that made the book possible.

In the process of writing the book, I incurred additional indebtedness to individuals as well as groups. It is hard to measure the degree to which I have been inspired and influenced by conversations with Wendy Cadden, Monica Green, Katharine Park, and Linda Williams. Specialized assistance with languages and texts, in the form of advice and exchanges, came from experts, including Caroline Boucher, Gregory Clark, Gijs Coucke, Noah Guynn, Everett

Rowson, and David Traill. And once a manuscript had taken shape, I profited immeasurably from the great wisdom and frankness of colleagues and friends who read all or part of it. There is no adequate way to thank Beverly Bossler, Mark Jordan, Krystyna von Henneberg, Ruth Mazo Karras, Joseph Ziegler, and Curt Zimansky. When I had done my imperfect best to take advantage of their invaluable help, the team from the University of Pennsylvania Press, including Jerry Singerman, Caroline Winschel, Noreen O'Connor–Abel, and Otto Bohlmann, did all it could to turn the result into an actual book.